PRINCIPLES
OF
INSTRUCTIONAL
DESIGN

PRINCIPLES
OF
INSTRUCTIONAL
DESIGN
Third Edition

Robert M.Gagné
Leslie J. Briggs
Walter W. Wager

Florida State University

HOLT,
RINEHART
AND WINSTON, INC

*New York   Chicago
San Francisco   Philadelphia
Montreal   Toronto   London
Sydney   Tokyo*

Library of Congress Cataloging-in-Publication Data

Gagné, Robert Mills, 1916–
    Principles of instructional design / Robert M. Gagné, Leslie J.
Briggs, Walter W. Wager.—3rd ed.
        p.   cm.
    Includes bibliographies and index.
    1. Instructional systems—Design. 2. Learning. I. Briggs,
Leslie J. II. Wager, Walter W., 1944–   . III. Title.
LB1028.35.G33 1988
371.3—dc19   87-18618
                                                       CIP

ISBN 0-03-011958-8

Requests for permission to make copies of any part of the work should be mailed to:
Permissions
Holt, Rinehart and Winston, Inc.
111 Fifth Avenue
New York, NY 10003

PRINTED IN THE UNITED STATES OF AMERICA

8 9 0 118 9 8 7 6 5 4 3 2

Holt, Rinehart and Winston, Inc.
The Dryden Press
Saunders College Publishing

Dedicated to the memory of
our friend and colleague
LESLIE J. BRIGGS

CONTENTS

PREFACE

We are glad to be able to continue the presentation of ideas of the Gagné–Briggs collaboration in a third edition of this book. The sad event of Leslie Briggs's death has left a void in the teaching of the concepts of instructional design, an effort for which he was so deeply admired. The influence of Briggs's ideas can, we hope, be extended broadly, with the help of his colleague Walter Wager, who worked closely with him for several years in the teaching of instructional design principles. Both authors of this revision have learned a great deal from their association with Leslie Briggs and hope that their memories of his teachings will continue to inform their work.

This third edition has the aim of including several varieties of up-to-date material and contemporary references. We have corrected or elaborated particular points and changed the sequence of presentation somewhat in the interest of clarification. We have added a number of sections to chapters to reflect recent progress in research and theory. Chapter 6 is new and deals with learner characteristics. Suggestions about how these learner characteristics should influence instructional design are made in subsequent chapters. We have taken a new look at procedures for instructional media selection in an effort to reflect modern work. In this revised edition, we expand our treatment of instructional procedures aimed at learner strategies and metacognitive skills.

Our purpose, however, remains one of describing a rationally consistent basis for instructional design. The procedures we suggest are more accurately viewed as "what to do" rather than specifically "how to do it." This approach reflects the belief that instructional design efforts must meet intellectually convincing standards of quality and that such standards need to be based on scientific research and theory in the field of human learning. Methods of instructional design should go as far as possible in defining the learning purposes of each design step. At that point, however, other considerations take over. The designer must ultimately deal with the details of print, pictures, and sound, and these elements have their own technologies. We trust that the designer who follows the principles described here can be assured that the details of instruction, however arrived at, will have a sound foundation in research and theory.

It is our expectation that this third edition will fill a need in courses of education for teachers and instructors at both undergraduate and graduate levels. With its current scope and emphasis, the book may find a place in courses that deal with instructional systems, including those that approach the selection of media as a part of educational technology. Instructors of courses in teaching methods, instructional planning, curriculum theory, and classroom techniques should find the book useful. In graduate-level education, the text may be of use in these same areas, as well as in courses on learning and educational psychology.

This book has four parts, and we have attempted to make the transitions between them apparent. Chapter 1, an introductory chapter, is followed by a chapter outlining the nine stages of instructional design procedure, beginning with the identification of instructional goals and continuing through the conduct of evaluation.

In Part Two, the major classes of learning outcome are described, including intellectual skills, cognitive strategies, verbal information, attitudes, and motor skills. An account is given of the conditions of learning applicable to the acquisition of these capabilities, and the implications of those conditions for the design of instruction are pointed out. The knowledge, skills, and abilities of learners, and how the differences among learners affect instructional planning constitute other topics covered in this section.

Part Three, comprising Chapters 7 through 13, deals directly and intensively with the procedures of design. These begin with an account of instructional objectives, their analysis, and classification. Procedures are identified for determining desirable sequences of instruction, for deriving the events of instruction for the single lesson, and for relating events to the selection of appropriate media. Chapter 13 describes procedures for assessing student performance through the use of criterion-based and norm-based measures.

Delivery systems for instruction is the general topic of Part Four. This includes the application of design products and principles to group instruction and to varieties of individualized instruction. The closing chapter, Chapter 16, deals with procedures for evaluating instructional programs.

Tallahassee, Florida
R. M. G.
W. W. W.

PRINCIPLES
OF
INSTRUCTIONAL
DESIGN

PART ONE

INTRODUCTION TO INSTRUCTIONAL SYSTEMS

1 INTRODUCTION

Instruction is a human undertaking whose purpose is to help people learn. Although learning may happen without any instruction, the effects of instruction on learning are often beneficial and usually easy to observe. When instruction is designed to accomplish a particular goal of learning, it may or may not be successful. The general purpose of this book is to describe what characteristics instruction must have to be successful, in the sense of aiding learning.

Instruction is a set of events that affect learners in such a way that learning is facilitated. Normally, we think of these events as being external to the learner—events embodied in the display of printed pages or the talk of a teacher. However, we also must recognize that the events that make up instruction may be partly internal when they constitute the learner activity called "self-instruction."

Why do we speak of *instruction* rather than *teaching*? It is because we wish to describe *all* of the events that may have a direct effect on the learning of a human being, not just those set in motion by an individual who is a teacher. Instruction may include events that are generated by a page of print, by a picture, by a television program, or by a combination of physical objects, among other things. Of course, a teacher may play an essential role in the arrangement of any of these events. Or, as already mentioned, the learners may be able to manage instructional events themselves. Teaching, then, may be considered as only one form of instruction, albeit a signally important one.

Considered in this comprehensive sense, instruction must be planned if it is to be effective. In detail, of course, a teacher may not have much time to plan instruction on a moment-to-moment basis. Each new event of the classroom requires one or more decisions on the part of a teacher. However, instruction is usually planned, which means that it is designed in some systematic way. Despite varying moment-to-moment decisions, a teacher follows the plan of a lesson design. The lesson is part of the larger design involved in the presentation of a topic (a course segment), and this topic in turn makes up part of a still more comprehensive design of the course or curriculum.

The purpose of designed instruction is to activate and support the learning of the individual student. This aim is characteristic of instruction wherever it occurs, whether between a tutor and a single student, in a school classroom, in an adult interest group, or in an on-the-job setting. Instruction for the support of learning must be something that is planned rather than haphazard. The learning it aids should bring all individuals closer to the goals of optimal use of their talents, enjoyment of life, and adjustment to the physical and social environment. Naturally, this does not mean that the planning of instruction will have the effect of making different individuals more alike. On the contrary, diversity among individuals will be enhanced. The purpose of planned instruction is to help each person develop as fully as possible, in his or her own individual direction.

BASIC ASSUMPTIONS ABOUT INSTRUCTIONAL DESIGN

How is instruction to be designed? How can one approach such a task, and how begin it? There must surely be alternative ways. In this book we describe one way that we believe to be both feasible and worthwhile. This way of planning and designing instruction has certain characteristics that need to be mentioned at the outset.

First, we adopt the assumption that instructional design must be aimed at *aiding the learning of the individual*. We are not concerned here with "mass" changes in opinion or capabilities, nor with education in the sense of "diffusion" of information or attitudes within and among societies. Instead, the instruction we describe is oriented to the individual. Of course, we recognize that learners are often assembled into groups; but learning nevertheless occurs within each member of a group.

Second, instructional design has phases that are both *immediate and long-range*. Design in the immediate sense is what the teacher does in preparing a lesson plan some hours before the instruction is given. The longer range aspects of instructional design are more complex and varied. The concern will more likely be with a set of lessons organized into topics, a set of topics constituting a course or course sequence, or perhaps with an entire instructional system. Such design is sometimes undertaken by individual teachers as well as groups or teams of teachers, by committees of school people, by groups and organizations of curriculum planners, by textbook writers, and by groups of scholars representing academic disciplines.

The immediate and long-range phases of instructional planning are best performed as separate tasks, and not mixed together. The job of the teacher in carrying out instruction is highly demanding in terms of time, effort, and intellectual challenge. The teacher has a great deal to do in planning instruction on an immediate, day-to-day or hour-to-hour basis. Such a task can be greatly facilitated when the products of careful long-range instructional design are made available in the form of textbooks, teachers' guides, audiovisual aids, and other materials. Trying to accomplish both immediate and long-range instructional design, while teaching 20 or 30 students, is simply too big a job for one person, and it can readily lead to the neglect of essential teaching functions. This is not to suggest, however, that teachers cannot or should not undertake long-range instructional design, either on their own or as part of a larger team. Teachers have essential contributions to make to long-range instructional design, and such contributions are best made during nonteaching periods.

Our third assumption in this work is that *systematically designed instruction can greatly affect individual human development.* Some educational writings (e.g., Friedenberg, 1965; Barth, 1972) indicate that education would perhaps be best if it were designed simply to provide a nurturing environment in which young people were allowed to grow up in their own ways, without the imposition of any plan to direct their learning. We consider this an incorrect line of thinking. Unplanned and undirected learning, we believe, is very likely to lead to the development of many individuals who are in one way or another incompetent to derive personal satisfaction from living in society, present or future. A fundamental reason for instructional design is to insure that no one is "educationally disadvantaged," and that all students have equal opportunities to use their individual talents to the fullest degree.

The fourth idea to which we give prominence is that instructional design should be *conducted by means of a systems approach.* This is discussed more fully in Chapter 2. Briefly, the systems approach to instructional design involves the carrying out of a number of steps beginning with an analysis of needs and goals, and ending with an evaluated system of instruction that demonstrably succeeds in meeting accepted goals. Decisions in each of the individual steps are based upon empirical evidence, to the extent that such evidence allows. Each step leads to decisions that become "inputs" to the next step, so that the whole process is as solidly based as is possible within the limits of human reason. Furthermore, each step is checked against evidence that is "fed back" from subsequent steps to provide an indication of the validity of the system.

Our fifth and final point, to be expanded in Part II and throughout the book, is that *designed instruction must be based on knowledge of how hu-*

man beings learn. In considering how an individual's abilities are to be developed, it is not enough to state what they should be; one must examine closely the question of how they can be acquired. Materials for instruction need to reflect not simply what their author knows, but also how the student is intended to learn such knowledge. Accordingly, instructional design must take fully into account *learning conditions* that need to be established in order for the desired effects to occur.

SOME LEARNING PRINCIPLES

At this point, it seems appropriate to expand upon the idea of basing instructional design on knowledge of the *conditions of human learning.* What sort of knowledge of these conditions is needed in order to design instruction?

Changes in the behavior of human beings and in their capabilities for particular behaviors take place following their experience within certain identifiable situations. These situations stimulate the individual in such a way as to bring about the change in behavior. The process that makes such change happen is called *learning,* and the situation that sets the process into effect is called a *learning situation.*

When examined closely, it is apparent that a learning situation has two parts—one *external* to the learner, and the other *internal* to that person. The internal part of the learning situation, it appears, derives from the stored memories of the learner. A person may experience the external stimulation that conveys the information that presidential elections in the United States are held on the first Tuesday after the first Monday in November. If that fact is to be learned, however, it is evident that certain internal conditions, provided by memory from previous learning, must also be present as part of the situation. The learner must have access to knowledge from memory, such as (1) the meanings of *Monday, Tuesday*, and *November* as designations of times, (2) the meaning of *presidential election* as the identification of an event, and (3) the basic skills involved in comprehending an English sentence. The person who possesses these internal capabilities (and certain others we will mention later on), and who is presented with the statement about presidential elections in oral or printed form, is potentially in a learning situation and likely to learn from it. The person who experiences that statement as the external part of the learning situation, but who lacks the internal part, will not learn what is being presented.

The process of learning has been investigated by the methods of science (mainly by psychologists) for many years. As scientists, learning investigators are basically interested in explaining how learning takes place. In other words, their interest is in relating both the external and internal parts of a learning situation to the process of behavior change called learning. The relationships they have found, and continue to find, between the situation and the behavior change may be appropriately called the *conditions of learning* (Gagné, 1985). These are the conditions, both external and internal to the learner, that make learning occur. If one has the intention of making learning occur, as in planning instruction, one must deliberately arrange these external and internal conditions of learning.

In the course of pursuing knowledge about how learning takes place, *theories* are constructed about structures and events (generally conceived as occurring in the central nervous system) that could operate to affect learning. The effects of particular events on learning may be, and usually are, checked again and again under a variety of conditions. In this way a body of facts about learning are collected along with a body of principles that hold true in a broad range of situations. The aspects of learning theory that are important for instruction are those that relate to *controlled events and conditions*. If we are concerned with designing instruction so that learning will occur efficiently, we must look for those elements of learning theory that pertain to the events about which an instructor can do something.

Some Time-Tested Learning Principles

What are some of the principles derived from learning theory and learning research that may be relevant to instructional design? First, we mention some principles that have been with us for many years. Basically, they are still valid, but they may need some new interpretations in the light of modern theory.

Contiguity

This principle states that the stimulus situation must be presented simultaneously with the desired response. One has to think hard to provide an example of a violation of the principle of contiguity. Suppose, for example, one wants a young child to learn to print an *E*. An unskilled teacher might be tempted to do it as follows: First, give the verbal instruction, "Show me how you print an *E*." Following this, show the child a printed *E* on a page, to illustrate what it looks like, and leave the page on the child's table. The child then draws an *E*. Now, has the child learned to print an *E*? Referring to the principle of contiguity, one would have to say, probably not yet. What has been made contiguous in this situation is:

Stimulus situation: a printed *E*
Child's response: printing an *E*

whereas the intended objective of the lesson was:

Stimulus situation: "Show me how you print an *E*"
Child's response: printing an *E*

Somehow, in order for the principle of contiguity to exert its expected effect, the first set of events must be replaced by the second by a staged removal of the intervening stimulus (the printed *E*). In the first case, the verbal instructions were *remote from* the expected response, rather than contiguous with it.

Repetition

This principle states that the stimulus situation and its response need to be repeated, or practiced, for learning to be improved and retention made more certain. There are some situations where the need for repetition is very apparent. For example, if one is learning to pronounce a new French word such as *variété*, repeated trials certainly lead one closer and closer to an acceptable pronunciation. Modern learning theory, however, casts much doubt on the idea that repetition works by "strengthening learned connections." Furthermore, there are many situations in which repetition of newly learned ideas does not improve either learning or retention (cf. Ausubel, Novak, and Hanesian 1978; Gagné, 1985). It is perhaps best to think of repetition not as a fundamental condition of learning, but merely as a practical procedure (practice) which may be necessary to make sure that other conditions for learning are present.

Reinforcement

Historically, this principle has been stated as follows: Learning of a new act is strengthened when the occurrence of that act is followed by a satisfying state of affairs (that is, a reward) (Thorndike, 1913). Such a view of reinforcement is still a lively theoretical issue, and there is a good deal of evidence for it. For instructional purposes, however, one is inclined to depend on another conception of reinforcement that may be stated in this form: A new act (A) is most readily learned when it is immediately followed by an old act (B) that the individual likes to do and performs readily, in such a way that doing B is made contingent upon doing A (Premack, 1965). Suppose a young child is fond of looking at pictures of animals, and his parents desire that he learn how to make drawings of animals. The new capability of animal drawing, according to this principle, will be most readily learned if one connects it to looking at additional animal pictures. In other words, the opportunity to look at animal pictures is made *contingent* upon drawing one or more animals. In this form, the principle of reinforcement is a most powerful one.

THE CONDITIONS OF LEARNING

As the study of human learning has proceeded, it has gradually become apparent that theories must be increasingly sophisticated. Contiguity, repetition, and reinforcement are all good principles, and one of their outstanding characteristics is that they refer to controllable instructional events. The designer of instruction, and also the teacher, can readily devise situations that include these principles. Nevertheless, even when all these things are done, an efficient learning situation is not guaranteed. Something seems to be missing.

It appears that instruction must take into account a whole set of factors that influence learning, and that collectively may be called the *conditions of learning* (Gagné, 1985). Some of these conditions, to be sure, pertain to the stimuli that are *external* to the learner. Others are *internal* conditions, to be sought within the individual learner. They are states of mind that the learner brings to the learning task; in other words, they are *previously learned capabilities* of the individual learner. These internal capabilities appear to be a highly important set of factors in ensuring effective learning.

The Processes of Learning

In order to take account of the conditions of learning, both external and internal, we must begin with a framework, or *model*, of the processes involved in an act of learning. A model widely accepted by modern investigators that incorporates the principal ideas of contemporary learning theories, is shown in Figure 1-1. This model conceives of learning as information processing.

Stimulation from the learner's environment (at left in Figure 1-1) activates the receptors and is transmitted as information to the central nervous system. The information attains a brief registration in one of the *sensory registers* and is then transformed into recognizable patterns that enter the *short-term memory*. The transformation that occurs at this point is called *selective perception*, or *feature perception*. The visually presented marks on a page of print become the letters *a, b*, and so on, when they are stored in *short-term memory*: a set of particular angles, corners, horizontal and vertical lines becomes a *rectangle*.

Storage of information in the short-term memory (STM) has a relatively brief duration, less that 20 seconds, unless it is rehearsed. The familiar example is remembering a seven-digit telephone number long enough to dial it. Once it is dialled (or punched in), it disappears from STM; but if it must be remembered longer, this can be done by internal *rehearsal*. Another aspect of the short-term memory of considerable importance for learning is its limited *capacity*. Only a few separate items, perhaps as few as four, can be "held in mind" at one time. Since short-term storage is one stage of the process of learning, its capacity limits can strongly affect the difficulty of learning tasks. For instance, the process of mentally multiplying 29×3 requires that the intermediate operations (30×3; $90 - 3$) be held in STM. This requirement makes the learning of such a task considerably more difficult than, say, 40×3.

Information to be remembered is again transformed by a process called *se-*

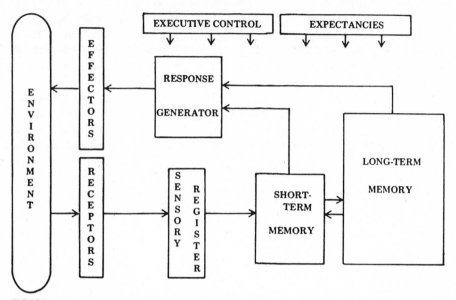

FIGURE 1-1 A basic model of learning and memory, underlying modern cognitive (information-processing) theories.
(From R. M. Gagné, *Essentials of learning for instruction*, copyright 1975 by Dryden Press—Holt, Rinehart and Winston, New York. Reprinted by permission.)

mantic encoding to a form that enters *long-term memory* (LTM). When encoded, information in long-term memory is meaningful; much of it has the form of propositions, that is, entities of language possessing sentencelike subjects and predicates. In this form, information may be stored for long periods of time. It may be returned to short-term memory by the process of *retrieval*, and it appears that such retrieved items may combine with others to bring about new kinds of learning. When functioning in this manner, the short-term memory is often referred to as a *working memory*.

Information from either the working memory or the long-term memory, when retrieved, passes to a *response generator* and is transformed into action. The message activates the effectors (muscles), producing a performance that can be observed to occur in the learner's environment. This action is what enables an external observer to tell that the initial stimulation has had its expected effect. The information has been "processed" in all of these ways, and the learner has indeed learned.

Control Processes

Two important structures shown in Figure 1-1 are *executive control* and *expectancies*. These are processes that activate and modulate the flow of information during learning. For example, learners have an expectancy of what they will be able to do once they have learned, and this in turn may affect how an external situation is perceived, how it is encoded in memory, and how it is transformed into performance. The *executive control* structure governs the use of *cognitive strategies*, which may determine how information is encoded when it enters long-term memory, or how the process of retrieval is carried out, among other things (see Chapter 4 for a fuller description).

The model in Figure 1-1 introduces the *structures* underlying contemporary learning theory and implies a number of processes made possible thereby. All of these processes compose the events that occur in an act of learning. In summary, the internal *processes* are:

1. *Reception* of stimuli by *receptors,*
2. *Registration* of information by *sensory registers,*
3. *Selective perception* for storage in short-term memory (STM),
4. *Rehearsal* to maintain information in STM,
5. *Semantic encoding* for storage in long-term memory (LTM),
6. *Retrieval* from LTM to *working memory* (STM),
7. *Response generation* to *effectors,*
8. *Performance* in the learner's environment, and
9. *Control* of processes through *executive strategies.*

Events outside the learner can be made to influence the processes of learning, particularly those numbered 3 through 6. These internal processes can be enhanced by events that take place in the learning environment. For example, the selective perception of the features of a plant can be aided by emphasizing them in a diagram. The semantic encoding of a prose passage can be more readily done if the passage opens with a topic heading.

Instruction and Learning Processes

If instruction is to bring about effective learning, it must somehow be made to influence the internal processes of learning implied by the information flow depicted in Figure 1-1. As the previous examples indicate, external events can affect these processes in a variety of ways, some of which are *supportive* of learning. If it is possible, then, to discover what kinds of events can provide such support, it should also be possible to select and put into effect those events that will do the most effective job. This is what instruction attempts to do. Instruction, then, may be conceived as *a deliberately arranged set of external events designed to support internal learning processes.*

We shall have occasion throughout this book to refer to the *events of instruction* (Gagné, 1985). When instruction is designed, it is these events that are being considered, chosen, and represented in the communications and other stimulation offered to the learner. These events, individually and collectively, are what constitute the external conditions of learning. Their purpose is to bring about the kinds of internal processing that will lead to rapid, obstacle-free learning.

The events of instruction are more fully described in Chapter 9. In brief, these events involve the following kinds of activities in roughly this order, relating to the learning processes previously listed:

1. Stimulation to *gain attention* to assure the reception of stimuli;
2. *Informing learners of the learning objective*, to establish appropriate expectancies;
3. *Reminding* learners of previously learned content for retrieval from LTM;
4. *Clear and distinctive presentation* of material to assure selective perception;
5. *Guidance of learning* by suitable semantic encoding;
6. *Eliciting performance*, involving response generation;
7. *Providing feedback* about performance;
8. *Assessing the performance*, involving additional response feedback occasions; and
9. *Arranging variety of practice* to aid future retrieval and transfer.

These events will be more fully and precisely described in a later chapter. They are presented here in this form to give a general impression of their relation to the processes of learning.

The Contributions of Memory

Besides the external events of instruction, the conditions of learning include the presence in working memory of certain *memory contents*. As previously noted, these are retrieved from LTM during the learning episode, when the learner is reminded of (or asked to recall) the contents learned on previous occasions. For example, learners who are acquiring new knowledge about the presidential election of 1980 will recall prior general knowledge about elections—when they are held, what events they include, and so on. Learners who are acquiring the skills for writing effective sentences will recall the skills they learned previously for spelling, word sequence, and punctuation.

The contents of LTM, when retrieved to working memory, become essential parts of the conditions of learning. These contributors to new learning are of many kinds and have many sorts of specific relationships with whatever is involved in the new learning. Our point of view, as reflected in subsequent chapters, is that these contents of memory can best be differentiated into five general categories. These are five classes of previously learned content, which can be exhibited in a corresponding five kinds of performance outcomes. On account of this latter quality, they may best be referred to as *five kinds of previously learned capabilities*. They are memory contents that make the learner *capable* of performing in ways implied by their titles.

Obviously, the capabilities that were previously learned fall into the same categories as those which are to be newly learned. Chapters 3, 4, and 5 describe in detail the five categories of learned capabilities and the conditions of learning that relate to them. Briefly, however, the five kinds of learned capabilities with which this book deals are:

1. *Intellectual skills* that permit the learner to carry out symbolically controlled procedures;
2. *Cognitive strategies*, by means of which learners exercise control over their own learning processes;
3. *Verbal information*—the facts and organized "knowledge of the world" stored in the learner's memory;
4. *Attitudes*—internal states that influence the personal action choices a learner makes; and
5. *Motor skills*—the movements of skeletal muscles organized to accomplish purposeful actions.

Concentrating instruction on any one type of these capabilities alone, or any two in combination, is insufficient. Verbal information, in and of itself, represents a highly inadequate instructional goal. Learning intellectual skills leads to practical competence. Yet these, too, are insufficient for the totality of new learning because such learning makes use of verbal knowledge as well. Furthermore, the learning of intellectual skills does not by itself equip learners with the cognitive strategies they need to become independent self-learners. Cognitive strategies themselves cannot be learned or progressively improved without the involvement of verbal information and skills—they must, in other words, have "something to work on." Attitudes, too, require a substrate of information and intellectual skills to support them. Finally, motor skills, although a somewhat specialized area of school learning, are nevertheless of recurring relevance to human development. In sum, *multiple aims* for instruction must be recognized. The human learner needs to attain several varieties of learned capabilities.

Intellectual Skills as Building Blocks for Instruction

For purposes of planning instruction in scope ranging from entire systems to individual lessons, intellectual skills have a number of desirable characteristics as components of a design framework (Gagné, 1985). An intellectual skill cannot be learned by being simply looked up or provided to the learner by a verbal communication. It must be learned, recalled, and put into use at the proper time. As an example, consider the intellectual skill of spelling words containing a long *a* sound. When the learner possesses this skill, he is able to perform such spelling rapidly and without the necessity of looking up a set of rules. His performance shows that he is able to recall such rules and put them immediately into effect. At the same time, learning the necessary rules for spelling words with long *a* is not something that takes a period of many months to accomplish (as seems to be true of cognitive strategies). Essentially flawless performance calling upon such an intellectual skill can be established in a short time.

There are other advantages to intellectual skills as a major framework for instruction and instructional design. Such skills come to be highly interrelated to each other and to build elaborate internal intellectual structures of a cumulative sort (Gagné, 1985). The learning of one skill aids the learning of other "higher-order" skills. Suppose an individual has learned the skill of substituting specific numerical values for letters in a symbolic expression such as the following:

$$\sigma^2 = E(X - m)^2.$$

Such a skill will aid the learning of many kinds of advanced skills, not simply in mathematics, but in many areas of science and social studies. Intellectual skills are rich in transfer effects, which bring about the building of increasingly complex structures of intellectual competence.

Another advantage of intellectual skills as a primary component in instruction is the relative ease with which they can be reliably observed. When a learner has attained an intellectual skill, as, for example, "representing quantitative increases graphically," it is relatively easy to show that the skill has indeed been learned. One would provide the person with numerical values of any increasing variable, and ask him to construct a graph to show the changes in that variable. An intellectual skill can always be defined in operational terms; that is, it can always be related to a class of human performances—to something the successful learner *is able to do.*

The choice of intellectual skills as a primary point of reference in the design of instruction, then, is based mainly upon practical considerations. In contrast to factual information, skills cannot be simply looked up or made available by "telling," but must be learned. In contrast to cognitive strategies, intellectual skills are typically learned over relatively short time periods and do not have to be refined and sharpened by months or years of practice. Intellectual skills build upon each other in a cumulative fashion to form increasingly elaborate intellectual structures. Through the mechanism of transfer of learning, they make possible an ever broadening intellectual competence in each individual. And finally, such skills can be readily observed, so that it is easy to tell that they have been learned.

THE RATIONALE FOR INSTRUCTIONAL DESIGN

The design of instruction must be undertaken with suitable attention to the conditions under which learning occurs—conditions which are both external and internal to the learner. These conditions are in turn dependent upon *what* is being learned.

To design instruction systematically you must first establish a rationale for what is to be learned. This requires going back to the initial sources that have given rise to the idea of employing instruction to meet a recognized need. A *system* of instruction may then be constructed step by step, beginning with a base of information that reflects identified goals.

The planning of instruction in a highly systematic manner, with attention to the consistency and compatibility of technical knowledge at each point of decision, is usually termed the "systems approach." This kind of design uses various forms of information, data, and theoretical principles as input at each planning stage. Further, the prospective outcomes of each stage are checked against whatever goals may have been adopted by those who manage the system as a whole. It is within this systems framework that we seek to apply what is known about the conditions of human learning to instructional design.

The Derivation of an Instructional System

The rational steps in the derivation of an instructional system, which we will describe more completely in the next chapter, may be outlined briefly as follows:

1. The *needs* for instruction are investigated as a first step. These are then carefully considered by a responsible group to arrive at agreements on the *goals* of instruction. The *resources* available to meet these goals must also be carefully weighed, along with those circumstances that impose *constraints* on instructional planning.
2. Goals of instruction may be translated into a framework for a *curriculum*, and for the individual courses contained in it. The goals of individual courses may be conceived as *target objectives*, and grouped to reflect a rational organization.
3. *Objectives* of courses are achieved through learning. In this book the lasting effects of learning are defined as the acquisition of various *capabilities* by the learner. As outcomes of instruction and learning, human capabilities are usually specified in terms of the classes of *human performance* which they make possible. We need to consider what kinds of capabilities may be learned. We shall describe the varieties of human performance made possible to the learner by each kind of learned capability—intellectual skills, cognitive strategies, verbal information, attitudes, and motor skills.
4. The identification of *target objectives* and the *enabling objectives* that support them and contribute to their learning makes possible the grouping of these objectives into units of comparable types. These can then be systematically arranged to form the course.
5. The determination of types of capabilities to be learned, and the inference of necessary learning conditions for them, makes possible the planning of *sequences of instruction*. This is so because the information and skills that need to be recalled for any given learning task must themselves have been previously learned. For example, learning the intellectual skill of using an adverb to modify a verb requires the recall of "subordinate" skills of constructing adverbs from adjectives, identifying verbs, identifying adjectives, and classifying action modifications. Thus by tracing backwards from the outcome of learning for a particular topic, one can identify a sequence of intermediate (or prerequisite) objectives that must be met to make possible

the desired learning. In this way, instructional sequences may be specified that are applicable to topics or to courses.

6. Continuation of instructional planning proceeds to the design of units of instruction that are smaller in scope and thus more detailed in character. Consideration of target objectives and the skills and verbal information that support them leads to a requirement for the delineation of precisely defined objectives called *performance objectives*. These identify the expected or planned outcomes of learning, and thus fall into the categories of *learned capabilities* previously mentioned. They represent instances of human performance that can be reliably observed and assessed as outcomes of learning.

7. Once a course has been designed in terms of target objectives, detailed planning of instruction for the individual *lesson* can proceed. Here again the first reference for such planning is the performance objectives that represent the outcomes of the lesson. Attention centers on the arrangement of *external conditions* that will be most effective in bringing about the desired learning. Consideration must also be given to the *characteristics of the learners* since these will determine many of the *internal conditions* involved in the learning. Planning the conditions for instruction also involves the choice of appropriate media and combinations of media that may be employed to promote learning.

8. The additional element required for completion of instructional design is a set of procedures for *assessment* of what students have learned. In conception, this component follows naturally from the definitions of instructional objectives. The latter statements describe domains from which items are selected. These in turn may be teacher observations or may be assembled as tests. Assessment procedures are designed to provide *criterion-referenced* measurement of learning outcomes (Popham, 1981). They are intended as direct measures of what students have learned as a result of instruction on specfic objectives. This kind of assessment is sometimes called *objective-referenced*.

9. The design of lessons and courses with their accompanying techniques of assessing learning outcomes makes possible the planning of entire systems. Instructional systems aim to achieve comprehensive goals in schools and school systems. Means must be found to fit the various components together by way of a management system, sometimes called an "instructional delivery system." Naturally, teachers play key roles in the operation of such a system. A particular class of instructional systems is concerned with *individualized instruction*, involving a set of procedures to ensure optimal development of the individual student. It is instructive to contrast these methods with others that characterize *group instruction*.

Finally, attention must be paid to *evaluation*. Procedures for evaluation are first applied to the design effort itself. Evidence is sought for needed revisions aimed at the improvement and refinement of the instruction (*formative evaluation*). At a later stage, *summative evaluation* is undertaken to seek evidence of the learning effectiveness of what has been designed.

WHAT THIS BOOK IS ABOUT

The design of instruction, the background of knowledge from which its procedures are derived, and the various ways in which these procedures are carried out are described in the 16 chapters of this book, arranged as follows:

Introduction to Instructional Systems

Chapter 1, the introduction, outlines our general approach to instruction, and includes an account of some principles of human learning that form the bases of instructional design.

Chapter 2 introduces the reader to instructional systems and the systems approach to the design of instruction. The stages of instructional system design are described, to be developed further in subsequent chapters.

Basic Processes in Learning and Instruction

Chapter 3 introduces the reader to the five major categories of instructional outcomes—the human capabilities that are learned with the aid of instruction. The varieties of human performance that these capabilities make possible are described and distinguished.

Chapter 4 enters into an intensive description of the characteristics and conditions of learning for two of these categories of learning outcomes—intellectual skills and cognitive strategies.

Chapter 5 extends this description of learned capabilities to the three additional categories, with definitions and examples of information, attitudes, and motor skills.

Chapter 6 gives an account of the human learner, with emphasis on the characteristics the individual learner and various groups of learners bring to the learning situation. Planning instruction requires that provisions be made for learner differences in these qualities.

Designing Instruction

Chapter 7 deals with the derivation and description of specific instructional objectives (performance objectives). These are related on the one hand to the categories of objectives previously defined, and on the other to the particular learned capabilities that are the focus of interest for instruction.

Chapter 8 describes procedures for task analysis, beginning with a consideration of purposes and goals of instruction. The aim of analysis is the classification of objectives for use in instructional planning. Prerequisites are identified for various kinds of learning outcomes.

Chapter 9 describes procedures for constructing sequences of lessons in making up larger units of instruction such as topics, modules and courses.

Chapter 10 deals with sequences of instructional events within single lessons and shows how these may be related to the stages of information processing involved in learning.

Chapter 11 discusses the important matter of media selection and provides a systematic procedure for conducting such a step as a part of instructional design.

Chapter 12 gives an account of the design of individual lessons, including the placement of parts of lessons in sequence, the arrangement of effective conditions of learning, and the use of media for instructional delivery.

Chapter 13 deals with methods of assessing student performances as outcomes of instruction, describing appropriate uses of criterion-referenced and norm-referenced tests.

Delivery Systems for Instruction

Chapter 14 opens this part of the book by discussing the special features of design needed when instruction is to be delivered to groups of various sizes.

Chapter 15 presents a complementary account of how systematic procedures may be designed to accomplish individualized instruction.

Chapter 16 describes the basic logic for evaluating designed products and procedures, from lessons to systems.

SUMMARY

Instruction is planned for the purpose of supporting the processes of learning. In this book, we describe methods involved in the design of instruction aimed at the human learner. We assume that planned instruction has both short-range and long-range purposes in its effects on human development.

Instructional design is based upon some principles of human learning, specifically, the conditions under which learning occurs. Some time-tested principles of contiguity, repetition, and reinforcement indicate some of the conditons external to the learner that can be incorporated into instruction. A model of information processing that identifies a number of internal processes underlies contemporary theories of learning. These processes bring about several successive stages in the transformation of information on its way to storage in the long-term memory. The purpose of instruction is to arrange external events that support these internal learning processes.

An act of learning is greatly influenced by previously learned material retrieved from the learner's memory. The effects of prior learning on new learning is seen in the acquiring of verbal information, intellectual skills, cognitive strategies, attitudes, and motor skills. These human capabilities, established by learning, will be described in the chapters to follow. These varieties of learned capabilities and the conditions for their learning constitute the basis for instructional planning. Derived from these principles is the rationale for a set of practical procedures for the design of instruction.

Students who use this book will find it possible to follow up the ideas derived from research on human learning with further exploration and study of the references at the end of each chapter. Those who are interested in becoming skillful in designing instruction will need to undertake practice exercises that exemplify the procedures described. Because of the anticipated variety of particular courses and educational settings in which this book might be used, it has been our general expectation that such exercises would be supplied by a course instructor. Examples and exercises of particular relevance for such a purpose are provided in a previously published volume by Briggs and Wager (1981).

References

Ausubel, D. P., Novak, J. D., & Hanesian, H. (1978). *Educational psychology: A cognitive view* (2nd ed.). New York: Holt, Rinehart and Winston.

Barth, R. S. (1972). *Open education and the American school.* New York: Agathon Press.

Briggs, L. J., & Wager, W. W. (1981). *Handbook of procedures for the design of instruction.* Englewood Cliffs, NJ: Educational Technology Publications.

Friedenberg, E. Z. (1965). *Coming of age in America: Growth and acquiescence.* New York: Random House.

Gagné, R. M. (1985). *The conditions of learning* (4th ed.). New York: Holt, Rinehart and Winston.

Popham, W. J. (1981). *Modern educational measurement.* Englewood Cliffs, NJ: Educational Technology Publications.

Premack, D. (1965). Reinforcement theory. In D. Levine (Ed.), *Nebraska symposium on motivation.* Lincoln, NE: Univeristy of Nebraska Press.

Thorndike, E. L. (1913). *The psychology of learning: Educational psychology* (Vol. 2.). New York: Teachers College Press.

2 DESIGNING INSTRUCTIONAL SYSTEMS

An instructional system may be defined as an arrangement of resources and procedures used to promote learning. Instructional systems have a variety of particular forms and occur in many of our institutions. Public schools embody the most widely known forms of instructional systems. The military services have, perhaps, some of the largest instructional systems in the world. Businesses and industries have instructional systems that are often referred to as training systems. Any institution that has the express purpose of developing human capabilities may be said to contain an instructional system.

Instructional systems design is the *systematic* process of planning instructional systems, and instructional development is the process of implementing the plans. Together, these two functions are components of what is referred to as *instructional technology*. Instructional technology is a broader term than instructional systems and may be defined as the systematic application of theory and other organized knowledge to the task of instructional design and development. Instructional technology also includes the quest for new knowledge about how people learn and how best to design instructional systems or materials (Heinich, 1984).

It should be evident that instructional systems design can occur at many different levels. One could imagine a nationwide effort at planning and developing instructional systems, as was the case with the Biological Sciences Curriculum Study and the Intermediate Science Curriculum Study funded by the National Science Foundation. These efforts centered on developing materials within a subject area. It is also worthy of note that some programs for individualized instruction in several subject areas have been undertaken. These systems, Project PLAN (Program for Learning in Accordance with Needs), IPI (Individually Prescribed Instruction) and IGE (Individually Guided Instruction) are described in a book edited by Weisgerber (1971).

Instructional designers don't always have a chance to work on projects of national scope. They generally design smaller instructional systems such as courses, units within courses, or individual lessons. Despite the differences in size and scope, the process of designing an instructional system has features in common at all levels of the curriculum. Instructional systems design of the smaller components is simply referred to as instructional design since the focus is the piece of instruction itself, rather than the total instructional system.

INSTRUCTIONAL DESIGN

Several models are suitable for the design of instruction of course units and lessons. One widely known model is the Dick and Carey (1985) model presented in Figure 2-1. All the stages in any instructional systems model can be categorized into one of three functions: (1) identifying the outcomes of the instruction, (2) developing the instruction, and (3) evaluating the effectiveness of the instruction. We shall focus on the activities of instructional design that occur within the nine stages shown in Figure 2-1.

Stage I: Instructional Goals

A goal may be defined as a desirable state of affairs. For example, at a national level a desirable goal is that every adult at least be literate at a sixth-grade reading level. Notice that this is also an instructional goal. An example of a noninstructional goal might be that every adult have adequate medical care. This latter goal is not obtainable by instruction. Global instructional goals must be made more specific before systematic instruction can be designed to attain them. One responsibility of an instructional designer is to recognize which goals are instructional goals and which are not. This is especially true in industrial or vocational instructional courses where the goal may be related to employee motivation or job satisfaction. At this stage the instructional designer must ask, "What goals will represent a desirable state of affairs?"

After goals have been stated, the designer may conduct a *needs analysis*. Recent writers (Burton and Merrill, 1977; Kaufman, 1976) have defined a need as a discrepancy or gap between a desired state of affairs (a goal) and the present state of affairs. Therefore, needs can be determined after the stating of goals and the analysis of the present state of affairs. In the case of public schools, the desired state of affairs is usually established by tradition—a

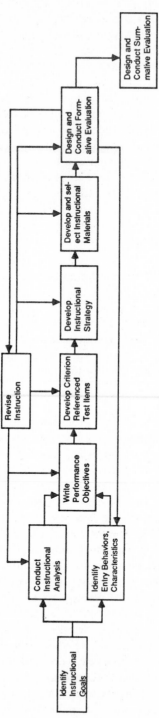

FIGURE 2-1   The systems approach model for designing instruction.
(From W. Dick & L. Carey, *The systematic design of instruction*, 2nd ed., copyright 1985 by Scott, Foresman & Co., Glenview, Illinois. Reprinted by permission.)

consensus on what school students ought to be learning and how well. Any gap between the students' achievement and the school's expectations identifies a need. For example, for a group of seniors at a particular high school, the mean score on the math portion of the SAT might be an indicator of how well the instructional system at that school was meeting its needs.

Training needs in business or industry may be derived from a job analysis or from data on the productivity of a particular department. Again, a discrepancy between desirable performance and present performance identifies a need (Branson, 1977). Other definitions of need include perceived or *felt* needs. These needs are not the result of any documented gap. Nevertheless, they sometimes are the basis for curricular decisions. As an example, parents may decide that their children should learn computer programming in elementary school. This felt need is not usually determined by an analysis of goal deficiencies. The prevailing view is that the general public should be involved in the process of determining instructional goals, and these are often expressed as needs. Needs and goals are further refined in stages 2 and 3 of the design process, *instructional analysis* and *learner analysis*.

Stage 2: Instructional Analysis

Stages two and three in the model of Figure 2-1 can occur in either order, or simultaneously. We have chosen to discuss instructional analysis first. The purpose of instructional analysis is to determine the skills involved in reaching a goal. For example, if the goal happens to be that every healthy adult will be able to perform cardiopulmonary resuscitation, an instructional analysis would reveal what component skills would have to be learned. In this case the designer would use a *task analysis* (or *procedural analysis*), the product of which would be a list of the steps and the skills used at each step in the procedure (Gagné, 1977).

Another kind of instructional analysis is an *information-processing analysis*, which is designed to reveal the mental operations used by a person who has learned a complex skill. This analysis makes possible inferences regarding the internal processes involved in an intended skill. It might be necessary to have a learner "talk through" a procedure used in solving a problem to determine whether appropriate skills and strategies are being applied. An important estimate to be made for each decision and action revealed by an information-processing analysis is whether the intended learners enter with these capabilities or whether they must be taught them (stage three).

An important outcome of instructional analysis is *task classification*. Task classification is the categorization of the learning outcome into a domain or subdomain of types of learning, as described in Chapters 3, 4, and 5. Gagné (1985) describes five major types of learning outcomes and some subtypes. Task classification can assist instructional design in several ways. Classifying the target objectives makes it possible to check whether any intended purpose of an instructional unit is being overlooked. Briggs and Wager (1981) have presented examples of how target objectives may be classified and then grouped into course units in the form of instructional curriculum maps. The resulting maps can then be reviewed to check whether necessary verbal information, attitudes, and intellectual skills are included in the instructional unit. Learning outcome classification also provides the information necessary to assure that the instructional treatment embodies the conditions most effective for different types of learning outcomes.

The final type of analysis to be mentioned is *learning-task analysis*. A learning-task analysis is appropriate for objectives of instruction that involve intellectual skills. If the aim is that fourth graders will be able to make change for a dollar, a learning-task analysis would reveal the subordinate skills needed for adding, subtracting, aligning decimals, carrying, and other skills that are related to this skill. The purpose of a learning-task analysis is to reveal the objectives that are *enabling* and for which teaching sequence decisions need to be made. One possible product of a learning task analysis is an *instructional curriculum map* (ICM) similar to the one shown in Figure 2-2. This ICM shows the targeted objectives and their subordinate objectives for an instructional unit on word processing.

A designer may need to apply any or all of these types of analysis in designing a single unit of instruction. Chapter 7 expands our description of the different types of analysis and the techniques for performing them.

Stage 3: Entry Behaviors and Learner Characteristics

As previously indicated, this step is often conducted in parallel with Stage 2. The purpose is to determine which of the required enabling skills the learners

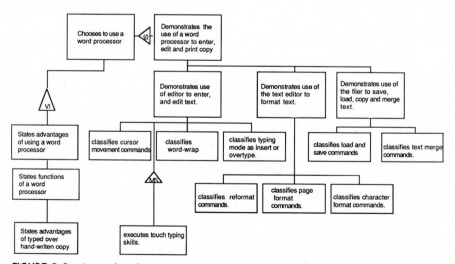

FIGURE 2-2 A unit-level instructional curriculum map (ICM) on word processing.

bring to the learning task. Some learners will know more than others, so the designer must choose where to start the instruction, knowing that it will be redundant for some, but necessary for others. The designer must also be able to identify those learners for whom the instruction would not be appropriate, so that they may be given instruction that remediates. The lack of understanding of a target audience can sometimes be seen in instructional design products. It is usually not sufficient for a designer to guess what the skills of an intended audience will be. A better procedure is to interview and test the skills of the target population until you know enough about them to design the instruction appropriately. Chapter 6 discusses the analysis of learner characteristics in more detail.

In addition to learner qualities such as intellectual skills, which are clearly learned, the designer of instruction may find it desirable to make some provision for learner *abilities* and *traits*, which are usually considered to be less readily alterable through learning. Abilities include such qualities as verbal comprehension and spatial orientation, for example. Instruction designed for learners who are low in verbal comprehension would best de-emphasize verbal presentations (such as printed texts). Instruction designed for learners who score high in spatial orientation ability might be able to use this ability to advantage in a course in architecture.

Traits of personality are another aspect of learner capability that may need to be considered in instructional design. Students who score high on the trait of anxiety, for example, may be better able to learn from instruction that is leisurely paced and that permits learners to choose optional next steps. As will be indicated in Chapter 6, learner traits and abilities may affect some of the general qualites of instruction, such as its employment of particular media and its pacing. In this respect, abilities and traits contrast with such learner characteristics as the possession of particular skills and verbal knowledge; the latter has quite specific effects on the content of effective instruction.

Stage 4: Performance Objectives

At this stage it is necessary to translate the needs and goals into performance objectives that are sufficiently specific and detailed to show progress toward the goals. There are two reasons for working from general goals to increasingly specific objectives. The first is to be able to communicate at different levels to different persons. Some people (for example, parents or a board of directors) are interested only in goals, and not in details, whereas others (teachers, students) need detailed performance objectives to determine what they will be teaching or learning.

A second reason for increased detail is to make possible planning and development of the materials and the delivery system. One thesis of this book is that different types of learning outcomes require different instructional treatments. To design effective instructional materials and choose effective delivery systems, the designer must be able to properly determine the conditions of learning necessary for acquisition of new information and skills. Specification of performance objectives facilitates this task. Once objectives are stated in performance terms, the curriculum can be analyzed in terms of sequence and completeness and the requirements of prerequisite skills. This work facilitates the planning of an effective delivery system. The size of the system needed can be estimated and development schedules can be planned to coordinate the work of the design team, the teachers, the media production team, and the trainers of teachers.

The final reason for eventually stating all objectives in terms of performance (rather than content outlines or teacher activities) is to be able to measure student performance to determine when the objectives have been reached. Objectives are of such central importance to the design process that an entire chapter of this book (Chapter 7) is devoted to their construction.

Performance objectives are statements of observable, measurable behaviors. Prior to this stage, the designer has given much thought to how the needs and goals may be translated into instructional plans at the course or unit level. It is likely that there have been many drafts of instructional objectives, objective groupings, and unit structures before this stage is reached. These modifications enable the designer to define the performance objectives that are to guide all the later work in developing lesson plans (or modules) and the measures to be used in monitoring student progress and evaluating the instruction.

The functions of performance objectives are to: (1) provide a means for determining if the instruction relates to the accomplishment of goals, (2) provide a means for focusing the lesson planning upon appropriate conditions of learning, (3) guide the development of measures of learner performance, and (4) assist learners in their study efforts. Thus the intimate relationships among *objectives, instruction, and evaluation* are emphasized. Briggs (1977) referred to these three aspects of instructional design as the "anchor points" in planning, and he emphasized the need to make certain that the three are in agreement with one another.

It is apparent that the objectives should guide the instruction and evaluation, not the other way around. Therefore, the objectives should be determined before the lesson plans or the evaluation instruments. Almost all instructional design models follow this sequence. Practices differ with regard to the step following the development of performance objectives. The model shown in Figure 2-1 places the development of test items before the development of instructional strategies. Briggs (1977) also placed the design of assessment instruments before lesson development, on the grounds that: (1) the novice is more likely to stray from the objectives in developing tests than in preparing lessons, and (2) a designer who had just finished developing lesson material might inadvertently focus on content rather than performance in constructing tests. The experienced designer, however, might choose to develop lessons before developing performance measures.

Stage 5: Criterion-Referenced Test Items

Since the design of measures of learner performance is discussed later on in Chapter 13, we need here only to summarize the purpose of this design stage. There are many uses for performance measures. First, they can be used for diagnosis and placement within a curriculum. The purpose of diagnostic testing is to assure that an individual possesses the necessary prerequisites for learning new skills. Test items allow the teacher to pinpoint the needs of individual students in order to concentrate on the skills that are lacking and to avoid unnecessary instruction.

Another purpose is to check the results of student learning during the progress of a lesson. Such a check makes it possible to detect any misunderstandings the student may have and to remediate them before continuing. In addition, performance tests given at the conclusion of a lesson or unit of instruction can be used to document student progress for parents or administrators.

These levels of performance assessment can be useful in evaluating the instructional system itself, either lesson-by-lesson, or in its entirety. Evaluations designed to provide data, whereby instruction is to be improved, are called *formative evaluations*. They are usually conducted while the instructional materials are still being formed and reformed. When no further changes are planned, and it is time to determine the success and worth of the course in its final form, *summative evaluations* are conducted. Types of performance measures suitable for these various purposes are discussed extensively in Chapter 13.

Some planning of performance measures may well be undertaken before the development of lesson plans and instructional materials because one wishes the tests to focus on the performance objectives (what the learner must be able to do) rather than on what the learner has read or what the teacher has done. Thus the performance measures are intended to determine if students have acquired the desired skill, not to determine if they merely remember the instructional presentation. Early determination of performance measures helps to focus on the goal of student learning and on the instruction needed to facilitate that learning.

Stage 6: Instructional Strategy

Our use of the term strategy is nonrestrictive. We do not intend to imply that all instruction must be self-contained instructional modules or mediated materials. Teacher-led or teacher-centered instruction can also benefit from instructional systems design. By instructional strategy we mean a plan for assisting the learners with their study efforts for each performance objective. This may take the form of a lesson plan (in the case of teacher-led instruction) or a set of production specifications for mediated materials. The purpose of developing the strategy before developing the materials themselves, is to outline how instructional activities will relate to the accomplishment of the objectives.

When teacher-led, group-paced instruction is planned, teachers use the instructional design process to produce a guide to help implement the intent of the lesson plan without necessarily conveying its exact content to the learners. The teacher gives directions, refers learners to appropriate materials, leads or directs class activities, and supplements existing materials with direct instruction. On the other hand, when a learner-centered, learner-paced lesson is planned, a *module* is typically presented to the learner. It usually presents a learning objective, an activity guide, the material to be viewed or read, practice exercises, and a self-check test for the learner. In this case the instructions or activity guide in the module is written for the student rather than for the teacher. Designers can use the process for both teacher-led and modular materials.

The purpose of all instruction, according to the view presented in this book, is to *provide the events of instruction*. These events are discussed in Chapter 10. They include such widely recognized functions as *directing attention, informing the learner of the objective, presenting the stimulus material,* and *providing feedback*. It matters little whether these events are performed by teachers or materials, so long as they are successfully performed. It may be noted further that these events of instruction are applicable to all domains of learning outcomes, although the details of how they are implemented imply somewhat different sets of conditions for learning (see Chapters 4 and 5 of this text; Gagné, 1985). You will notice that there are common elements across all instructional events and unique features to be selected for each type of desired outcome.

What will become evident as this book progresses is that different instructional media have different capabilities for providing the various events of instruction. For example, teachers are unbeatable for providing learning guidance and feedback; however, videotape can be used to present stimulus situations (e.g., a tour of Florence) that would be difficult for the classroom teacher to present in any other way. It may now be appreciated that the *choice of the delivery system* indicates a general preference for emphasizing certain agents to accomplish instructional events; within such a general preference (such as for individualized, learner-paced modules) specific agents or media can be assigned, event-by-event, objective-by-objective. That is what we mean by developing a strategy for instruction.

The planning of an instructional strategy is an important part of the instructional design process. It is at this point that the designer must be able to combine knowledge of learning and design theory with his experience of learners and objectives. Needless to say, creativity in lesson design will enhance this other knowledge and experience. Perhaps it is this component of creativity that separates the art of instructional design from the science of instructional design. It is clear that the best lesson designs will demonstrate knowledge about the learners, the tasks reflected in the objectives, and the effectiveness of teaching strategies. To achieve this combination, the designer most often functions as part of a team of teachers, subject-matter experts, script writers and producers, and perhaps others.

Stage 7: Instructional Materials

The word *materials* here refers to printed or other media intended to convey events of instruction. In most traditional instructional systems, teachers do not design or develop their own instructional materials. Instead, they are given materials (or they select materials) that they integrate into their lesson plans. In contrast, instructional systems design underscores the selection and development of materials as an important part of the design effort. Teachers can be hard-pressed to arrange instruction when there are no really suitable materials available for part of the planned objectives. Often, they improvise and adapt as best they can. Most often, however, teachers do find suitable materials. The danger is that teachers sometimes choose existing materials for convenience, in effect changing the objectives of the instruction to fit their available materials. In such circumstances, the student may be receiving information or learning skills that are unrelated to instructional goals.

The more well established are the objectives and hence the more precisely determined the content of the materials, the more likely it is that suitable materials will already be on the market. Nevertheless, such materials are more likely to be referenced by content than by objective (to say nothing of their failure to address the events of instruction they provide). It is possible that available materials will be able to provide some of the needed instruction. In this case a module could be designed to take advantage of the existing materials and could be supplemented with other materials to provide for the missing objectives. Materials production is a costly process, and it is desirable to take advantage of existing materials when possible.

A few general principles begin to emerge. First, the more innovative the objectives, the more likely it is that a greater portion of the materials must be developed since they are not likely to be available commercially. Secondly, developing materials for a particular delivery system is almost always more expensive than making a selection from those available. Third, it is possible to save development expenses by selecting available materials and integrating them into a module providing coverage of all the desired objectives of instruction. Fourth, the role of the teacher is affected by the choice of delivery system and the completeness of the materials because the teacher will have to provide whatever missing events may be needed by the learners.

Some new curricula and instructional systems have intentionally been planned from the outset either to develop all new materials or to make as much use as possible of existing materials. The reason in the first instance is probably to ensure that a central concept, method, theme, or body of content is carefully preserved. Since such programs are often recognized as experimental, the added development costs may be justified to preserve purity of the original concept. In the case of a decision to make maximum use of existing materials, cost is likely to be the primary consideration. An example of this latter kind of decision was that of Project PLAN (Flanagan, 1975). The design of that individualized system called for maximum use of available materials so that funds would be available for designing implementation plans, for performance measures to monitor student progress, and for computer costs to save teachers' time in scoring tests and keeping records.

It is beyond the scope of this book to describe how design teams operate to accomplish the various stages of instructional systems design including the development of materials. Carey and Briggs (1977) and Branson and Grow (1987) give a general account of the process, and Weisgerber (1971) gives some of the details for specific systems.

Stage 8: Formative Evaluation

Formative evaluation provides data for revising and improving instructional materials. Dick and Carey (1985) provide detailed procedures for a three-level process of formative evaluation. First, the prototype materials are tried *one-on-one* (one evaluator sitting with one learner) with learners representative of the target audience. This step provides a considerable amount of information about the structure and logistic problems the learners may have with the lessons. The designer can interview the learner or have him "talk through" the thoughts he has while going through the material. It has been estimated that the effectiveness of instructional materials could be improved 50 percent simply through the use of a few one-on-one evaluations. The second level is a small group tryout, in which the materials are given to a group of six to eight students. Here the focus is on how the students use the materials and how much help is requested. This information can be used to make the lesson more self-sufficient. It will also give the designer a better idea of the materials probable effectiveness in a large group, the mean scores of the students being more representative than the scores from the one-on-one student trials. The final step is a field trial in which the instruction, revised on the basis of the one-on-one and small group trials, is given to a whole class.

The purpose of formative evaluation is to revise the instruction so as to make it as effective as possible for the largest number of students. This stage in materials development is probably one of the most frequently overlooked because it comes late in the design process and represents a significant effort in planning and execution. However, the use of systems feedback to correct the system represents the essence of systems philosophy. Instructional design without formative evaluation is incomplete. The feedback loop in Figure 2-1 shows that formative evaluation data may call for the revision or review of products because of information derived from any of the previous stages of design.

Stage 9: Summative Evaluation

Studies of the effectiveness of a system as a whole are called *summative evaluations*, the basic form of which is described more fully in Chapter 16. As the term implies, a summative evaluation is normally conducted after the system has passed through its formative stage—when it is no longer undergoing point-by-point revision. This may occur at the time of the first field test or as much as five years later, when large numbers of students have been taught by the new system. If there is expectation that the system will be widely used in schools or classrooms throughout the country, summative evaluations need to be conducted under an equally varied range of conditions.

A national agency, the Joint Dissemination Review Panel (JDRP), conducts such reviews. The JDRP meets periodically to review evidence of effectiveness of educational products identified as potentially "exemplary" and suitable for dissemination. This is a form of summative evaluation, in which a team of evaluators audits a pilot project to judge evidence of its effectiveness. "The evidence must be shown to be valid and reliable, the effects must be of sufficient magnitude to have educational importance, and it should be possible to reproduce the intervention and its effects at other sites" (Tallmadge, 1977; p. 2). If the project passes the panel's scrutiny, it may qualify for funds to support dessemination from the National Diffusion Network.

EDUCATIONAL SYSTEM DESIGN

Many different models may be used to describe the process of instructional design as applied to total educational systems. Models for the most comprehensive level must include analyses of needs, goals, priorities, resources, and other environmental and social factors affecting the educational system. The model outlined in Table 2-1 lists 14 stages in the design of instruction for total systems of education.

In contrast to the nine-stage model we just described (Figure 2-1), Table 2-1 makes it apparent that additional factors and stages must be dealt with in planning instruction for large curriculum design efforts and for total educational systems. These include the analysis of resources, constraints, alternative delivery systems, teacher preparation, and the installation and diffusion of newly developed instruction.

Resources, Constraints, and Alternative Delivery Systems

Once needs and goals are identified, instructional planners need to consider issues such as: How will students learn the skills implied by the goals? From whom will they learn? Where will they find the resources, materials, or help

TABLE 2-1   Stages in Designing Instructional Systems

System Level
  1. Analysis of Needs, Goals, and Priorities
  2. Analysis of Resources, Constraints, and Alternate Delivery Systems
  3. Determination of Scope and Sequence of Curriculum and Courses; Delivery System Design

Course Level
  4. Determining Course Structure and Sequence
  5. Analysis of Course Objectives

Lesson Level
  6. Definition of Performance Objectives
  7. Preparing Lesson Plans (or Modules)
  8. Developing, Selecting Materials, Media
  9. Assessing Student Performance (Performance Measures)

System Level
  10. Teacher Preparation
  11. Formative Evaluation
  12. Field Testing, Revision
  13. Summative Evaluation
  14. Installation and Diffusion

they need? What resources will it take to teach the goals? Are the resources available? Do we want to spend that much? Can the present system do this? Will instructor training be needed? and, What alternative systems might be used? Once questions such as these are pursued, some alternative delivery systems suggest themselves.

A delivery system includes everything necessary to allow a particular instructional system to operate as it was intended and where it was intended. Thus a system can be designed to fit a particular physical plant or to require a new one. The basic decision about instructional delivery can directly affect the kind of personnel, media, materials, and learning activities that can be carried on to reach the goals. Can any of the resources or constraints be altered? This is a key question at several stages of planning, including this one.

Should the new set of goals appear out of reach of any of the available delivery systems, no further planning is possible until (1) some goals are changed, (2) some resources and constraints are changed, or (3) another delivery system can be conceived. Failure to do this may lead to piecemeal planning with generally unsatisfactory results. Lack of resolution of these issues may lead to various kinds of waste including: (1) equipment and materials sitting unused due to lack of supporting personnel, (2) laboratories not used because supplies were not budgeted for, (3) learning activities disrupted due to bad scheduling, (4) goals not achieved because essential prerequisite learning experiences were not provided.

Often the estimate of resources and constraints calls for the goals to be achieved within a currently existing delivery environment. In the case of schools, this generally means the teacher-led classroom. In industry, it could mean the use of videotaped instruction because the delivery system is already in place. What must be considered is whether the existing delivery system is capable of providing the environment needed for learning the new skills. Further discussion of this point is contained in later chapters.

Teacher Preparation

The term as used here does not refer to the initial education and training of new teachers, but rather to the special training of current teachers in the development and dissemination of new instructional systems. Teachers, as noted earlier, are generally important members of the design team. They assist in all the stages of design and become trainers of other teacher or demonstration teachers. If a new instructional system requires special skills beyond those already possessed by teachers in service, special training must be designed as part of the instructional systems design process to provide those new skills. Special workshops are one common mode for such training, but visits to schools where the system is first operating as a pilot test is an important alternative. The teachers need to perceive that the new system will work in their environment. Teachers are often skeptical of new approaches, and it is time-consuming to switch to new curricula and materials; accordingly, teachers must approach the task with a positive attitude toward the new system. In visits to schools adopting an individualized system of instruction, Briggs and Aronson (1975) discovered that most teachers felt they needed a year of experience beyond their initial training for them to prefer new systems of instruction over their prior practices.

The basic principle we want to stress is that teachers need to be prepared before materials are distributed in order for a new unit of instruction to be adopted. The more input teachers have along the way, the more likely new materials will fit into the existing system, and the more likely they will be adopted (Burkman, 1986).

Installation and Diffusion

This stage of instructional systems development was mentioned in some of the preceding discussion. After an acceptable degree of merit is shown in one or more summative evaluations, the new system (course, or curriculum) is ready for widespread adoption and regular use.

In the course of operational installation, a number of practical matters receive final attention or adjustment. For example, materials may have to be stored differently in some schools than in others, owing to differences in building design and available space. Time schedules for a new set of instruction may require modifications to fit within existing patterns for a particular school. There are inevitable logistical problems: the duplication and distribution of expendable materials, for example. Even more important, according to Heinich (1984), is the need to be aware of the nature of the system into which the innovation is to be introduced. New technology is often perceived as a threat to the existing system and is often blocked by those who should use it.

A frequent problem is securing enough adoptions of a new instructional system to amortize the costs of development, marketing, and maintenance (an often overlooked cost). Techniques relevant to the diffusion of educational systems and innovations have generated a great many research studies. It is beyond the scope of this book to discuss the merits of relevant techniques. As a follow-up to the JDRP, the U.S. Office of Education created the National Diffusion Network (NDN) in 1974, for the purpose of providing educators with information about exemplary programs. The NDN supports demonstration projects that provide training, materials, and technical assistance to those who adopt their programs. The NDN also has "facilitators" in each state, usually within the state's Department of Education; these are persons who help identify suitable NDN programs. NDN estimates that it presently supports over 400 programs in more than 15 thousand public schools. As a result of NDN's efforts, more than 50 thousand teachers and administrators have received in-service training, which may in turn have affected over 1.5 million students (National Diffusion Network, 1986).

If diffusion is one of the goals of a development project, it must be considered early in the design process. Collaboration with a publishing company is one approach, but the operating procedures of the company may put constraints on what the final product or delivery system can be. For example, a chosen delivery system may be unacceptable to a publisher, and the design team may have to accept a less desirable delivery system in order to achieve the adoption goals. This may require rethinking the instructional goals, needs, or system design objectives.

SUMMARY

The term *instructional systems design* was defined along with a general description of the design process. Stages of design are often presented as a flow diagram or model to be followed in the design of instructional materials. The instructional systems approach is a process of planning and developing instruction that makes use of research and learning theory and employs empirical testing as a means for the improvement of instruction.

The nine-stage model of design described in this chapter represents one of the possible ways of conceptualizing the process. All design models focus attention on the three "anchor points" of instruction: performance objectives, materials, and evaluation instruments. The purpose of lesson planning, as we see it, is to ensure that the necessary instructional events are provided to the learner. Key steps in the planning process include: (1) classifying the lesson objectives by learning type, (2) listing the needed instructional events, (3) choosing a medium of instruction capable of providing those events, and (4) incorporating appropriate conditions of learning into the prescriptions indicating how each event will be accomplished by the lesson. Some events may be executed by the learner, some by the materials, and some by the teacher.

The design process is iterative, and many of the earlier stages have to be revisited and the products reworked based on findings or new information uncovered during later stages. There is, then, much working back and forth as the total design work progresses. The entire design approach outlined here is considered to be internally consistent and in agreement with research findings on how learning takes place. The resulting designs are amenable to both formative and summative evaluations. Each design objective is stated in testable form so that the success of the design can be evaluated.

More comprehensive levels of systematic instructional design are encountered in efforts to develop courses or curricula for entire educational systems. At such levels, as many as fourteen stages of analysis and development may be involved. Procedures of design at this level usually include considerations of resources and constraints, requirements for teacher education, and techniques for installation and diffusion. Evaluation of an entire system involves assessing the effectiveness and viability of components of the system as a whole.

References

Branson, R. K. (1977). Military and industrial training. In L. J. Briggs (Ed.), *Instructional design: Principles and applications*. Englewood Cliffs, NJ: Educational Technology Publications.

Branson, R. K., & Grow, G. (1987). Instructional systems development. In R. M. Gagné (Ed.), *Instructional technology: Foundations*. Hillsdale, NJ: Erlbaum.

Briggs, L. J. (Ed.), (1977). *Instructional design: Principles and applications*. Englewood Cliffs, NJ: Educational Technology Publications.

Briggs, L. J. & Aronson, D. (1975). *An interpretive study of individualized instruction in the schools: Procedures, problems and prospects*. (Final Report, National Institute of Education, Grant No. NIE–G–740065). Tallahassee, FL: Florida State University.

Briggs, L. J. & Wager, W. W. (1981). *Handbook of procedures for the design of instruction*. Englewood Cliffs, NJ: Educational Technology Publications.

Burkman, E. (1986). Factors affecting utilization. In R. M. Gagné, (Ed.), *Instructional technology: Foundations*, Hillsdale, NJ: Erlbaum.

Burton, J. K. & Merrill, P. F. (1977). Needs assessment: Goals, needs, and priorities. In L. J. Briggs (Ed.), *Instructional design: Principles and applications*. Englewood Cliffs, NJ: Educational Technology Publications.

Carey, J. & Briggs, L. J. (1977). Teams as designers. In L. J. Briggs (Ed.), *Instructional design: Principles and applications*. Englewood Cliffs, NJ: Educational Technology Publications.

Dick, W. & Carey, L. (1985). *The systematic design of instruction* (2nd ed.) Glenview, IL: Scott, Foresman.

Flanagan, J. C., Project PLAN. In H. Talmage (Ed.). (1975), *Systems of individualized education*. Berkeley, CA: McCutchan.

Gagné, R. M. (1977). Analysis of objectives. In L. J. Briggs (Ed.), *Instructional design: Principles and applications*. Englewood Cliffs, NJ: Educational Technology Publications.

Gagné, R. M. (1985). *The conditions of learning* (4th ed.). New York: Holt, Rinehart and Winston.

Heinich, R. (1984). The proper study of instructional technology. *Educational Communications and Technology Journal, 32*(2), 67–87.

Kaufman, R. A. (1976). *Needs assessment: What it is and how to do it*. San Diego, CA.: University Consortium on Instructional Development and Technology.

National Diffusion Network. (1986). *Educational programs that work* (ed. 12). Longmont, CO: Sopris West.

Tallmadge, G. K. (1977). *The joint dissemination review panel IDEABOOK*. Washington, D.C.: U.S. Office of Education.

Weisgerber, R. A. (1971). *Developmental efforts in individualized instruction*. Itasca, IL: Peacock.

PART TWO

BASIC PROCESSES IN LEARNING AND INSTRUCTION

3 THE OUTCOMES OF INSTRUCTION

The best way to design instruction is to work backwards from its expected outcomes. Some ways of working backwards and the implications of these procedures for the content of instruction are described in this chapter. These procedures begin with the identification of human capabilities to be established by instruction. The instructional outcomes, introduced and defined here in terms of five broad categories, run throughout the book as the framework on which the design of instruction is built.

INSTRUCTION AND EDUCATIONAL GOALS

The basic reason for designing instruction is to make possible the attainment of a set of educational goals. The society in which we live has certain functions to perform in serving the needs of its people. Many of these functions—in fact, most of them—require human activities that must be learned. Accordingly, one of the functions of a society is to ensure that such learning takes place. Every society, in one way or another, makes provision for the education of people in order that the variety of functions necessary for its survival can be carried out. *Educational goals* are those human activities that contribute to the functioning of a society (including the functioning of an individual *in* the society) and that can be acquired through learning.

Naturally, in societies whose organization is simple—often called "primitive" societies—the goals of education and the means used to reach them are fairly easy to describe and understand. In a primitive society whose economy revolves around hunting animals, for example, the most prominent educational goals center upon the activities of hunting. The son of a hunter is educated in these activities by his father or, perhaps, by other hunters of the village to which he belongs. Fundamentally, educational goals have the same kind of origin in a modern complex society. Obviously, though, as societies become more complex, so must educational goals.

Every so often in our own society, we hold conferences, appoint committees, or establish commissions to study educational goals. One of the most famous of these bodies formulated a set of goals called the "Cardinal Principles of Secondary Education" (Commission on the Reorganization of Secondary Education, 1918). The key statement of this document was (p. 9):

> Education in a democracy, both within and without the school, should develop in each individual the knowledge, interests, ideals, habits, and powers whereby he will find his place and use that place to shape both himself and society toward ever nobler ends.

The composition of the "knowledge, interests, ideals, habits, and powers" was considered by this commission to fall into the seven areas of (1) health, (2) command of basic skills, (3) worthy home-membership, (4) pursuing a vocation, (5) citizenship, (6) worthy use of leisure, and (7) ethical character. You might suppose that these guidelines would lend themselves to more specific objectives for education. This sort of analysis, however, is an overwhelming task, so great, in fact, that it has never really been attempted for our society. Instead, we depend upon a number of different simplifications to specify educational goals in detail. These simplifying approaches condense information in several stages, therefore losing some information along the way.

Thus it has come about that we tend to structure education in terms of various kinds of "subject matters" that are actually gross simplifications of educational goals, rather than activities reflecting the actual functions of human beings in society. It is as though the activity of shooting a bear in a primitive society were to be transformed into a subject called marksmanship. We represent an educational goal with the subject-matter name of English rather than with the many different human activities that are performed with language. The formulation of educational goals within various subject matter fields has been carried out by the National Assessment of Educational Progress (Womer, 1970). Goals derived from analyses of contemporary educational needs are discussed in books by Boyer (1983) and Goodlad (1984).

Goals as Educational Outcomes

The reflection of societal needs in educational goals is typically expressed in statements describing categories of *human activity*. A goal is preferably stated, not as "health," but as "performing those activities that will maintain health." The goal, or goals, are most inadequately conveyed by the term "citizenship'; they are better reflected in a statement such as "carries out the activities of a citizen in a democratic society."

Although it has not yet been done, it would be helpful for educational scholars to define the array of human *capabilities* that would make possible the kinds of activities expressed in educational goals. It is these capabilities that represent the proximate goals of instruction. To carry out the activities required for maintaining health, the individual must possess certain kinds of capabilities (knowledge, skills, attitudes). In most cases these are learned through deliberately planned instruction. Similarly, to perform the various activities appropriate to being a citizen, the individual must have learned a variety of capabilities through instruction.

Educational goals are statements of the outcomes of education. They refer particularly to those activities made possible by learning, which in turn is often brought about by deliberately planned instruction. The rationale in our society is not different from that of a primitive society. In the latter, for example, the educational goal of becoming a hunter is achieved by a customary regime of instruction in the various component human capabilities (locating prey, stalking, shooting, etc.) that make possible the total activity of hunting. The difference, however, is an important one. In the more complex society, the *capabilities* required for one activity may be shared by a number of others. Thus, the human capability of "performing arithmetic operations" serves not only one educational goal (such as making a family budget) but several, including changing money and making scientific measurements.

To design instruction, one must seek a means of identifying the *human capabilities* that lead to the outcomes called educational goals. If these goals were uncomplicated, as in a primitive society, defining these human capabilities might be equally simple. But such is not the case in a highly differentiated and specialized society. Instruction cannot be adequately planned separately for each educational goal necessary to a modern society. One must seek, instead, to identify the human capabilities that contribute to a number of different goals. A capability such as reading comprehension, for example, obviously serves several purposes. The present chapter is intended to serve as an introduction to the concept of human capabilities.

Courses and Their Objectives

The planning of instruction is often carried out for a single *course*, rather than for larger units of a total curriculum. There is no necessary fixed length of a course or no fixed specification of "what is to be covered." A number of factors may influence the choice of duration or amount of content. Often the length of time available in a semester or year is the primary determining factor. In any case, a course is usually defined rather arbitrarily by the designation of some topics understood within the local environment of the school. A course may take on a general title such as "American History," "Beginning French," "English 1," and so on.

The ambiguity in meaning of courses with such titles is evident. Is "American History" in grade 6 the same as or different from the course of the same title in grade 12? Is "English 1" concerned with composition, literature, or both? These are by no means idle questions because they represent sources of difficulty for many students in many places, particularly when they are planning programs of study. It is not entirely uncommon, for example, for a student to choose a course such as "First-year French," only to find that he should have elected "Beginning French."

Ambiguity in the meaning of courses with title or topic designations can readily be avoided when courses are described in terms of the *objectives* (Mager, 1975; Popham and Baker, 1970). Examples of objectives in many subject areas are described by Bloom, Hastings, and Madaus (1971). Thus, if "English 1" has the objective of having the student be able to "compose a unified composition on any assigned single topic, in acceptable printed English, within an hour" it is perfectly clear to everyone what a portion of the course is all about. It will not help the student, in any direct fashion, to "identify imagery in modern poetry," nor to "analyze the conflicts in works of fiction." It will, however, if successful, teach him the basic craft of writing a composition. Similarly, if an objective of "Beginning French" is that the student be able to "conjugate irregular verbs," this is obviously fairly clear. It will not readily be confused with an objective that makes it possible for the student to "write French sentences from dictation."

As usually planned, courses often have several objectives, not just one. A course in social studies may have the intention of providing the student with several capabilities: "describing the context of (specified) historical events," "evaluating the sources of written history," and "showing a positive liking for the study of history." A course in science may wish to establish in the student the ability "to formulate and test hypotheses," to "engage in scientific problem solving," and also to "value the activities of scientists." Each of these kinds of objectives within a single course may be considered equally worthwhile. They may also be differentially valued by different teachers. The main point to be noted about them at this juncture, however, is that they are different. The most important difference among them is that each requires a different plan for its achievement. Instruction must be differentially designed to ensure that each objective is attainable by students within the context of a course.

Are there a great many specific objectives for which individual instructional planning must be done, or can this task be reduced in some manner? To answer this question, one has to think of what common categories there may be among all the different subject matter to be learned. For example, learning to describe the size and composition of the Washington Monument in some sense is not inherently different from learning to describe something else, such as the events of the siege of Vicksburg. Applying the rules of trigonometry to triangles is a task comparable to applying the rules of grammar to sentences. Instructional planning can be vastly simplified by assigning learning objectives to *five major categories of human capabilities* (Gagné, 1985). Such categories can be formed because each leads to a different class of human performance. (As will be seen later, each category also requires a different set of instructional conditions for effective learning.) Within each of these five categories, regardless of the subject matter of instruction, the same qualities of performance apply. Of course, there may be subcategories within each of the five categories. In fact, there *are* some subcategories that are useful for instructional planning, as the next chapter will show. But for the moment, in taking a fairly general look at instructional planning from the standpoint of courses, five categories provide the comprehensive view.

FIVE CATEGORIES OF LEARNING OUTCOMES

What are the categories of objectives expected as learning outcomes? A brief definition and description of each is given in the following paragraphs. The performances that may be observed as learning outcomes are considered to be made possible by internally stored states of the human learner, called capabilities. A fuller description of the usefulness of these capabilities will be given in a later section; conditions necessary for their learning are described in following chapters.

Intellectual Skills

Intellectual skills enable individuals to interact with their environment in terms of symbols or conceptualizations. Their learning begins in the early grades with the three Rs, and progresses to whatever level is compatible with the individual's interests and intellectual ability. They make up the most basic and the most pervasive structure of formal education. They range from elementary language skills such as composing a sentence to the advanced technical skills of science, engineering, and other disciplines. Examples of intellectual skills of the latter sort would be finding the stresses in a bridge or predicting the effects of currency devaluation. The five kinds of capabilities that are outcomes of learning are listed in Table 3-1 along with the examples of the intellectual skills of identifying a diagonal, and demonstrating the rule of using pronouns in the objective case following a preposition.

TABLE 3-1. Five Kinds of Learned Capabilities

Capability	Examples of Performance
Intellectual Skill	Identifying the diagonal of a rectangle. Demonstrating use of objective case of pronoun following a preposition.
Cognitive Strategy	Using an image link to learn a foreign equivalent to an English word. Rearranging a verbally stated problem by working backwards.
Verbal Information	Stating the provisions of the Fourth Amendment to the U.S. Constitution. Recounting the events of an automobile accident.
Motor Skill	Planing the edge of a board. Printing the letter *E*.
Attitude	Choosing to read science fiction. Choosing running as a regular form of exercise.

Learning an intellectual skill means learning *how to do* something of an intellectual sort. Generally, what is learned is called *procedural knowledge* (Anderson, 1985). Such learning contrasts with learning *that* something exists or has certain properties. The latter is *verbal information*. Learning how to identify a sonnet by its rhyme pattern is an intellectual skill, whereas learning what the sonnet says is an instance of verbal information. A learner may, of course, learn both, and often does, but it is perfectly possible for a person to learn how to do the first (identify a sonnet) without being able to do the second (state what a particular sonnet says). Likewise, as teachers know well, it is possible for a student to learn the second without being able to do the first. For these reasons, it is important to maintain this distinction between knowing *how* and knowing *that*, even while recognizing that a particular unit of instruction may involve both as expected learning outcomes.

Another example of an intellectual skill may be given here. A student of the English language learns at some point in his studies what a metaphor is. More specifically, if his instruction is adequate, he learns to use a metaphor. (In the next chapter, we identify this particular subcategory of intellectual skill as a *rule*.) In other words, it may be said that the student has learned to use a rule to show what a metaphor is; or that he has learned to apply a rule. This skill then has the function of becoming a component of further learning. That is to say, the skill of using a metaphor now may contribute to the learning of more complex intellectual skills, such as writing illustrative sentences, describing scenes and events, and composing essays.

If one wishes to know whether the student has learned this intellectual skill, one must observe a category of *performance*. Usually this is done by asking the student to "show what a metaphor is" in one or more specific instances. In other words, observations might be made to determine whether the student performed adequately when asked to use a metaphor to describe (1) the cat's movements, (2) a cloudy day, and perhaps (3) the moon's surface.

Cognitive Strategies

These are special and very important kinds of skills. They are the capabilities that govern the individual's own learning, remembering, and thinking behavior. For example, they control his behavior when he is reading with the intent to learn; and the internal methods he uses to "get to the heart of a problem." The phrase cognitive strategy is usually attributed to Bruner (Bruner, Goodnow, and Austin, 1956). Rothkopf (1971) has named them "mathemagenic behaviors"; Skinner (1968) "self-management behaviors." One expects that such skills will improve over a relatively long period of time as the individual engages in more and more studying, learning, and thinking. An example shown in Table 3-1 is the cognitive strategy of using images as links to connect words in the learning of foreign language vocabulary (Atkinson, 1975).

Provided it has previously been learned, a cognitive strategy may be selected by a learner as a *mode* of solving a novel problem. Often, for example, newly encountered problems can be efficiently approached by working backwards in stages beginning with the goal to be achieved by a solution. This "working backwards" approach is an example of a cognitive strategy. Intellectual skills (such as basic arithmetic operations) frequently have to be recalled by the learner and brought to bear upon a problem. But although these skills are essential, they are not sufficient. A *mode* of seeking a solution must also be used by the learner, a *cognitive strategy* that he has practiced in the past, perhaps many times in a variety of situations.

The most commonly occurring cognitive strategies are *domain-specific*. For example, there are strategies for retaining information from reading; for aiding the solution of word problems in arithmetic; for helping the composition of effective sentences; and many others that focus on particular domains of learning tasks. However, some cognitive strategies are more general, like the process called *inference* or *induction*. Suppose that a student has become acquainted with magnetic attraction in a bar magnet—noting that a force is exerted by each pole of the magnet on certain kinds of metal objects. Now the student is given some iron filings to sprinkle on a piece of paper placed over the magnet. When the paper is tapped, the filings exhibit "lines of force" around each pole of the magnet. The student proceeds to verify this observation in other situations, perhaps using other magnets and other kinds of metal objects. These observations, together with other knowledge, may lead the induction of the idea of a magnetic field of force surrounding each pole of the magnet. It is important to note in this example that the student has not been told of the magnetic field beforehand, nor given instruction in "how to induce." But this kind of mental operation is carried out.

Learning a cognitive strategy such as induction, however, is apparently not done on a single occasion. Instead, this kind of capability develops over fairly long periods of time. Presumably, the learner must have a number of experiences with induction in widely different situations for the strategy to become dependably useful.

When a learner becomes capable of induction, this strategy may be used in a great variety of other situations. Provided other requisite intellectual skills and information have been learned, an induction strategy may be used to arrive at an explanation of what makes smoke rise in the air, why pebbles in a stream are rounded and smooth, or what intention a writer had in composing an editorial essay. In other words, the cognitive strategy of induction may be put to use in a great many situations of thinking and learning—situations that are enormously varied in their describable properties. In fact, the performances that the learner is able to exhibit in these situations may be seen to resemble each other only in the respect that they involve induction. And this, of course, is the basic reason for believing that such a cognitive strategy exists—it is by an act of induction that one arrives at the presence of the cognitive strategy of induction in other people.

Verbal Information

This is the kind of knowledge we are able to *state*. It is *knowing that*, or *declarative knowledge*. All of us have learned a great deal of verbal information or verbal knowledge. We have readily available in our memories many commonly used items of information such as the names of months, days of the week, letters, numerals, towns, cities, states, countries, and so on. We also have a great store of more highly organized information, such as many events of U.S. history, the forms of government, the major achievements of science and technology, and the components of the economy. The verbal information we learn in school is in part "for the course only" and in part the kind of knowledge we are expected to be able to recall readily as adults.

The learner usually acquires a great deal of information from formal instruction. Much is also learned in an incidental fashion. Such information is stored in the learner's memory, but it is not necessarily "memorized," in the sense that it can be repeated verbatim. Something like the *gist* of paragraph-long passages is stored in memory and recalled in that form when the occasion demands. The example given in Table 3-1 refers to the performance of telling what the Fourth Amendment says. A second example is a learner's description of a set of events, such as might have taken place in an automobile accident. Students of science learn much verbal information, just as students do in other fields of study. They learn the properties of materials, objects, and living things, for example. A large number of "science facts" may not constitute a defensible primary goal of science instruction. Nevertheless, the learning of such facts is an essential part of the learning of science. For example, a student may learn that "the boiling point of water is 100°C." One major function of such information is to provide the learner with directions for how to proceed in further learning. Thus, in learning about the change of state of materials from liquid to gaseous form, the learner may be acquiring an intellectual skill (that is, a *rule*) that relates atmospheric pressure to vaporization. In working with this relationship, a student may be asked to apply the rule to a situation that describes the temperature of boiling water at an altitude of nine thousand feet. At this juncture the *information* given in the example must be recalled in order to proceed with the application of the rule. One may be inclined to say this information is not particularly important—rather, the learning of the *intellectual skill* is the important thing. There is no disagreement about this point. However, the *information is essential* to these events. The learner must have such information available to learn a particular application.

Information may also be of importance for the transfer of learning from one situation to another. For example, a student of government may hit upon the idea that the persistence of bureaucracy bears some resemblance to the growth of an abscess in the human body. If he or she has some information about abscesses, such an analogy may make it possible to think of causal relationships pertaining to bureaucracies that would not otherwise be possible. A variety of cognitive strategies and intellectual skills may now be brought to bear on this problem by the student, and new knowledge thereby generated. The initial transfer in such an instance is made possible by an "association of ideas," in other words, by the possession and use of certain classes of information.

Finding out whether students have learned some particular facts or some particular organized items of information is a matter of observing whether they can communicate them. The simplest way to do this, of course, is to ask for a statement of the information either orally or in writing. This is the basic method commonly employed by a teacher to assess what information has been learned. In the early grades, assessing the communications children can make may require the use of simple oral questions. Pictures and objects that the child can point to and manipulate may also be employed.

Motor Skills

Another kind of capability we expect human beings to learn is a motor skill (Fitts and Posner, 1967; Singer, 1980). The individual learns to skate, to ride a bicycle, to steer an automobile, to use a can opener, to jump rope. There are also motor skills to be learned as part of formal school instruction, such as printing letters (Table 3-1), drawing a straight line, or aligning a pointer on a dial face. Despite the fact that school instruction is so largely concerned with intellectual functions, we do not expect a well-educated adult to be lacking in certain motor skills (such as writing) that may be used every day. A motor skill is one of the most obvious kinds of human capabilities. Children learn a motor skill for each printed letter they make with a pencil on paper. The function of the skill, as a capability, is simply to make possible the motor performance. Of course, these motor performances may themselves enter into further learning. For example, students employ the skill of printing letters when they are learning to make (and print) words and sentences. The acquisition of a motor skill can be reasonably inferred when the students can perform the act in a variety of contexts. Thus, if youngsters have acquired the skill of printing the letter *E* they should be able to perform this motor act with a pen, a pencil, or a crayon, on any flat surface, constructing letters with a range of sizes. Obviously one would not want to conclude that the skill has been learned from a single instance of an *E* printed with pencil on a particular piece of paper. But several *E*s, in several contexts, observably distinct from *F*s or *H*s, provide convincing evidence that this kind of capability has been learned.

Attitudes

Turning now to what is often called the *affective domain* (Krath-Bloom, and Masia, 1964), we identify a class of learned capabilities called *attitudes*. All of us possess attitudes of many sorts towards various things, persons, and situations. The effect of an attitude is to amplify an individual's positive or negative reaction toward some person, thing, or situation. The strength of people's attitudes toward some item may be indicated by the frequency with which they *choose* that item in a variety of circumstances. Thus, an individual with a strong attitude toward helping other people will offer help in many situations, whereas a person with a weaker attitude of this sort will tend to restrict offers of help to fewer situations. The schools are often expected to establish socially approved attitudes such as respect for other people, cooperativeness, personal responsibility, as well as positive attitudes toward knowledge and learning, and an attitude of self-efficacy. A student learns to have preferences for various kinds of activities, preferring certain people to others, showing an interest in certain events rather than others. One infers from a set of such observations that the student has *attitudes* toward objects, persons, or events that influence the choice of courses of action towards them. Naturally, many such attitudes are acquired outside of the school, and there are many that schools cannot appropriately consider relevant to their instructional function. As one possibility, though, school instruction may have the objective of establishing positive attitudes toward subjects being studied (e.g., Mager, 1968). Often, too, school learning is successful in modifying attitudes toward activities that provide esthetic enjoyment. One of the examples of Table 3-1 is a positive attitude toward reading a particular kind of fiction.

Considered as a human capability, an attitude is a persisting state that modifies the individual's choices of action. A positive attitude toward listening to music makes the student *tend* to choose such activity over others, when such choices are possible. Of course, this does not mean he or she will always be listening to music, under all circumstances. Rather, it means that when there is an opportunity for leisure (as opposed to other pressing concerns) the probability of a choice to listen to music is noticeably high. If one were able to observe the student over an extended period of time, one would be able to note that the choice of this activity was relatively frequent. From such a set of observations, it could be concluded that the student had a positive attitude toward listening to music.

In practice, of course, making such a set of observations about a single student, not to mention a class of students, would be an exceedingly time-consuming and therefore expensive undertaking. As a result, inferences about the possession of attitudes are usually made on the basis of "self-reports." These may be obtained by means of questionnaires that ask students what choices of action they would make (or in some cases, *did* make) in a variety of situations. There are, of course, technical problems in the use of self-reports for attitude assessment. Since their intentions are rather obvious, students can readily make self-reports of choices that do not reflect reality. However, when proper precautions are taken, such reports make possible the inference that a particular attitude has been learned or modified in a particular direction.

Thus, the performance that is affected by an attitude is the *choice of a course of personal action*. The tendency to make such a choice, towards a particular class of objects, persons, or events, may be stronger in one student than in another. A change in an attitude would be revealed as a change in the probability of choosing a particular course of action on the part of the student. Continuing the previous example, over a period of time or as a result of instruction, the probability of choosing to listen to music may be altered. The observation of such change would give rise to the inference that the student's attitude toward listening to music had changed, that is, had become "stronger" in the positive direction.

Human Capabilities as Course Goals

A single course of instruction usually has objectives that fit into several categories of human capability. The major categories, which cut across the "content" of courses, are the five we have described. From the standpoint of the expected outcomes of instruction, the major reason for distinguishing these five categories is that they *make possible different kinds of human performance*.

For example, a course in elementary science may foresee as general objectives such learning outcomes as (1) solving problems of velocity, time, and acceleration; (2) designing an experiment to provide a scientific test of a stated hypothesis; or (3) valuing the activities of science. Number one obviously names *intellectual skills* and therefore implies some performances involving intellectual operations the student can show he can do. Number two pertains to the use of *cognitive strategies* since it implies that the student will need to exhibit this complex performance in a novel situation, where little guidance is provided in the selection and use of rules and concepts he has previously learned. Number three has to do with an *attitude*, or possibly with a set of attitudes, that will be exhibited in behavior as choices of actions directed toward science activities.

The human capabilities distinguished in these five categories also differ from each other in another highly important way. They each require *a different set of learning conditions* for their efficient learning. The conditions necessary for learning these capabilities efficiently, and the distinctions among these conditions, constitute the subjects of the next two chapters. There, we give an account of the conditions of learning that apply to the acquisition of each of these kinds of human capability, beginning with intellectual skills and cognitive strategies and following with the remaining three categories.

DESIGNING INSTRUCTION USING HUMAN CAPABILITIES

The point of view presented in this chapter is that instruction should always be designed to meet accepted educational goals. When goals are matched with societal needs, the ideal condition exists for the planning of a total program of education. Were such an undertaking to be attempted, the result would be, as a first step, a list of human activities, each of which would have associated with it an estimate of its importance in meeting the needs of the society.

When human activities derived from societal needs are in turn analyzed, they yield a set of *human capabilities*. These are descriptions of what human adults in a particular society ought to *know*, and particularly what they ought to *know how to do*. Such a set of capabilities would probably not bear a close resemblance to the traditional subject matter categories of the school curriculum. There would, of course, be a relationship between human capabilities and the subjects of the curriculum, but it would probably not be a simple correspondence.

Most instructional design, as currently carried out, centers upon *course* planning and design. We shall use such a framework in this book. However, we shall continue to maintain an orientation toward the *goals* of instruction. Learning outcomes cannot always be well identified, it appears, by the topical titles of courses. They *can* be identified as the varieties of learned human capabilities that make possible different types of human performances. Accordingly, the present chapter has provided an introduction to the five major categories of capabilities, which will serve throughout the book as the basic framework of instructional design.

If the instructional designer thinks "These five categories are all well and good, but all I'm *really* interested in is producing creative thinkers," he is fooling himself. With the exception of motor skills, *all* of these categories are likely to be involved in the planning of any course. One cannot have a course without information, and one cannot have a course that doesn't affect attitudes to some degree. And most importantly, one cannot have a course without intellectual skills.

There are a couple of reasons why intellectual skills play a central role in designing the structure of a course of *study*. First, they are the kinds of capabilities that determine what the student can *do*, and thus are intimately bound up with the description of a course in terms of its learning outcomes. A second reason is that intellectual skills have a *cumulative* nature—they build upon each other in a predictable manner. Accordingly, they provide the most useful model for the sequencing of course structure. In the next chapter, we begin to look more closely at intellectual skills—what kinds are there, how can they be learned, and how does one know when they are learned?

SUMMARY

This chapter has shown that the defining of goals for education is a complex problem. In part, this is because so much is expected of education. Some persons would like education to emphasize the importance of understanding the history of mankind; some would like it to perpetuate the present culture or present academic disciplines; some would stress the need to help children and young adults adjust to a rapidly changing society; and others would hope that education could prepare students to become agents improving themselves and the society in which they live.

One source of complexity in defining educational goals arises from the need to translate goals from the very general to the increasingly specific. Many layers of such goals would be needed to be sure that each topic in the curriculum actually moves the learner a step closer to the distant goal. Probably this mapping has never been done completely for any curriculum. Thus there tend to be large gaps from general goals to the specific objectives for courses in the curriculum. A major problem then remains—the need to define course objectives in the absence of an entire network of connections between the most general goals and the specific course objectives.

Despite the involved nature of this problem, means are available for classifying course objectives into categories, that then make it possible to examine the scope of types of *human capabilities* the course is intended to develop. One purpose of such taxonomies (sets of performance categories) is to evaluate the objectives themselves in their entirety. The taxonomy presented in this chapter contains the following categories of learned capabilities:

1. Intellectual skills
2. Cognitive strategies
3. Verbal information

4. Motor skills
5. Attitudes

The usefulness of learning each of these types of capabilities has been discussed and will be treated in greater detail in later chapters.

Uses of such a taxonomy, in addition to the evaluation of the variety of capabilities a course is intended to produce in the learner, include the following:

1. The taxonomy can help to group specific objectives of a similar nature together, and thus reduce the work needed to design a total instructional strategy.
2. The groupings of objectives can aid in determining the sequence of segments of a course of study.
3. The grouping of objectives into types of capabilities can then be utilized to plan the internal and external conditions of learning estimated to be required for successful learning.

Each performance objective of a course defines a unique performance expected as an outcome of the instruction. By grouping objectives into the five categories of capabilities which have been described, one also can assess the adequacy of coverage in each category, while capitalizing upon the fact that the conditions of learning are the same for each objective within that category. Identification of the conditions of learning for each type of human capability is the main topic of the next two chapters.

REFERENCES

Anderson, J. R. (1985). *Cognitive psychology and its implications* (2nd ed.). San Francisco: Freeman.

Atkinson, R. C. (1975). Mnemotechnics in second language learning. *American Psychologist, 30,* 821–828.

Bloom, B. S., Hastings, J. T., & Madaus, G. F. (1971). *Handbook on formative and summative evaluation of student learning.* New York: McGraw-Hill.

Boyer, E. L. (1983). *High school.* New York: Harper & Row.

Bruner, J. S., Goodnow, J. J., & Austin, G. A. (1956). *A study of thinking.* New York: Wiley.

Commission on the Reorganization of Secondary Education. (1918). *Cardinal principles of secondary education.* Washington, D.C.: Department of the Interior, Bureau of Education.

Fitts, P. M., & Posner, M. I. (1967). *Human performance.* Belmont, CA: Brooks/Cole.

Gagné, R. M. (1985). *The conditions of learning* (4th ed.). New York: Holt, Rinehart and Winston.

Goodlad, J. I. (1984). *A place called school.* New York: McGraw-Hill.

Krathwohl, D. R., Bloom, B. S., & Masia, B. B. (1964). *Taxonomy of educational objectives. Handbook II: Affective domain.* New York: McKay.

Mager, R. F. (1968). *Developing attitude toward learning.* Belmont, CA: Fearon.

Mager, R. F. (1975). *Preparing objectives for instruction* (2nd ed.). Belmont, CA: Fearon.

Popham, W. J., & Baker, E. L. (1970). *Establishing instructional goals.* Englewood Cliffs, NJ: Prentice-Hall.

Rothkopf, E. Z. (1971). Experiments on mathemagenic behavior and the technology of written instruction. In E. Z. Rothkopf & P. E. Johnson (Eds.), *Verbal learning research and the technology of written instruction.* New York: Teachers College.

Singer, R. N. (1980). *Motor learning and human performance* (3rd ed.). New York: Macmillan.

Skinner, B. F. (1968). *The technology of teaching.* New York: Appleton.

Womer, F. G. (1970). *What is national assessment?* Denver: Education Commission of the States.

4 VARIETIES OF LEARNING: INTELLECTUAL SKILLS and STRATEGIES

When one begins to think about the application of learning principles to instruction, there is no better guide than to ask the question, *what* is to be learned? We have seen that the answer to this question may in any given instance fall into one of the general classes (1) intellectual skills, (2) cognitive strategies, (3) information, (4) motor skills, or (5) attitudes. In this chapter, we intend to consider the conditions affecting the learning of *intellectual skills*, which are of central importance to school learning and which in addition provide the best structural model for instructional design. It is a reasonable step to proceed then to a consideration of *cognitive strategies*, which are a special kind of intellectual skill deserving of a separate categorization. In the following chapter, we will consider the learning requirements for the remaining three classes of human capabilities.

An intellectual skill makes it possible for an individual to respond to his environment through symbols. Language, numbers, and other kinds of symbols represent the actual objects of the person's environment. Words "stand for" objects. They also represent relations among objects, such as *above, behind, within*. Numbers represent the quantity of things in the environment, and various symbols are used to represent relations among these quantities (+, =, and so forth). Other kinds of symbols, such as lines, arrows, and circles, are commonly used to represent spatial relations. Individuals communicate aspects of their experience to others by using such symbols. Symbol using is one of the major ways people remember and think about the world in which they live. We need to provide here an expanded description of these intellectual skills. What kinds of intellectual skills may be learned, and how are they learned?

TYPES OF INTELLECTUAL SKILLS

The intellectual skills learned by the individual during school years are many, surely numbering in the thousands. One can appreciate this fact by thinking of a single domain—language skills. Even topics of instruction such as oral reading, expressive reading, sentence composition, paragraph construction, conversing, and persuasive speaking contain scores of specific intellectual skills that must be learned. This is also true of skills of number and quantification within the various fields of mathematics. Many skills of spatial and temporal patterning form a part of such subjects as geometry and physics. In dealing with intellectual skills, one must be prepared to look at the "fine-grained" structure of human intellectual functioning.

In whatever domain of subject matter they occur, intellectual skills can be categorized by *complexity*. This means the intricacy of the mental process that may be inferred to account for the human performance. For example, suppose that a learner is shown two novel and distinctive-looking objects, and told to learn how to tell them apart when they are brought back at a later time. The kind of mental processing required is not very complex. One can infer that what has been learned in this situation and can later be recalled is a "discrimination."

Quite a different level of complexity is indicated by the following example: Following instruction the learner is able to comprehend adjectives in the German language that he has never before encountered, constructed by adding the suffix *lich* (as with *Gemüt–gemütlich*). This kind of performance is often referred to as *rule-governed* because the kind of mental processing it requires is "applying a rule." It is not necessary for the learner to state the rule, or even for him to be able to state it. He is, however, performing in a way that implies he must have learned an internal capability that makes his behavior regular or rule-governed. What he has learned is called a *rule*. Obviously, such a process is more complex than the discrimination referred to in the previous paragraph.

Different levels of complexity of mental processing, then, make possible the classification of intellectual skills. Such categories cut across and are independent of types of subject matter. How many levels of complexity of intellectual processing can be distinguished or need to be? For most instructional purposes, the useful distinctions among intellectual skills are as shown in Figure 4-1.

Learning affects the intellectual development of the individual in the manner suggested by the diagram. In solving problems for which instruction has

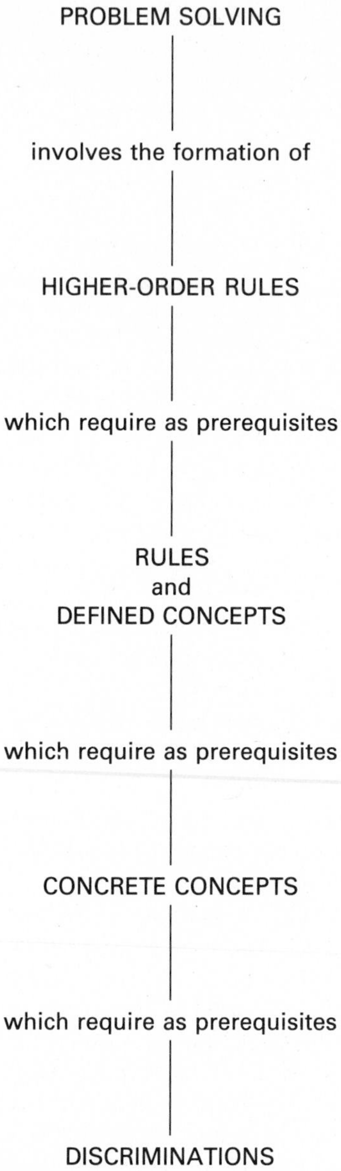

PROBLEM SOLVING

|

involves the formation of

|

HIGHER-ORDER RULES

|

which require as prerequisites

|

RULES
and
DEFINED CONCEPTS

|

which require as prerequisites

|

CONCRETE CONCEPTS

|

which require as prerequisites

|

DISCRIMINATIONS

FIGURE 4-1 Levels of complexity in intellectual skills.

prepared them, learners are acquiring some *higher order rules* (that is, *complex rules*). Problem solving requires that they recall some simpler, previously learned *rules* and *defined concepts*. To acquire these rules, learners must have learned some *concrete concepts*, and to learn these concepts, they must be able to retrieve some previously learned *discriminations*. For example, the reader who is confronted with the problem of inferring the pronunciation of an unfamiliar printed word must bring to bear on this problem some previously learned rules (decoding skills), whose learning has in turn required the prerequisite of identifying the word components called *phonemes* (defined concept) and printed letters (concrete concepts). The child who is learning to identify a letter such as a printed *E* must have previously learned to distinguish three horizontal lines $\equiv$ from two $=$; that is, this discrimination must have been acquired as a prerequisite. Of course, the teacher who is designing instruction to get children to identify *E* may find it possible to assume that they already know two lines from three. If this assumption is not correct, it may be necessary to design instruction so that the learners "catch up" with specific capabilities that reflect the simpler forms of intellectual skills.

Discriminations

A discrimination is the capability of making different responses to stimuli that differ from each other along one or more physical dimensions. In the simplest case, the person indicates by responding that two stimuli are the same or different. Examples in secondary and adult education may occur with stimuli encountered in art, music, foreign languages, and science. Industrial examples pertain to the discrimination of differences in woods, metals, textiles, papers, forms of printing, and a host of others.

Discrimination is often a regular part of instruction for children in early school grades. Here children are asked to distinguish between two "pictures," one having vertical lines and another horizontal lines; or one having a circle and the other a square. Matching to a sample is another variant form of the discrimination task. The child may be asked to "select the block that has the same color as this one" from a group of blocks of various colors. In beginning music instruction, the child may be asked to learn to discriminate which of the two tones is louder or which of two tone-pairs contains tones that are the same or different in pitch.

Discrimination is a very basic kind of intellectual skill. Deliberate instruction in discrimination is undertaken most frequently for young children and for the mentally retarded. As far as most school learning is concerned, relevant discriminations are usually assumed to have been learned early in life. Every once in a while, however, one is surprised to realize that certain elementary discriminations may not have been learned and cannot be assumed. Does the learner of the French uvular and frontal *r* actually hear this distinction (that is, has it been learned as a discrimination)? Has the student microscopist actually seen the distinction between the bright and dark boundary that will later be identified as a cell wall? Can the untrained adult discriminate between the wood grains of cherry and maple?

In describing the characteristics of a discrimination, as well as other types of intellectual skills to follow, we need to account for three components of the learning situation:

1. The *performance* that is acquired or to be acquired. What is it that the learner will be able to do after learning that he was not able to do before?
2. The *internal conditions* that must be present for the learning to occur. These consist of capabilities that are recalled from the learner's memory and that then become integrated into the newly acquired capability.
3. The *external conditions* that provide stimulation to the learner. These may be visually present objects, symbols, pictures, sounds, or meaningful verbal communications.

For discrimination learning, these characteristics are described below.

Performance

There must be a response which indicates that the learner can distinguish stimuli that differ on one or more physical dimensions. Often this is an indication of *same* or *different*.

Internal Conditions

On the sensory side, the physical difference must give rise to different patterns of brain activity. Otherwise, the individual must have available only the responses necessary to indicate that the difference is detectable, as in saying *same* and *different*. The required responses may be as simple as pointing, making a checkmark, or drawing a circle around a pictured object.

External Conditions

The learning of discriminations involves external conditions reflected in some of the most generally applicable learning principles. *Contiguity* is necessary in that the response must follow the stimulus within a short time. *Reinforcement* is of particular importance to discrimination learning and is made to occur *differently* for right and wrong responses. A response indicating a correct distinction between same or different stimuli is followed by a pleasant familiar activity (for example, circling *other* figures of the same sort), whereas an incorrect response is not followed by such activity. When reinforcement occurs in this manner, the discrimination will soon be learned. *Repetition* also plays a particular role. The situation may need to be repeated several times, in order that the correct stimulus difference is selected. Sometimes this may happen in one trial, but often a few repetitions may be necessary to permit reinforcement to take its effect. Additional repetitions become necessary when *multiple* discriminations are being learned, as when several different object shapes must be distinguished at one time.

Concrete Concepts

A concept is a capability that makes it possible for an individual to identify a stimulus as a member of a class having some characteristic in common, even though such stimuli may otherwise differ from each other markedly. A concrete concept identifies an *object property* or object attribute (color, shape, etc.). Such concepts are called "concrete" because the human performance they require is recognition of a concrete object.

Examples of object properties are round, square, blue, three, smooth, curved, flat, and so on. One can tell whether a concrete concept has been learned by asking the individual to identify, by "pointing to," two or more members belonging to the same object-property class; for example, by pointing to a penny, an automobile tire, and the full moon as round. The operation of pointing may be carried out practically in many different ways; it is a matter of choosing, checking, circling, or grasping. Frequently, pointing is carried out by naming (labeling). Thus, the particular *response* made by the individual is of no consequence, so long as it can be assumed that he knows how to do it.

An important variety of concrete concept is *object position*. This can be conceived as an object property, since it can be identified by pointing. It is clear, however, that the position of an object must be in relation to that of another object. Examples of object positions are above, below, beside, surrounding, right, left, middle, on, in front of. Obviously, one can ask that such positional characteristics be "pointed to" in some manner or other. Thus, object positions qualify as concrete concepts.

The distinction between a discrimination and a concept is easy to appreciate: the first is "responding to a difference", the second, identifying something by name or otherwise. A person may have learned to tell the difference between a triangle and a rectangle drawn on a piece of paper. These may be seen as different figures, by choosing, pointing, or otherwise responding differentially. Such a performance permits only the conclusion that the person can discriminate between these particular figures. To test whether the *concept* triangle has been learned, however, one would need to ask the person to identify several figures exhibiting this property—figures that otherwise differ widely in their other qualities such as size, color, border thickness, and so on. In other words, acquiring a concrete concept means that the individual is capable of identifying the *class* of object properties.

The capability of identifying concrete concepts is fundamentally significant for more complex learning. Many investigators have emphasized the importance of "concrete learning" as a prerequisite to "the learning of abstract ideas." Piagét (1950) made this distinction a key idea in his theory of intellectual development. The acquisition of *concepts by definition* (to be described next) requires that the learner be able to identify the referents of the words used in such definitions. Thus, to acquire the concept *rim* by way of the definition "the edge of a round thing," the learner must have as prerequisites the concrete concepts "edge" and "round." If he is not able to identify these concepts concretely, it will not be possible for him in any true or complete sense to "know the meaning" of *rim*.

Performance

The student identifies a class of object properties, including object positions, by "pointing to" two or more members of the class. The "pointing" may be done in any of a number of ways (checking, circling, and so on) equivalent only in the sense that identification occurs. Examples: (1) display a set of objects made of metal, plastic, and wood, and ask for identification of those made of wood; (2) given a model of Old English type, ask for identification of this type in several samples of print.

Internal Conditions

In acquiring a concrete concept, discriminations must be recalled. Thus, an individual who is learning the concept *two* must be able to discriminate a variation in object quality like this: $|\,|$, from one like this: $|\,|\,|$. For the concept *o* to be learned, the difference between O and other physical-forms of O must have been discriminated. Examples in the previous paragraph require discriminations of (1) surface appearances of wood, plastic, and metal; and (2) the same letters printed in different type.

External Conditions

Instances of this class are presented, varying widely in their nonrelevant characteristics, and the individual is asked to identify each by pointing or picking out from a group. For example, a concept like *two* may be identified by objects as vastly different in other characteristics as two dots on a page, two persons, two buildings, or two baseballs. The concept *wooden* may be displayed in a variety of material objects of different sizes, colors, and shapes. Negative instances of the concept (objects of metal or plastic) are included, in order to reveal rejection of their choice as an indicator of correct performance.

Defined Concepts

An individual is said to have learned a defined concept when he can demonstrate the meaning of some particular class of objects, events, or relations. For example, consider the concept *alien*, a citizen of a foreign country. An individual who has learned such a concept will be able to classify a particular person in accordance with the definition, by showing that that person is currently in a country of which he is not a citizen, and that he is a citizen of some other country. The demonstration may involve verbal reference to the definition, and this is an adequate demonstration when one assumes that the individual knows the meaning of the words *citizen, other,* and *country*. Should it be the case that such knowledge cannot be assumed, it might be necessary to ask for the demonstration in other terms, perhaps involving pointing to pictures of people and maps of countries. *Demonstration* of the meaning distinguishes this kind of mental processing from the kind involved in memorized verbal information such as the statement "An alien is a citizen of a foreign country."

Many concepts can only be acquired as defined concepts and cannot be identified by "pointing to" them, as can concrete concepts. Familiar examples are *family, city,* and abstractions like *justice*. However, some defined concepts have corresponding concrete concepts that carry the same name and possess certain features in common. For example, many young children learn the basic shape of a triangle as a concrete concept. Not until much later in studying geometry do they encounter the defined concept of triangle, "a closed plane figure formed by three line segments that intersect at three points." The concrete and defined meanings of *triangle* are not exactly the same, yet they overlap considerably.

An example of a defined concept is *boundary line,* the definition of which may be stated as "a line marking where an area ends." This concept must be demonstrated for an external observer to know that it has been learned. Such a demonstration by the learner would consist, essentially, of (1) identifying an area, either by pointing to a piece of ground, a map, or by drawing one on paper; (2) identifying a *line* that shows the limits of the area; and (3) demonstrating the meaning of *end,* by showing that passage of a moving object is brought to a stop at the line.

Why doesn't one just ask the question, what does *boundary line* mean? Why describe this elaborate procedure? As mentioned previously—only by assuring that the individual is capable of operations identifying the *referents* of the words can one be confident that the meaning of a defined concept has been learned. In practice, of course, the procedure of obtaining verbal answers to verbal questions is often used. But such a procedure is always subject to the ambiguity that the learner may be repeating a verbalization, and so may not know the meaning of the concept after all. It is for this reason that we use the phrase *demonstrate a concept* rather than a simpler phrase like *state a definition* or *define.* We want to imply that the learner has a genuine understanding of a defined concept, rather than the superficial acquaintance indicated by reeling off a string of words.

Performance

The learner demonstrates the concept by identifying instances of concepts that are components of the definition, and showing an instance of their relation to one another. In the example just given, the concepts *area* and *line* are first identified, followed by the meaning of the verb *end* (bringing a moving object to a stop), which is the relation to be demonstrated. Example: A definition of the physical concept *mass* is "the property of a body that determines how much *acceleration* it would attain when subjected to a particular *force*." Ideally, a performance demonstrating the possession of this concept would identify several bodies of different *mass*, indicated by differing *accelerations* resulting from the same *force*. In quantitative terms, of course, a proportional relation between *mass* and *acceleration* should be demonstrated, in accordance with the relation $a = f/m$.

It is instructive to note that the demonstration that a defined concept has been attained goes far beyond repeating a definition in words. Obviously a student can learn to *state* a definition such as "Mass is the property that determines the amount of acceleration imparted to a body by a particular force", without having an understanding of the concept. To assess the idea properly, one must devise questions that require identification of the component concepts, including the relation between them. Thus a question like the following does not do the job required: "A body is acted on by a force of 1000 dynes and accelerates to 20 cm/sec/sec. What is its mass?" A question like the following would be a better indicator: "A body is observed to accelerate at 20 cm/sec/sec. Describe two different sets of conditions that could bring about this acceleration."

Internal Conditions

To acquire a concept by definition, the learner must retrieve all of the component concepts included in the definition, including the concepts that represent relations among them (such as *end* in the case of *boundary line*, or *acts upon* in the case of *force* and *mass*).

External Conditions

A defined concept may be learned by having the learner watch a demonstration. Laboratory exercises in elementary physics typically include a demonstration of the concept *mass*. Most frequently, though, a defined concept is "demonstrated" by means of a verbally stated definition. Thus, the concept *scum* may be communicated by the statement "a filmy covering floating on a liquid." Provided the learners' internal conditions are met, such a statement is sufficient to induce learning of the concept. What must be retrieved by the learners are the concepts *filmy*, *covering*, *liquid*, and *floating on*; not just the words.

Rules

A rule has been learned when it is possible to say with confidence that the learner's performance has a kind of "regularity" over a variety of specific situations. In other words, the learner shows that he is able to respond with a *class* of relationships among *classes* of objects and events. When a learner shows that he can sort cards marked X into a bin marked A, and cards marked Y into a bin marked B, this is insufficient evidence that his behavior is "rule-governed." (He may simply be exhibiting learning of the concrete concepts X and Y.) But suppose he has learned to put each X card into any bin two positions away from his last choice, and each Y card one position away from his last choice. In that case, he has learned a rule. He is responding to classes of objects (X and Y cards) with classes of relationships (one position away, two positions away). His behavior cannot be described in terms of a *particular* relation between the stimulus (the card) and his sorting response to a bin.

There are many common examples of rule-governed behavior. In fact, most behavior of human beings falls into this category. When we make a sentence using a given word such as *girl*, as in "The girl rode a bicycle," we are using a number of rules. For example, we begin the sentence with *The*, not with *girl*, employing a rule for the use of the definite article. The subject of the sentence is followed with a predicate, a verb coming next in order—that is, we say "The girl rode," and not "Rode the girl." The verb is followed in turn by the object *bicycle*, which, according to one rule, is placed in a particular order, and according to another is preceded (in this case) by the indefinite article *a*. Finally, we complete the sentence by bringing it to a close, which in written form involves a rule for the use of a period. Since we have acquired each of these rules, we are able to construct *any* sentence of the same structure, with *any* given words as subject and object.

Principles learned in science courses are exhibited by the learner as rule-using behavior. For example, we expect students who have learned Ohm's law, $E = I \times R$, to *apply* the rule embodied in this statement. A question like the following may be asked: "Assuming that an electric circuit has a resistance of 12 ohms, if the current is increased from 20 amps to 30 amps, what change is required in the voltage?"

Obviously, possessing the capability called a *rule* does not mean being able to state it verbally. The student who can state "voltage equals current times resistance" cannot necessarily apply the rule to a specific concrete problem. The child performs the behavior of constructing oral sentences long before learning grammatical rules. The observer of learning behavior may have to "state the rule" being learned, in explanation of what is being talked about. There are many instances, however, in which learners are quite unable to state a rule, even though their performance indicates that they "know" it.

Now that we have indicated what a rule is, we can admit that a defined concept, as previously described, is actually not formally different from a rule, and is learned in much the same way. In other words, a defined concept is a particular type of rule whose purpose it is to classify objects and events; it is a *classifying rule*. Rules, however, include many other categories besides classifying. They deal with such relationships as equal to, similar to, greater than, less than, before, after, and many others.

Performance

The rule is demonstrated by showing that it applies to one or more concrete instances. Examples: (1) The rule relating electrical resistance to cross-sectional area of a conductor can be demonstrated by the decrease in ohms when wire of larger diameter is selected for an electric circuit. (2) The rule governing the case of pronouns following prepositions can be demonstrated by a correct choice of pronoun in completing the sentence: "The secret was strictly between *(she) (her)* and *(I) (me)*." (3) The rule for multiplying fractions can be shown by application to an example such as $\frac{5}{6} \times \frac{2}{3}$.

Internal Conditions

In learning a rule the learner must retrieve each of the component concepts of the rule, including the concepts that represent relations. The instructor needs to assume that these concepts have been previously learned and can readily be recalled. In the example of resistance of a wire conductor, the learner must be able to retrieve such concepts as "cross section," "area," "conductor," and "decrease."

External Conditions

Usually, the external conditions for learning rules involve the use of verbal communications. The rule may be communicated to the learner verbally, although not necessarily in a precise manner. The purpose of such verbal statements is to *cue* the arrangement of concepts in a correct order by the learner. They are not to teach the learner a formal verbal proposition representing the rule. Suppose, for example, a teacher intends to impart a particular rule in the decoding of printed words (the rule for pronouncing words having consonants followed by a final *e*). The teacher may say, "Notice that the letter *a* has a long sound when followed by a consonant in a word that ends in *e*. This is true in words that you know like *made, pale, fate*. When the word does not end in *e*, the letter *a* has a short sound, as in *mad, pal, fat*. Now tell me how to pronounce the following words: *dade, pate, kale*."

The basic reasons for the verbal communication are (1) to remind the learner of component concepts to be recalled (such as "long vowel sounds," "consonants"); and (2) to get the learner to arrange component concepts in the proper order (that is, "consonant followed by final *e*," not "vowel followed by final consonant," nor "vowel followed by final *e*," nor "consonant followed by final vowel," nor any other incorrect ordering).

It is evident that the verbal communication used in rule learning may be more or less lengthy. Accordingly, more or less of the actual rule construction may be left up to the learner. Another way to say this is that the external conditions for instruction in a rule may provide different amounts of *learning guidance*. When minimal amounts of learning guidance are provided, instruction is said to emphasize *discovery* on the part of the learner (Bruner, 1961; Shulman and Keislar, 1966). Conversely, discovery is de-emphasized when the amount of learning guidance provided is large, as tends to be true in more detailed verbal communications. Studies of "discovery learning" suggest that small amounts of learning guidance have advantages for retention and transfer of the rules that are learned (cf. Worthen, 1968). Often techniques to bring about learning by discovery incorporate the use of pointed questioning of the learner. These questions lead to the discovery of proper ordering of component concepts.

Higher Order Rules—Problem Solving

Sometimes, the rules we learn are complex combinations of simpler rules. Moreover, it is often the case that these more complex, or "higher order," rules are *invented* for the purpose of solving a practical problem or class of problems. The capability of problem solving is, naturally, a major aim of the educational process—most educators agree that the school should give priority to teaching students "how to think clearly." When students work out the solution to a problem that represents real events, they are engaging in the behavior of thinking. There are, of course, many kinds of problems, and an even greater number of possible solutions to them. In attaining a workable solution to a problem, students also achieve a new capability. They learn something that can be generalized to other problems having similar formal characteristics. This means they have acquired a new rule or perhaps a new set of rules.

Suppose that a small car has been parked near a low brick fence and is discovered to have a flat tire on one of its front wheels. No jack is available, but there is a ten-foot two-by-four and a piece of sturdy rope. Can the front of the car be raised? In this situation, a possible solution might be found by using the two-by-four as a lever, the wall as a fulcrum, and the rope to secure the end of the lever when the car is in a raised position. This solution is invented to meet a particular problem situation. It is evident that the solution represents a "putting together" of certain rules that may not have been applied to previous similar situations by the individual who is solving the problem. One rule pertains to the application of force on an end of the car to achieve a lifting of that end. Another rule pertains to the use of the wall as a fulcrum that will bear an estimated weight. And still another, of course, is the rule regarding use of the two-by-four as a lever. All of these rules, in order to be used in an act of problem solving, must be recalled by the individual, which means they must have been previously learned. (Note once again that the rules to which we refer cannot necessarily be verbalized by the problem solver; nor have they necessarily been learned in a physics course.) These previously acquired rules are then brought together by the individual to achieve the solution to the problem. And once solved, the individual has learned a new rule, more complex than those used in combination. The newly learned rule will be stored in the memory and used again to solve other problems.

The invention of a complex rule can be illustrated with a problem in mathematics. Suppose a student has learned to add monomials such as $2x$ and $5x$, $3x^2$ and $4x^2$, $2x^3$ and $6x^3$. Now he is shown a set of polynomials, such as:

$$2x + 3x^2 + 1$$
$$2 + 3x + 4x^2.$$

The student is asked, "What do you suppose is the sum of these two expressions?" This question asks for the solution of a new problem, which (we assume) has not been previously encountered. Possibly, the student may make some false starts that could be corrected. The chances are, however, that previously learned subordinate rules will permit her to think out the solution to this problem (for example, the rule that a variable a added to the variable a^2 results in the sum $a + a^2$, also the rule for adding monomials, such as $2a^2 + 3a^2 = 5a^2$. It is probably not a difficult problem for the student, therefore, to devise the complex rule: Add variables with the same exponents; express the sum as a set of terms connected by the $+$ sign. Again in this example, the problem solver has remembered and "combined" simpler rules into a more complex rule to solve the problem.

The essential condition that makes this sort of learning a problem-solving event is the *absence* of any learning guidance, whether in the form of a verbal communication or in some other form. The solution has been "discovered," or invented. The learning guidance is provided by the problem solver alone, not by a teacher or other external source. One may guess that some problem-solving *strategies*, learned in quite different situations, are probably brought to bear. But in any case, relevant rules are recalled and combined to form a new higher order rule.

Rules play an essential role in problem solving. It is impossible for a learner to acquire all the rules needed for every situation. Concepts and rules must be synthesized into new complex forms for the learner to cope with new problem situations. Problem solving for a particular class of tasks is facilitated by adding to a student's repertoire of intellectual skills relevant to those tasks. Problem solving should be thought of as a human activity that combines previously acquired concepts and rules, and not as a generic skill. The ability to solve problems in mathematics does not automatically transfer to solving the mechanical problems of an automobile.

Performance

Performance requires the invention and use of a complex rule to achieve the solution of a problem novel to the individual. When the higher order rule has been acquired, it should also be possible for the learner to demonstrate its use in other physically different, but formally similar situations. In other words, the new complex rule exhibits *transfer of learning.* Here are two examples of higher order rules developed in problem-solving situations: (1) To create a potting soil mixture for greenhouse plants in a region having average humidity of 65% requires the combining of component rules about soil components, plant nutrients, and evaporation times; (2) in the absence of a conversion chart, the problem of determining proper proportions for a substitute sweetener in a recipe calling for sugar may require retrieval and combining of previously learned rules about "sweetness quotients" in the combining of specific ingredients.

Internal Conditions

In solving a problem, the learner must retrieve relevant subordinate rules and relevant information. It is assumed that these capabilities have been previously learned.

External Conditions

The learner is confronted with an actual or a represented problem situation not previously encountered. Cues in the form of verbal communication are at a minimum or may be absent entirely. In general, learners engage in discovery learnings; they invent solutions that embody a *higher order rule.*

COGNITIVE STRATEGIES

A very special kind of intellectual skill, of particualr importance to learning and thinking, is the *cognitive strategy.* In terms of modern learning theory, a cognitive strategy is a *control process,* an internal process by which learners select and modify their ways of attending, learning, remembering, and thinking (Gagné, 1985). Several of Bruner's writings (1961, 1971) describe the operation and usefulness of cognitive strategies in problem solving. More recently, many different strategies have been identified that relate to the entire range of cognitive processes of the learner (O'Neil and Spielberger, 1979).

Varieties of Learner Strategies

Although specific strategies may be used by the learner in dealing with all conceivable kinds of learning tasks, it is convenient to classify them into a few categories that indicate their control functions. The following categories are suggested by Weinstein and Mayer (1986).

Rehearsal Strategies

By means of these strategies, learners conduct their own practice of the material being learned. In simplest form, the practice is simply repeating to themselves the names of items in an ordered list (for instance, the U.S. presidents or the states). In the case of more complex learning tasks, such as learning the main ideas of a printed text, rehearsal may be accomplished by underlining the main ideas or by copying portions of the text.

Elaboration Strategies

In using the techniques of elaboration, the learner deliberately associates the item to be learned with other readily accessible material. In learning foreign language vocabulary, for example, the foreign word may be associated with a mental image of an English word that forms an "acoustic link" with the word having the correct meaning (Atkinson, 1975; Levin, 1981). When applied to learning from prose texts, elaboration activities include paraphrasing, summarizing, note taking, and generating questions with answers.

Organizing Strategies

Arranging material to be learned into an organized framework is the basic technique of these strategies. Sets of words to be remembered may be arranged by the learner into meaningful categories. Relations among facts may be organized into a table, making possible the use of spatial arrangement cues to recall the material. Outlining the main ideas in prose passages and generating new organizations for the ideas is another method. Learners are able to acquire strategies that organize passages of text into several particular kinds of relations among ideas, such as "comparison," "collection," and "description" (Meyer, 1981).

Comprehension Monitoring Strategies

These strategies, sometimes referred to as *metacognitive strategies* (Brown, 1978), pertain to the student's capability of setting goals for learning, estimating the success with which the goals are being met, and selecting alternative strategies to meet the goals. These are strategies having the function of *monitoring*, the presence of which becomes evident in reading for understanding (Golinkoff, 1976). Students have been taught to develop their own statements and questions to be used in guiding and controlling their performance in the comprehension of prose (Meichenbaum and Asarnow, 1979).

Affective Strategies

These techniques may be used by learners to focus and maintain attention, to control anxiety, and to manage time effectively. Such strategies can be taught by making students aware of their operation and providing ways for them to practice their use (Dansereau, 1985; McCombs, 1982).

Learning Cognitive Strategies

A cognitive strategy is a cognitive skill that selects and guides the internal processes involved in learning and thinking. Notice that it is the *object* of the skill that differentiates cognitive strategies from other intellectual skills. The latter, concepts and rules, are oriented toward environmental objects and events, such as sentences, graphs, or mathematical equations. In contrast, cognitive strategies have as their objects the *learner's own cognitive processes*. Undoubtedly, the efficacy of an individual's cognitive strategies has a crucial effect upon the quality of information processing. A learner's cognitive strategies may determine, for example, how readily he learns, how well he recalls and uses what has been learned, and how fluently he thinks.

Statements of educational goals often give the highest priority to cognitive strategies. Many statements of goals for school learning give a prominent place to "teaching students how to think." Although it would be difficult to find disagreement with the importance of such a goal, it seems wise to temper one's enthusiasm with a few facts concerning the feasibility of reaching it. First, one should realize that genetic factors, not amenable to the influence of education, are likely to play an important part in the determination of creative thought (cf. Tyler, 1965; Ausubel, Novak, and Hanesian, 1978, Chap. 16). In other words, there are bound to be enormous differences in intellectual capacity among people, which can never be completely overcome by environmental influences such as education. Second, the internally organized nature of cognitive strategies means that the conditions of instruction can have only an indirect effect upon their acquisition and improvement. In the case of other types of intellectual skills, one can plan a sequence of learning events so as to increase the probability of certain internal events; and these in turn determine the learning of the cognitive strategy. Accordingly, the design of instruction for cognitive strategies has to be done in terms of "favorable conditions." Generally, the favorable conditions are those that *provide opportunities for development and use* of cognitive strategies. In other words, to "learn to think," the student needs to be given opportunities to think.

Derry and Murphy (1986) describe a system of learning-strategies training that begins with direct instruction of such strategies as reading comprehension monitoring, problem solving, and control of affect. Following the initial training, the same strategies are practiced in a variety of learning situations over an extended period, using distinctive cues that remind the learner in each instance of an appropriate useful strategy. The method thus incorporates the idea of spaced and varied practice, which is considered to be desirable for the learning of these higher order strategies.

Performance

The performance of cognitive strategies cannot be observed directly but must be inferred from performances calling for the use of other intellectual skills. Investigators usually discover strategies by asking learners to "think aloud" while they are learning, remembering, or solving problems (Ericsson & Simon, 1980). However, inferences about the quality of reading comprehension reveal the use of comprehension strategies; inferences about the quality of learning new rules of trigonometry (for example) reveal the use of organizing strategies; and inferences about the quality of problem solving reveal the use of thinking strategies.

Internal Conditions

Prior knowledge (that is, intellectual skills and verbal information) relevant to the subject matter to be learned or thought about must be retrievable, just as is true for other intellectual skills. However, it should be noted that cognitive strategies often intrinsically possess a simple structure (for example, "underline main ideas," "divide the problem into parts").

External Conditions

Strategies may often be suggested to learners by verbal communications or demonstrated to them in simple form. Even young children, for example, can respond appropriately to the suggestion that they use an organizing strategy like classifying lists of words into meaningful categories. In other instances, cognitive strategies result from learning by discovery. Favorable conditions always depend upon the provision of opportunities for practice.

METACOGNITION

The internal processing that makes use of cognitive strategies to monitor and control other learning and memory processes is known generally as *metacognition* (Flavell, 1979). In confronting problems to be solved, learners are able to select and regulate the employment of relevant intellectual skills and bring to bear task-oriented cognitive strategies. Such *metacognitive strategies*, which govern the use of other strategies, are also spoken of as "executive" or "higher level." Learners may also become aware of such strategies and may be able to describe them, in which case they are said to possess *metacognitive knowledge* (Lohman, 1986). Planning models for direct training in metacognitive knowledge are involved in many schemes for study skills and general problem solving.

Broadly speaking, there are two different viewpoints about the origins of metacognitive strategies (Derry and Murphy, 1986). One is that they may be acquired by learners through the communication of metacognitive knowledge (that is, via verbal information) followed by practice in their use. This approach is exemplified by courses in problem-solving strategies, such as that described by Rubinstein (1975). The second view proposes that metacognitive strategies arise from the generalization of a number of specific task-oriented strategies, usually after a considerable variety of problem-solving experiences by the learner. This latter view appears to be supported by the weight of evidence (Derry and Murphy, 1986) and we have therefore adopted it in our discussion.

Strategies for Problem Solving

Often of particular interest in instructional design are those cognitive strategies called into play when the learner defines and thinks out a solution to a highly novel problem. Although such strategies are often of primary interest in educational programs, our knowledge of how to assure their learning is weakest (Gagné, 1980; Polson and Jeffries, 1985). A number of strategies employed by adults in solving verbally stated problems are described by Wickelgren (1974). These include: (1) inferring transformed conceptions of the "givens"; (2) classifying action sequences, rather than randomly choosing them; (3) choosing actions at any given state of the problem that get closer to the goal ("hill climbing"); (4) identifying contradictions that prove the goal cannot be attained from the givens; (5) breaking the problem into parts; and (6) working backwards from the goal statement. Strategies like these are obviously applicable to "brain teaser" problems of an algebraic or geometric sort.

Programs designed to teach problem solving strategies are critically reviewed by Polson and Jeffries (1985). They point out the existence of three different models of problem solving that rest upon different assumptions, and are currently irreconcilable with each other. Model 1 makes the assumption that general problem solving skills (such as those mentioned previously) can be directly taught and will exhibit generalization to other situations. Model 2 maintains that general problem solving skills can be taught, but not directly. Instead, general strategies are most likely to develop indirectly by generalization from task-specific strategies. There is, of course, dependable evidence that the latter kinds of strategies (such as strategies for solving mechanical problems or for constructing geometric proofs) can readily be acquired. Model 3 argues that direct instruction in general problem-solving strategies is effective in establishing only weak strategies that help problem solving very little, even though they are broadly generalizable ("break the problem into parts" is an example). Consequently, although the strategies viewed by Model 3 are teachable, they are not very useful.

In estimating the value of general problem-solving strategies for an instructional program, one should take into consideration the findings of studies contrasting the capabilities of experts with those of novices, in various fields (Gagné and Glaser, 1986). In general, these studies indicate that experts do not necessarily use better problem-solving strategies than do novices; but that they approach problems with larger and better organized knowledge bases. The organized knowledge of the expert includes *verbal information*, as well as *intellectual skills*.

VARIETIES OF INTELLECTUAL SKILLS IN SCHOOL SUBJECTS

The range of human capabilities called intellectual skills includes the varieties of discriminations, concrete concepts, defined concepts, rules, and the higher order rules often acquired in problem solving. An additional category of internally organized skills is cognitive strategies, which govern the learner's behavior in learning and thinking, and thus determine its quality and efficiency. These varieties of learning are distinguishable by: (1) the class of *performance* they make possible; (2) the internal and external *conditions* necessary for their occurrence; and (3) the *complexity* of the internal process they establish in the individual's memory.

Any school subject may at one time or another involve any of these types of learned capabilities. However, the frequency with which they are encountered in various school subjects varies widely. Examples of discriminations can be found in such elementary subjects as printing letters and reading music. In contrast, there are few examples of this type, and many more of defined concepts, in a history course. However, quite a few examples of discriminations also occur in the beginning study of a foreign language, which may be undertaken in the ninth grade. In the same grade, the writing of compositions very frequently involves defined concepts and rules but seems not to require the learning of discriminations or concrete concepts. In this case, the necessary learning of these simpler skills has been accomplished years ago.

Any school subject *can* be analyzed to reveal the relevance of *all* of these kinds of learning. But this is not always a practical course of action because the presentation of the subject in a particular grade may begin with the assumption that simpler kinds of learning have already been accomplished. Thus, discrimination of ·· from · is certainly relevant to the study of algebra. But one doesn't begin the study of algebra with the learning of discriminations because it is possible to assume these discriminations have been previously learned. In science, however, certain discriminations, such as those involved in using a microscope or spectrophotometer, may have to be newly acquired. Such simple skills must be learned before the student can progress to the concepts, rules, and problem solving that may represent the major aims of the course.

Adult education in technical and professional subjects sometimes exhibits objectives representing a limited range of intellectual skills, sometimes the entire range. Personal counseling, for example, partakes of few of the simpler intellectual skills except those involved in reading and the understanding of language. In contrast, the processing of lumber and wood products requires training that may need to begin with discriminations of wood textures and proceed to concrete concepts of wood grain before instruction on the characteristics and uses of various woods can reach an advanced stage.

Is there, then, a *structure* of intellectual skills that represents the "path of greatest learning efficiency" for every subject in the curriculum? In theory, yes. Do we know what this structure is? Only vaguely, as yet. After all, teachers, curriculum specialists, and textbook writers *try* to represent structure in their lesson and curriculum plans. Nevertheless, on the whole their efforts must be characterized as partial and inadequate. The purpose of this book is to describe a systematic method for approaching the problem, as free of culs-de-sac as possible. Such a method will also be subject to empirical verification, revision, and refinement. The application of the method to be described can lead to descriptions of the "learning structure" of any subject taught in the school. This structure may be represented as a kind of *map* of the terrain to be covered in progressing from one point in human development to any other point.

The mapping of learning structures does not lead to "routinization" or "mechanization" of the process of learning. A map indicates starting points, destinations, and alternative routes in between; it does not tell how to make the journey. Making the "learning journey" requires a different set of internal events for each and every individual. In a fundamental sense, there are as many learning "styles" as there are individuals. Describing the learning structures for a progression of objectives within any school subject does not lead to prescribing how the individual student must learn. On the contrary, learning structures are simply descriptions of the accepted goals, or *outcomes* of learning, together with subordinate steps along the way.

SUMMARY

Starting with the need to identify goals as the desired outcomes of the educational system, Chapter 3 proposed that in attempting to design specific courses, topics, and lessons, there is a need to classify performance objectives into broad categories: intellectual skills, cognitive strategies, verbal information, motor skills, and attitudes. Doing so, it was shown, facilitates (1) review of the adequacy of the objectives; (2) determination of the sequencing of instruction; and (3) planning for the conditions of learning needed for successful instruction.

The present chapter has begun the account of the *nature* of the performance capabilities implied by each of the five categories of learned capabilities, beginning with intellectual skills and cognitive strategies. For each of these two domains, this chapter has (1) presented examples of learned performances in terms of different school subjects; (2) identified the kinds of internal conditions of learning needed to reach the new capability; and (3) identified the external conditions affecting its learning.

For intellectual skills, several subcategories were identified: discriminations, concrete and defined concepts, rules, and the higher order rules often learned by problem solving. Each represents a different class of performance, and each is supported by different sets of internal and external conditions of learning. Cognitive strategies were not broken down into subcategories, as intellectual skills were. Research in the future may suggest that this can and should be done.

An important distinction is made between cognitive strategies relating to specific domains of knowledge (such as geometry or poetry) and those that are more general in their relevance. The latter are sometimes called *executive* or *metacognitive strategies* because their function is to govern the use of other strategies, and they apply generally to information processing independently of specific knowledge domains. It may be that metacognitive strategies can be directly taught; more likely, they are generalized by learners from their experience with a variety of specific task-oriented strategies.

The next chapter gives a corresponding kind of treatment to the remaining kinds of learned capabilities: information, attitudes, and motor skills. The purpose of Chapters 4 and 5 is to move one more step toward specification of an orderly series of steps to be used in the actual design of instruction for a lesson, a unit, a course, or an entire instructional system. Specifically, these chapters identify the appropriate internal and external conditions of learning for each kind of learned capability. They lead to suggestions of how to proceed with two aspects of instructional design: (1) how to take account of the *prior* learning assumed to be necessary for the learner to undertake the new learning; and (2) how to plan for the *new* learning in terms of the appropriate *external conditions* needed for the attainment of each type of learning outcome. In later chapters, these conditions will be translated into guidelines for instructional planning.

REFERENCES

Atkinson, R. C. (1975). Mnemotechnics in second language learning. *American Psychologist, 30,* 821–828.

Ausubel, D. P., Novak, J. D., & Hanesian, H. (1978). *Educational psychology: A cognitive view* (2nd ed.). New York: Holt, Rinehart, and Winston.

Brown, A. L. (1978). Metacognitive development and reading. In R. J. Spiro, B. C. Bruce, & G. W. F. Brewer (Eds.), *Theoretical issues in reading comprehension.* Hillsdale, NJ: Erlbaum.

Bruner, J. S. (1961). The act of discovery. *Harvard Educational Review, 31,* 21–32.

Bruner, J. S. (1971). *The relevance of education.* New York: Norton.

Dansereau, D. F. (1985). Learning strategy research. In J. Segal, S. Chipman, & R. Glaser (Eds.), *Thinking and learning skills* (Vol. 1). Hillsdale, NJ: Erlbaum.

Derry, S. J., & Murphy, D. A. (1986). Designing systems that train learning ability: From theory to practice. *Review of Educational Research, 56,* 1–39.

Ericsson, K. A., & Simon, H. A. (1980). Verbal reports as data. *Psychological Review, 87,* 215–251.

Flavell, J. H. (1979). Metacognition and cognitive monitoring: A new area of psychological inquiry. *American Psychologist, 34,* 906–911.

Gagné, R. M. (1980). Learnable aspects of problem solving. *Educational Psychologist, 15* (2), 84–92.

Gagné, R. M. (1985). The conditions of learning (4th ed.). New York: Holt, Rinehart and Winston.

Gagné, R. M., & Glaser, R. (1986). Foundations in research and theory. In R. M. Gagné (Ed.), *Instructional technology: Foundations.* Hillsdale, NJ: Erlbaum.

Golinkoff, R. A. (1976). A comparison of reading comprehension processes in good and poor comprehenders. *Reading Research Quarterly, 11,* 623–659.

Levin, J. R. (1981). The mnemonic '80s: Keywords in the classroom. *Educational Psychologist, 16(2),* 65–82.

Lohman, D. F. (1986). Predicting mathemathanic effects in the teaching of higher-order thinking skills. *Educational Psychologist, 21,* 191–208.

McCombs, B. L. (1982). Transitioning learning strategies research into practice: Focus on the student in technical training. *Journal of Instructional Development, 5(2),* 10–17.

Meichenbaum, D, & Asarnow, J. (1979). Cognitive-behavior modification and metacognitive development: Implications for the classroom. In P. C. Kendall & S. D. Hollon (Eds.), *Cognitive-behavioral intervention: Theory, research and procedures.* New York: Academic Press.

Meyer, B. J. F. (1981). Basic research on prose comprehension: A critical review. In D. F. Fisher & C. W. Peters (Eds.), *Comprehension and the competent reader.* New York: Praeger.

O'Neil, H. G., Jr., & Spielberger, C. D. (Eds.). (1979). *Cognitive and affective learning strategies.* New York: Academic Press.

Piaget, J. (1950). *The psychology of intelligence.* New York: Harcourt, Brace & Jovanovich.

Polson, P. G., & Jeffries, R. (1985). Instruction in general problem-solving skills: An analysis of four approaches. In J. Segal, S. Chipman, & R. Glaser (Eds.), *Thinking and learning skills* (Vol. 1). Hillsdale, NJ: Erlbaum.

Rubinstein, M. F. (1975). *Patterns of problem solving.* Englewood Cliffs, NJ: Prentice-Hall.

Shulman, L. S. & Keislar, E. R. (1966). *Learning by discovery: A critical appraisal.* Chicago: Rand McNally.

Tyler, L. E. (1965). *The psychology of human differences* (3rd ed.). New York: Appleton.

Weinstein, C. E., & Mayer, R. E. (1986). The teaching of learning strategies. In M. C. Wittrock (Ed.), *Handbook of research on teaching* (3rd ed.). New York: Macmillan.

Wickelgren, W. A. (1974). *How to solve problems.* San Francisco: Freeman.

Worthen, B. R. (1968). Discovery and expository task presentation in elementary mathematics. *Journal of Educational Psychology, Monograph Supplement, 59(1),* Part 2.

5

VARIETIES OF LEARNING:
Information, Attitudes, and Motor Skills

In this chapter, we need to continue our description of the varieties of human capabilities that may be learned. Courses and lessons are, of course, not always aimed at developing intellectual skills or cognitive strategies, as discussed in the previous chapter. Furthermore, a topic, course of study, or even an individual lesson may have more than one class of objective as a learning outcome. Instruction is typically designed to encompass several objectives in any given unit of instruction and to achieve a suitable balance among them.

We shall be describing here the conditions applicable to these outcomes: the learning of *verbal information,* the establishment or changing of *attitudes,* and the acquisition of *motor skills.* These three outcomes of learning are obviously very different from each other and of differing importance within particular instructional programs. For each of them, as in the previous chapter, we need to consider three aspects of the learning situation:

1. the *performance* to be acquired as a result of learning;
2. the *internal conditions* that need to be present for learning to occur; and
3. the *external conditions* that bring essential stimulation to bear upon the learner.

VERBAL INFORMATION (KNOWLEDGE)

Verbal information is also called verbal knowledge; according to theory, it is stored as networks of propositions (Anderson, 1985; E. D. Gagné, 1985) that conform to the rules of language. Another name for it, intended to emphasize the performance capability it implies, is *declarative knowledge*.

A great deal of information is learned and stored in memory as a result of school instruction. Of course, an enormous amount is acquired outside of school as well, from the reading of books, magazines, newspapers, and by way of radio and television programs. From this very fact, it is apparent that special means of instruction do not have to be provided for a large amount of learning to occur. The communications provided by the various media bring about learning in many people, provided of course that those who hear or see or read these communications possess the basic intellectual skills for interpreting them.

In school learning, however, there are many circumstances in which one desires greater certainty of learning than can ordinarily be expected from various extraschool communications. The literate individual may gain much information from a radio lecture on modern developments in chemistry. The amount of information learned by this means may vary greatly among different individuals, depending on their interests and previous experience. A formally planned course in chemistry, in contrast, may have the aim of teaching all students certain information deemed essential for further study of the subject, such as the names of elements, the physical states of matter, and so on. Similarly, the purpose of a course in U.S. government may be to teach all students the content of the articles of the Constitution. Planned instruction in school subjects is undertaken because of this need for certainty of learning particular bodies of information.

Two primary reasons exist for desiring a high degree of certainty in the learning of information. As previously mentioned, particular information may be needed for a learner to continue learning a topic or subject. Of course, some of the necessary detailed information may be looked up in a book or other source. A great deal of it, however, may need to be recalled and used again and again in pursuing study within a subject. Thus there is typically a body of information that is basic in the sense that future learning will be more efficiently conducted if it is acquired and retained.

A second reason for learning information is that much of it may be continually useful to the individual throughout life. All people need to know the names of letters, numerals, common objects, and a host of facts about themselves and their environment in order to receive and give communications. A great deal of such factual information is acquired informally without any formal planning. In addition, an individual may acquire unusual quantities of factual information in one or more areas of particular interest (a mass of facts about flowers or automobiles or the game of baseball). The problem faced in designing school curricula in any area is one of distinguishing between information that is more or less essential. Some may be used by the individual for communication throughout his lifetime. Other information may be personally interesting but not essential. The former category is one for which the standard of certainty of learning becomes the concern of formal education. There seems to be no reason to limit the information a person wishes to learn because of particular interests or desire for further learning.

When information is organized into bodies of meaningfully interconnected facts and generalizations, it is usually referred to as *knowledge*. Obviously, the information possessed by individuals within their own particular field of work or study is usually organized as a body of knowledge. Thus we expect a chemist, for example, to have learned and stored a specialized body of knowledge about chemistry; and similarly, we expect a cabinetmaker to possess a body of knowledge about woods and joints and tools. Besides these masses of specialized knowledge, one must face the question of whether there is a value to acquiring knowledge that may be called *general*. It may be noted that most, if not all, human societies have answered this question affirmatively. In one way or another, means have been found to pass on the accumulated knowledge of the society from one generation to the next. Information about the origins of the society, tribe, or nation, its development through time, its goals and values, its place in the world, is usually considered a body of knowledge desirable for inclusion in the education of each individual.

In the past within our own society, there was a body of general knowledge, fairly well agreed upon, that was considered desirable for the "educated class" (those who went to college) to learn. It was composed of historical information about Western culture extending back to early Greek civilization, along with related information from literature and the arts. As mass education progressively replaced class education, there was for a while an accompanying reduction in the amount of general cultural knowledge considered desirable for all students to learn. More recently, however, the "back-to-basics" movement has revived interest in the learning of general cultural information as a contributor to societal stability.

What function does general cultural knowledge serve in the life of the individual? Evidently, such knowledge serves the purpose of communication, particularly in those aspects of life pertaining to citizenship. Knowing the facts about the community, the state, and the nation and the services they provide, as well as the responsibilities owed them, enables the individual to participate as a citizen. Cultural and historical knowledge may also contribute to the achievement and maintenance of the individual's "identity", or sense of self-awareness of his origins in relation to those of the society to which he belongs.

A much more critical function of general knowledge can be proposed, although evidence concerning it is incomplete. This is the notion that knowledge is the *vehicle* for thought and problem solving. In the previous chapter we have seen that thinking in the sense of problem solving requires certain prerequisite intellectual skills, as well as cognitive strategies. These are the tools individuals possess that enable them to think clearly and precisely. How does one think broadly? How can a scientist, for example, think about the social problem of the isolation of aged people? Or how can a poet capture in words the essential conflict of youthful rebellion and alienation? It is probable that these individuals need to possess bodies of knowledge shared by many other people to solve these problems. The thinking that takes place is "carried" by the associations, metaphors, and analogies of language within these bodies of knowledge. The importance of a "knowledge background" for creative thought has been discussed by many writers, including Polanyi (1958), and more recently by Glaser (1984).

In summary, it is evident that a number of important reasons can be identified for the learning of information, whether this is conceived as facts, generalizations, or as organized bodies of meaningful knowledge. Factual information is needed in learning the increasingly complex intellectual skills of a subject or discipline. Such information may in part be looked up, but is often more conveniently stored in memory. Certain types and categories of factual information must be learned because it is necessary for communication pertaining to the affairs of everyday living. Information is often learned and remembered as organized bodies of knowledge. Specialized knowledge of this sort may be accumulated by the individual learner while pursuing a field of study or work. General knowledge, particularly that which reflects the cultural heritage, is often considered desirable or even essential in making possible the communications necessary for functioning as a citizen of a community or nation. In addition, however, it seems likely that such bodies of general knowledge become the carriers of thought for the human being engaged in reflective thinking and problem solving.

THE LEARNING OF VERBAL INFORMATION

Verbal information may be presented to the learners in various ways. It may be delivered to their ears in the form of oral communications or to their eyes in the form of printed words and illustrations. There are many interesting research questions relating to the effectiveness of communication media (Bretz, 1971; Clark and Salomon, 1986), and some of their implications for instructional design will be discussed in a later chapter. At this point, however, we attend to a different set of dimensions, which cut across those of communication media.

Verbal information presented for learning may vary in amount and in organization. Some variations along these dimensions appear to be more important than others for the design of instruction. From this point of view, it seems desirable to distinguish three kinds of learning situations. The first concerns the learning of *labels,* or *names.* A second pertains to the learning of isolated or single *facts,* which may or may not be parts of larger meaningful communications. The third is the learning of *organized information.* The latter two types are often referred to as *declarative knowledge.*

Learning Labels

To learn a label simply means to acquire the capability of making a consistent verbal response to an object or object class in such a way that it is "named." The verbal response itself may be of almost any variety—"$x - 1$," "petunia," "pocket dictionary," or "spectrophotometer." Information in this form is simply a short *verbal chain*. Reference to the substantial body of research on the learning of verbal paired associates may be found in many texts (for example, Hulse, Egeth, and Deese, 1980; Kausler, 1974).

Learning the name of an object in the sense of a label is quite distinct from learning the *meaning* of that name. The latter phrase implies the acquisition of a *concept*, which has been described previously. Teachers are well acquainted with the distinction between "knowning the name of something" and "knowing what the name means." A student knows a label when he can simply supply the name of a specific object. To know that object as a concept (that is, know its meaning) he or she must be able to identify examples and nonexamples that serve to define and delimit the class.

In practice, a name for a concept is often learned at the time the concept itself is learned, or just prior to that time. Although the task of name learning may be easy for one or two objects at a time, difficulty increases rapidly when several different names for several objects or many names for many objects must be learned at once. Such a situation arises in school learning when students are asked to acquire the names of a set of trees, a set of leaves, or the set of members in a president's cabinet. Students engaged in such tasks may accurately be said to be memorizing the names, but there is scarcely harm in that, and students often enjoy doing it. In any case, label learning is a highly useful activity. Among its other uses, it establishes the basis for communication between the learner and the teacher, or between the learner and a textbook.

Learning of sets of names can often be aided by the use of *mnemonic techniques,* most of which have been known for many years. In associating words in pairs, such as *car–wolf*, the learner may be encouraged to make up a sentence, "The *car* chased the *wolf*", and to treat each pair similarly with other sentences. This strategy usually brings about remarkable improvements in paired associate learning (Rohwer, 1970). The learning of foreign language vocabulary is another example of a task in which mnemonic techniques can be put to good use (Pressley, Levin, and Delaney, 1982). The strategy called the *keyword method* involves the use of learner-generated images to cue the retrieval of English equivalents to foreign words. For example, for the Spanish word *carta*, the keyword *cart* can be used as part of an image serving as a link: "The cart was used to deliver the *letter*."

Learning Facts

A fact is a verbal statement that expresses a relation between two or more named objects or events. An example is, "That book has a blue cover." In normal communication, the relation expressed by the fact is assumed to exist in the natural world. Thus, the words that made up the fact have *referents* in the environment of the learner. The words refer to those objects and to the relation between them. In the example given, the objects are *book* and *blue cover* and the relation is *has*. It is of some importance to emphasize that a fact, as employed here, is defined as the *verbal statement* and not the referent or referents to which it refers. (Alternative meanings of a common word like *fact* may readily be found in other contexts.)

Students learn a host of facts in connection with their studies in school. Some of these are isolated in the sense of being unrelated to other facts or bodies of information. Others form a part of a connected set, related to each other in various ways. For example, children may learn the fact "the town siren is sounded at noontime," and this may be a fairly isolated fact that is well remembered, even though not directly related to other information. Isolated facts may be learned and remembered for no apparent reason; in studying history, a student may learn and remember that Charles G. Dawes served as vice-president in the administration of Calvin Coolidge, and at the same time learn the names of other vice-presidents. Most frequently, though, a specific learned fact is related to others in a total set, or to a larger body of information. For example, a student may learn a number of facts about Mexico that are related to each other in the sense that they pertain to aspects of Mexico's geography, economy, or culture. Such facts may also be related to a larger body of information including facts about the culture, economy, and geography of other countries, including the student's native country.

Whether isolated or connected with a larger set, learned facts are of obvious value to the student, for two major reasons. The first is that they may be essential to everyday living. Examples are: the fact that many stores and banks are closed on Sunday; or the fact that molasses is sticky; or the fact that the student's birthday is the tenth of February. The second and more obvious reason for the importance of learned facts to students is that they are used in further learning. To find the circumference of a circle, for example, one needs to know the value of *pi*. To complete a chemical equation, the student may need to know the valence of the element sodium.

With regard to the function of facts as elements in the learning of skills or additional information, it is evident that such facts *can* be looked up in convenient reference books or tables when this further learning is about to take place. There are many instances when looking-up may be proper and desirable. The alternative is for the student to learn the facts and store them in his memory, so that he may then retrieve them whenever he needs them. This alternative is often chosen as a matter of convenience and efficiency. Facts that are likely to be used again and again might as well be stored in memory—the student would likely find the constant looking-up a nuisance. The designer of instruction, however, has the obligation of deciding which of a great many facts in a given course are (1) of such infrequent usage that they had better be looked up; (2) of such relatively frequent reference that learning them would be an efficient strategy; or (3) of such fundamental importance that they ought to be remembered for a lifetime.

Performance

The performance that indicates a fact has been learned consists of *stating*, either orally or in writing, relations that have the syntactic form of a sentence.

Internal Conditions

For acquisition and storage, an organized network of declarative knowledge needs to be accessed in memory, and the newly acquired fact must be related to this network (E. D. Gagné, 1985). For example, to learn and remember that Mount Whitney is the highest peak in the continental United States, an organized network of propositions (which may differ for each individual learner) needs to be accessed, that may include a classification of mountain peaks and ranges, a set of categories of mountains in the United States, and information about the range of mountains that includes Mount Whitney. The new fact is associated by the learner with a number of other facts within this larger network.

External Conditions

Externally, a verbal communication, picture, or other cue is presented to remind the learner of the larger network of organized knowledge with which the new fact will be associated. The new fact is then presented, usually by means of a verbal statement. The communication may also suggest the association to be acquired, as in conveying the idea that Mount Whitney "sticks up highest" in the Sierra Nevada range. *Imagery* of the mountain and its name may help retention of the fact. *Elaboration* of the new fact may be advantageous by presenting other facts relating Mount Whitney to mountain ranges. Provision also needs to be made for the rehearsal of the new fact or repetition of it in the form of a spaced review.

Learning Organized Knowledge

Larger bodies of interconnected facts, such as those pertaining to events of history or to categories of art, science, or literature, may also be learned and remembered. As is the case with the learning of single facts, the networks of propositions that constitute the new knowledge become linked to the larger propositional networks already existing in memory. Larger bodies of knowledge are organized from smaller units so that they become meaningful wholes.

The key to remembering bodies of knowledge appears to be one of having them *organized* in such a way that they can be readily retrieved. Organizing verbal information appears to require generating new ideas that relate sets of information already stored in memory. Such organizing, when carried out during learning, aids in the later retrieval of information by providing effective cues to retrieval (E. D. Gagné, 1985). The periodic table of chemical elements, for example, besides having a theoretical rationale, also helps students of chemistry to remember the names and properties of a large number of elements. Similarly, students of U.S. history may have acquired a framework of historical periods into which many individual facts can be fitted for learning and remembering. The more highly organized this previously acquired information, the easier it is for a student to acquire and retain any given new fact that can be related to this organized structure.

Performance

The substance of paragraphs or longer passages of connected prose appear to be learned and retained in a way that preserves their *meaning* but not necessarily the detailed component facts contained in them (Reynolds and Flagg, 1977). The more general ideas appear to be recalled better than the more specific ones (Meyer, 1975). Details are often "constructed" by the learner, apparently in accordance with a general *schema* (Spiro, 1977) that represents the gist of a story or passage.

Internal Conditions

As in the case of individual facts, the learning and storage of larger units of organized verbal information occurs within the context of a network of interconnected and organized propositions previously stored in the learner's memory. It may be that newly learned knowledge is *subsumed* into larger meaningful structures (Ausubel, Novak, and Hanesian, 1978), or it may be that the new information is *linked* to a network of propositions already in the learner's memory (R. M. Gagné, 1985).

External Conditions

The external conditions that favor the learning and retention of organized sets of verbal information pertain primarily to the provision of *cues*. Such cues enable the learner to search successfully for the information at a later time, and thus to retrieve it for use.

Cues need to be as *distinctive* as possible, in order to avoid interference among stored propositions. Distinctiveness of cues can be assured by introducing readily memorable stimuli (such as rhymes) within the material to be learned, and by reducing the confusion with other, similar ideas. Organizing component ideas into tables or spatial arrays is another method of increasing the distinctiveness of cues (Holley and Dansereau, 1984). *Elaboration* is another technique that enchances retrieval. By adding related information to the new ideas to be learned, the learner adds more cues for retrieval. The cues may be within the learner's environment, as when parts of a room are used to recall the sequence of ideas in a speech. More frequently, though, the cues are themselves retrieved from the learner's memory as words, phrases, or images.

Another external condition that plays a part in the retention of meaningful prose is the adoption of an *attentional strategy* by the learner. Suggestions of "what to look for" or "what to remember" may be made to a learner before learning begins; these may have the effect of activating a cognitive strategy for learning. A suggestion may be given directly or indirectly via questions inserted in a text (cf. Frase, 1970). Another method involves the use of an *advance organizer* (Ausubel, Novak, and Hanesian, 1978; Mayer, 1979), a brief passage given before the text to be learned, which has the effect of orienting the learner to what is to be remembered in a subsequent passage.

Repetition has long been known to have a marked effect on the remembering of information, and this is true whether one is dealing with isolated facts or with larger bodies of information. The effective employment of repetition, however, is in providing *spaced* occasions for the learner to recall the information he has learned. The processes put into effect when information is retrieved from memory are apparently the most important factors in the remembering of such information.

ATTITUDES

We now shift gears again to consider the achievement of a very different kind of learning outcome, one which partakes not so much of knowledge as it does of *emotion* and *action*. This is the acquired state of the learner called *attitude*.

It would be difficult to overemphasize the importance of attitudes in school learning. In the first place, as is so evident to those who teach, students' attitudes toward attending school, toward cooperating with instructor and fellow students, toward giving attention to the communications made to them and toward the act of learning itself are all of great significance in determining how readily the students learn.

A second large class of attitudes are those that institutions (such as schools) aim to establish as a result of instruction. Attitudes of tolerance and civility towards other people are often mentioned as goals of education in the schools. Positive attitudes towards the seeking and learning of new skills and knowledge are usually stated as educational goals of far-reaching importance. More specific likings for various subjects of instruction, whether science, literature, salesmanship, or labor negotiations, are usually conceived as objectives of high value within each subject area. And finally, there are attitudes of broad generality, often called *values,* to which schools and other societal institutions are expected to contribute and influence. These are attitudes pertaining to such social behaviors as are implied by the words fairness, honesty, charitableness, and the more general term *morality* (cf. R. M. Gagné, 1985, pp. 226–228).

Regardless of the great variety exhibited by the content of these types of attitudes, one must expect that they all resemble each other in their formal properties. That is to say, whatever the particular content of an attitude, it functions to affect "approaching" or "avoiding." In so doing, an attitude influences a large set of specific behaviors of the individual. It is reasonable to suppose, then, that there are some general principles of learning that apply to the acquisition and changing of attitudes.

Definition of Attitude

Attitudes are complex human states that affect behavior towards people, things, and events. Many investigators have studied and emphasized in their writings the conception of an attitude as a system of beliefs (Festinger, 1957). These views serve to point out the *cognitive* aspects of attitudes. Other writers deal with their *affective* components, the feelings they give rise to or that accompany them, as in liking and disliking. Learning outcomes in the "affective domain" are described by Krathwohl, Bloom, and Masia (1964).

There are several varieties of theory concerning the nature and origin of attitudes. A comprehensive review of the major theories and their implications for instruction is presented by Martin and Briggs (1986). These authors describe procedures for instructional design that integrate affective and cognitive objectives.

For a number of reasons, including practical ones, it seems desirable in the present context to give emphasis to the aspect of attitudes relating to *action.* Acknowledging that an attitude may arise from some complex of beliefs, and that it may be accompanied and invigorated by emotion, the important question would appear to be, what action does it support? The general answer to this question is that an attitude influences *a choice of personal action* on the part of the individual. A definition of attitude, then, is an internal state that affects an individual's choice of personal action toward some object, person, or event (Gagné, 1985).

Portions of this definition require some comments. An attitude is an *internal state,* inferred from observations (or often, from reports) of the individual's behavior; it is not the behavior itself. If one observes a person depositing a gum wrapper in a waste basket, an inference cannot be made from that single instance alone that the person has a positive attitude towards disposing of personal trash, or a negative attitude towards pollution, and certainly not an attitude towards gum wrappers. A number of instances of behavior of this general class, however, may occur in a number of different situations, making possible an inference about that person's attitude. The inference is that some internal state affects a whole class of specific instances, in each of which the individual is making a *choice*.

This choice, which is inferred to be affected by the attitude, is of a *personal action*. Thus a person may choose to throw away a gum wrapper or to hold it until a trash basket is handy. Another person may choose to vote for or against a presidential candidate—the choice indicates the attitude. A white student may choose to speak in a friendly manner to a black classmate or may not so speak—again, an indicator (along with other instances) of the student's attitude. In following this definition, one cannot answer the question, "What is this person's attitude toward black Americans," because that is altogether too general a question to be answered sensibly. Instead, one asks, "What is this person's attitude toward *working* with black people or *living near* black people or *sitting beside* a black person?" It is the choice of a personal action in each case that is affected by an attitude. In connection with school learning, one may be interested in a student's attitude toward *reading* books, *doing* scientific experiments, *writing* stories, or *constructing* an art object.

This definition implies that attitudes should be measured in terms of the personal actions chosen by the individual. In some instances, such measurement can be done by observation over a period of time. For example, a teacher may record observations of an elementary pupil over a weekly period, recording the number of times that pupil helps his classmates as opposed to interfering with their activities (cf. Mager, 1968). A proportion of this sort, recorded over several such periods, can serve well as a measure of the student's "attitude toward helping others." Of course, such direct indicators of choice cannot always be obtained. For example, the teacher would find it difficult to obtain behavioral measures of the pupil's "attitude towards listening to classical music," or her "attitude towards reading novels" because many choices in these areas are made outside of the school environment. Attitude measures are therefore frequently based upon *self-reports* of choices in situations described in questionnaires. Typical questions, for example, may ask the student to use a 10-point scale to answer a variety of questions, such as: "When choosing a book from the public library to read on a summer afternoon, how likely are you to pick a novel about adventure on the seas?" This method of attitude measurement, emphasizing choices of action, has been described in the work of Triandis (1964).

Attitude Learning

The conditions favoring the learning of attitudes and the means of bringing about changes in attitudes are rather complex matters, about which much is yet to be discovered. A number of contrasting views on the effectiveness of attitude-change methods are reviewed by Martin and Briggs (1986). Certainly the methods of instruction to be employed in establishing desired attitudes differ considerably from those applicable to the learning of intellectual skills and verbal information (Gagné, 1985).

How does the individual acquire or modify an internal state that influences his choices in a particular area of action? One way that this is *not* done, according to a great deal of evidence, is solely by the use of persuasive communication (McGuire, 1969). Perhaps most adults would recognize the ineffectiveness of repeated use of such maxims as "Be kind to others," or "Learn to appreciate good music," or "Drive carefully." Even more elaborate communications, however, often have equally poor effects, such as those that make emotional appeals or those that are developed by a careful chain of reasoning. Apparently, one must seek more sophisticated means than these for changing attitudes, and more elaborately specified conditions for attitude learning.

Direct Methods

There are direct methods of establishing and changing attitudes, which sometimes occur naturally and without prior planning. On occasion, such direct methods can also be employed deliberately. At least, it is worthwhile to understand how attitude change can come about by these means.

A conditioned response of the classical sort (cf. Gagné, 1985, pp. 24–29) may establish an attitude of approach or avoidance toward some particular class of objects, events, or persons. Many years ago, Watson and Rayner (1920) demonstrated that a child could be conditioned to "fear" (that is, to shrink away from) a white rat he previously had accepted and petted. This type of response was also made to other small furry animals. The unconditioned stimulus used to bring about this marked change in the child's behavior was a sudden sharp sound made behind the child's head, when the animal (the conditioned stimulus) was present. Although this finding may not have specific pedagogical usefulness, it is important to realize that attitudes can be established in this way, and that some of the attitudes students bring to school with them may be dependent upon earlier conditioning experiences. A tendency to avoid birds or spiders or snakes, for example, may be instances of attitudes having their origin in a prior event of conditioning. In theory, almost any attitude might be established in this way.

Another direct method of attitude learning having more usefulness for school situations is based upon the idea of arranging *contingencies of reinforcement* (Skinner, 1968). If a new skill or element of knowledge to be learned is followed by some preferred or rewarding activity, in such a way that the latter is contingent upon achieving the former, this general situation describes the basic prototype of learning, according to Skinner. In addition, the student who begins with a *liking* for the second activity (called a *reinforcer*) will, in the course of this act of learning, acquire a liking for the first task. Following this principle, one might make a preferred activity for an elementary pupil, such as examining a collection of pictures, contingent upon her asking to see the pictures by means of a complete sentence ("May I look at the pictures?") as opposed to asking by blurting out a single word ("Pictures?"). Continuation of this practice in a consistent way and in a variety of situations will likely result in the child's using complete sentences when making a request. The child will also come to enjoy the newly learned way of asking for things because of experiencing success in doing so. In other words, an attitude toward "using complete sentences" will take a positive turn.

Generalizing somewhat from this learning principle of reinforcement contingencies, it would appear that *success* in some learning accomplishment is likely to lead to a positive attitude toward that activity. Young people acquire definitely positive attitudes toward ice skating when they achieve some success at it. Students develop positive attitudes toward listening to classical music when they realize that they are able to recognize the forms and themes the music contains.

An Important Indirect Method

A method of establishing or changing attitudes of great importance and widespread utility is *human modeling* (Bandura, 1969, 1977). We regard this method as indirect because the chain of events that constitutes the procedure for learning is longer than that required for more direct methods. Furthermore, as its name implies, this method operates through the agency of another human being, real or imagined.

Students can observe and learn attitudes from many sorts of human models. In early years, one or both parents serve as models for actions that are instances of fairness, sympathy, kindness, honesty, and so on. Older siblings or other members of the family may play this role. During school years, one or more teachers may become models for behavior, and this remains a possibility from kindergarten through graduate school. But the varieties of human modeling are not confined to the school. Public figures may become models, as may prominent sports people, or famous scientists, or artists. It is not essential that people who function as human models be seen or known personally—they can be seen on television or in movies. In fact, they can even be read about in books. This latter fact serves to emphasize the enormous potential that printed literature has for the determination of attitudes and values.

The human model must, of course, be someone whom the learners *respect;* or as some writers would have it, someone with whom they can *identify.* In addition, desirable characteristics of the model are to be perceived as *credible* and *powerful* (Gagné, 1984). Observation must be made of the model making the desired kinds of choices of personal action; as in exhibiting kindness, rejecting harmful drugs, or cleaning up litter. A teacher model may be seen to dispense praise consistently and impartially. Having perceived the action, whatever it may be, the learner also must see that such action leads to satisfaction or pleasure on the part of the model. This step in the process is called *vicarious reinforcement* by Bandura (1969). A sports figure may receive an award or display his pleasure in breaking a record; a scientist may exhibit his satisfaction in discovering something new, or even in getting closer to such a discovery. The teacher may show that she is glad to have helped a slow-learning-child to acquire a new skill.

The essential characteristics of attitudes and the conditions for their learning, using human modeling, are summarized in the following paragraphs.

Performance

An attitude is indicated by the choice of a class of personal actions. These actions can be categorized as showing either a positive or negative tendency toward some objects, events, or persons.

Internal Conditions

An attitude of respect for or identification with the human model must preferably be already present in the learner. If it is not, it needs to be established as a first step in the process. Intellectual skills and knowledge related to the behavior exhibited by the model must have been previously acquired for this behavior to be imitated. For example, an attitude of rejection of harmful drugs must be preceded by knowledge of the common names of the drugs and the situations in which they may be available. It may be noted, however, that such prerequisite knowledge does not in itself engender the attitude.

External Conditions

These may be described as the following sequence of steps.

1. presentation of the model and establishment of the model's appeal and credibility
2. recall by the learner of knowledge of the situations to which the attitude applies
3. communication or demonstration by the model of the desired choices of personal action
4. communication or demonstration that the model obtains pleasure or satisfaction with the outcome of the behavior. This step is expected to lead to vicarious reinforcement on the part of the learner.

Of course, modification of these steps is possible when the human model is not directly seen and when the desired performance cannot be directly observed. The essential conditions may still be present when the learner is viewing television or reading a book. Other variations on the human modeling theme may also be noted (Gagné, 1985, p. 239). One form that human modeling can take is *role playing,* in which the actor models action choices after those of an imagined rather than an actual person. A student playing the role of a fair-minded work supervisor is influenced by the choices made by that (imaginary) person. Human modeling also takes place during *class discussions* bearing upon social or personal problems. In such situations, more than one point of view toward personal action choices is likely to be presented. Discussion leaders become, in effect, the human models that influence the establishment of attitudes.

The modification of attitudes undoubtedly takes place all the time in every portion of a student's daily life. Models with whom the student comes in contact bear a tremendous responsibility for the determination of socially desirable attitudes and the development of *moral behavior* (Gagné, 1985, pp. 226*ff*). Teachers obviously need to appreciate the importance of the human modeling role if for no other reason than an appreciation of the large proportion of time spent by students in their presence. It is likely that those teachers the student later remembers as "good teachers" are the ones who have modeled positive attitudes.

MOTOR SKILLS

Sequences of unitary motor responses are often combined into more complex performances called *motor skills*. Sometimes these are referred to as "perceptual-motor skills" or "psychomotor skills", but these phrases appear to carry no useful added meaning. They imply, of course, that the performance of motor skills involves the senses and the brain as well as the muscles.

Characteristics of Motor Skills

Motor skills are learned capabilities that underlie performances whose outcomes are reflected in the rapidity, accuracy, force, or smoothness of bodily movement. In the school, these skills are interwoven throughout the curriculum at every age, and include such diverse activities as using pencils and pens, writing with chalk, drawing pictures, painting, using a variety of measuring instruments, and of course, engaging in various games and sports. Basic motor skills such as printing numerals on paper are learned in early grades and assumed to be present thereafter. In contrast, a motor skill like tying a bowline knot may not previously have been learned by a fifth-grader, and so would conceivably constitute a reasonable objective for instruction at that age or later.

As motor skills are practiced, they appear to give rise to a centrally organized *motor program* that controls the skilled movements without feedback from the senses (Keele, 1973). However, this is not the whole story. The increased smoothness and timing that result from practice of a motor skill are considered by Adams (1977) to be dependent on feedback that is both internal and external. Internal feedback takes the form of stimuli from muscles and joints that make up a *perceptual trace,* a kind of motor image that acts as a reference against which the learner assesses error on successive practice trials. External feedback is often provided by *knowledge of results,* an external indication to the learner of the degree of error. As practice proceeds, improvement in the skill comes to depend increasingly upon the internal type of feedback and to a lesser degree upon externally provided knowledge of results.

Usually, motor skills can be divided into *part skills* that constitute the total performance in the sense that they occur simultaneously or in a temporal order. Swimming the crawl, for example, contains the part skills of foot flutter and arm stroke, both of which are carried out at the same time; and also the part skill of turning the head to breathe, which occurs in a sequence following an arm stroke. Thus the total performance of swimming is a highly organized and precisely timed activity. Learning to swim requires the integration of part skills of various degrees of complexity, some of them as simple as motor chains. The *integration* of these parts must be learned, as well as the component part skills themselves.

Learning to integrate part skills that are already learned has been recognized by investigators of motor skills as a highly significant aspect of the total learning required. Fitts and Posner (1967) refer to this component as an *executive subroutine,* using a computer analogy to express its organizing function. Suppose, for example, an individual learning to drive an automobile has already mastered the part skills of driving backward, of turning the steering wheel to direct the motion of the car, and of driving (forward or backward) at minimal speed. What does such a person still need to learn in order to turn the car around on a straight two-lane street? Evidently, he needs to learn a procedure in which these part skills are combined in a suitable order, so that by making two or three backward and forward motions, combined with suitable turning, the car is headed in the other direction.

This instructive example of turning a car around shows the importance of the intellectual component of a motor skill. Obviously, the executive subroutine is not in itself "motor" at all, rather, it is a *procedure,* conforming to the qualities of a procedural *rule* as described in the previous chapter. The rule-governed aspect of the motor skill performance is what controls the sequence of action—which movement is executed first, which second, and so on (Gagné, 1985, pp. 202–212).

Swimming provides an interesting comparison. It, too, has an executive subroutine pertaining to the timing of flutter kicks, arm movements, and head turning to breathe. But in this case, the smooth performance of these part skills is usually being improved by practice at the same time that the executive routine is being exercised. Many studies have been performed to find out whether practicing the part skills of various motor skills first is more advantageous than practicing the whole skill (including the executive subroutine) from the outset (Naylor and Briggs, 1963). No clear answer has emerged from these studies, and the best one can say is that it depends on the skill; sometimes part skill practice is an advantage, and sometimes not. It is clear, however, that *both* the executive subroutine and the part skills must be learned. Practice on either without the other has many times been shown to be ineffective for the learning of the total skill.

Learning Motor Skills

The learning of motor skills is best accomplished by repeated practice. There is no easy way of avoiding practice if one seeks to improve the accuracy, speed, and smoothness of motor skills. In fact, it is interesting to note that practice continues to bring about improvement in motor skills over very long periods of time (Fitts and Posner, 1967; Singer, 1980), as performers in sports, music, and gymnastics are well aware.

Performance

The performance of a motor skill embodies the intellectual skill (procedure) that constitutes a movement sequence of muscular activity. When observed as a motor skill, the action meets certain standards (either specified or implied) of speed, accuracy, force, or smoothness of execution.

Internal Conditions

The executive subroutine that governs the procedure of the motor skill must be retrieved from prior learning or must be learned as an initial step. For example, the part skills of "backing" and "turning" in an automobile must be previously acquired and retrieved to enter into the skill of "turning the car around on a street." As for the part skills that will make up the total skill, they will depend upon the retrieval of individual responses or unitary motor chains.

External Conditions

For the learning of the *executive subroutine,* the instructor provides one of several different kinds of communications to the learner. Sometimes, verbal instructions are used ("Bend your knee and put the weight on your left foot"). In fact, any of the kinds of verbal communications intended for encoding of a procedure (see previous chapter) may be employed. A checklist showing the sequence of required movements can be presented, with the expectation that it will be learned as a part of practice. Pictures or diagrams may be used to show the required sequence. For the improvement of accuracy, speed, and quality of *part skills,* as well as the *total skill,* the learner engages in practice, repeating the movements required to produce the desired outcome in each case. The skill is improved by continued practice with the accompaniment of informative feedback (Singer, 1980).

SUMMARY

The present chapter has been concerned with a description of three different kinds of learning—verbal information, attitudes, and motor skills. Although they have some features in common, their most notable characteristic is that they are in fact different. They differ, first, in the kinds of outcome performances which they make possible:

1. verbal information—verbally stating facts, generalizations, organized knowledge
2. attitude—choosing a course of personal action
3. motor skill—executing a performance of bodily movement

As our analysis of the conditions of learning has shown, these three kinds of learning differ markedly from each other in the conditions necessary for their effective achievement. For verbal information, the key condition is the provision of external cues that relate the new information to a *network of organized knowledge* from prior learning. For attitude, although such a network of knowledge indicates the situational limits, one must either assure direct reinforcement of personal action choices, or depend upon *human modeling* to bring about vicarious reinforcement of the learner. And for the learning of motor skills, besides the early learning of the executive subroutine and provision for the integration of part skills, the important condition is *practice* with frequent information feedback to the learner.

The kinds of performances associated with these capabilities and the conditions for their learning are also obviously different from those described in the previous chapter pertaining to intellectual skills and cognitive strategies. Are the types of learning outcomes dealt with in this chapter in some ways less important than intellectual skills? Assuredly not. The storage and accessibility of verbal information, particularly in the form of organized knowledge, is a legitimate and desirable objective of instruction in both formal and informal educational settings. The establishment of attitudes is widely acknowledged to be a highly significant objective of many courses of study, and some would accord prosocial attitudes the highest importance of all. Motor skills, although they often appear to contrast with the cognitive orientation of schools, individually have their own justification as fundamental components of basic skills, of art and music, of science, and of sports.

The contrasting features of these kinds of learned capabilities, as compared with those of intellectual skills, do not reside in their differing importance for programs of instruction. Instead, these capabilities differ with respect to the internal conditions that must be assumed and the external conditions that must be arranged for instruction to be effective. Included in these differences are the *enabling prerequisite* relationships that intellectual skills have with each other, as described in the previous chapter, as compared with the *supportive* effects of prior learning on the varieties of learned capabilities covered in this chapter (cf. Gagné, 1985, pp. 286–272). This characteristic has particular implications for the determination of desirable *sequences of instruction,* as will be seen in a later chapter.

The system of instructional design being developed in this book is one that suggests that first priority be given to intellectual skills as central planning components. That is, the fundamental structures of instruction are designed in terms of what the student will be *able to do* when learning has occurred, and this capability is in turn related to what has previously been learned. This strategy of instructional design typically leads to the identification of intellectual skills as a first step, followed by analysis and identification of their prerequisites. To this basic sequence may then be added, at appropriate points, the cognitive strategies which these basic skills make possible. In other instances, the primary objective of instruction may be a particular body of verbal information, or the changing of an attitude, or the mastery of a motor skill. These cases also require analysis to reveal the supportive effects of prior learning. Most typical of all, as will be seen later, are instances of instruction that require provisions for multiple objectives.

Several of the following chapters directly address procedures for designing instruction. In large part, the techniques we describe are derived from the knowledge of varieties of learning outcomes which we have detailed, and represent direct applications of this knowledge.

References

Adams, J. A. (1977). Motor learning and retention. In M. H. Marx & M. E. Bunch (Eds.), *Fundamentals and applications of learning.* New York: Macmillan.

Anderson, J. R. (1985). *Cognitive psychology and its implications* (2nd ed.). New York: Freeman.

Ausubel, D. P., Novak. J. D., & Hanesian, H. (1978). *Educational psychology: A cognitive view* (2nd ed.). New York: Holt, Rinehart and Winston.

Bandura, A. (1969). *Principles of behavior modification.* New York: Holt, Rinehart and Winston.

Bandura, A. (1977). *Social learning theory.* Englewood Cliffs, NJ: Prentice-Hall.

Bretz, R. (1971). *A taxonomy of communication media.* Englewood Cliffs, NJ: Educational Technology Publications.

Clark, R. E. & Salomon, G. (1986). Media in teaching. In M. C. Wittrock (Ed.), *Handbook of research on teaching* (3rd ed.). New York: Macmillan.

Festinger, L. (1957) *A theory of cognitive dissonance.* New York: Harper & Row.

Fitts, P. M., & Posner, M. I. (1967). *Human performance.* Monterey, Calif.: Brooks/Cole.

Frase, L. T. (1970). Boundary conditions for mathemagenic behavior. *Review of Educational Research, 40,* 337–347.

Gagné, E. D. (1985). *The cognitive psychology of school learning.* Boston: Little, Brown.

Gagné, R. M. (1984). Learning outcomes and their effects. *American Psychologist, 39,* 377–385.

Gagné, R. M. (1985). *The conditions of learning* (4th ed.). New York: Holt, Rinehart and Winston.

Glaser, R. (1984). Education and thinking: The role of knowledge. *American Psychologist, 39,* 93–104.

Holley, C. D., & Dansereau, D. F. (1984). *Spatial learning strategies: Techniques, applications, and related issues.* New York: Academic Press.

Hulse, S. H., Egeth, H., & Deese, J. (1980). *The psychology of learning* (5th ed.) New York: McGraw-Hill.

Kausler, D. H. (1974). *Psychology of verbal learning and memory.* New York: McGraw-Hill.

Keele, S. W. (1973). *Attention and human performance.* Pacific Palisades, CA: Goodyear.

Krathwohl, D. R., Bloom, B. S., & Masia, B. B. (1964). *Taxonomy of educational objectives. Handbook II: Affective domain.* New York: McKay.

Mager, R. F. (1968). *Developing attitude toward learning.* Belmont, CA: Fearon.

Martin, B. L., & Briggs, L. J. (1986). *The affective and cognitive domains: Integration for instruction and research.* Englewood Cliffs, NJ: Educational Technology Publications.

Mayer, R. E. (1979): Can advance organizers influence meaningful learning? *Review of Educational Research, 49,* 371–383.

McGuire, W. J. (1969). The nature of attitudes and attitude change. In G. Lindzey & E. Aronson (Eds.), *Handbook of social psychology,* Vol. 3 (2nd ed.). Reading, MA: Addison-Wesley.

Meyer, B. J. F. (1975). *The organization of prose and its effects on memory.* New York: Elsevier.

Naylor, J. C., & Briggs, G. E. (1963). Effects of task complexity and task organization on the relative efficiency of part and whole training methods. *Journal of Experimental Psychology, 65,* 217–224.

Polanyi, M. (1958). *Personal knowledge.* Chicago: University of Chicago Press.

Pressley, M., Levin, J. R., & Delaney, H. D. (1982). The mnemonic keyword method. *Review of Educational Research, 52,* 61–91.

Reynolds, A. G., & Flagg, P. W. (1977) *Cognitive psychology.* Cambridge, MA: Winthrop.

Rohwer, W. D., Jr. (1970). Images and pictures in children's learning: Research results and educational implications. *Psychological Bulletin, 73,* 393–403.

Singer, R. N. (1980). *Motor learning and human performance* (3rd ed.). New York: Macmillan.

Skinner, B. F. (1968). *The technology of teaching.* New York: Appleton.

Spiro, R. J. (1977). Remembering information from text: The "state of the schema" approach. In R. C. Anderson, R. J. Spiro, & W. E. Montagne (Eds.), *Schooling and the acquisition of knowledge.* Hillsdale, NJ: Erlbaum.

Triandis, H. C. (1964). Exploratory factor analyses of the behavioral component of social attitudes. *Journal of Abnormal and Social Psychology, 68,* 420–430.

Watson, J. R., & Rayner, R. (1920). Conditioned emotional reactions. *Journal of Experimental Psychology, 3,* 1–14.

6   THE LEARNER

Whatever it is that gets learned can be identified as one or another of the capabilities described in the two previous chapters or as some combination of them. Learners may be expected to learn a set of intellectual skills involving mathematical operations, for example, and perhaps also a positive attitude toward the use of these operations. Or, learners may be asked to acquire some organized verbal information about the history of the zipper fastener plus some intellectual skills about how to assemble such a fastener. Young learners who encounter a zipper fastener for the first time will doubtless need to learn the motor skill required for its operation.

This range of learning tasks is undertaken by learners who themselves exhibit diversity that is enormous in scope and detail. The learners who approach new learning tasks are all quite different in their characteristics as learners. In the face of this almost bewildering diversity, the procedures of instructional design must do the following things:

1. Discover a rational means of reducing the great diversity of individual learner characteristics to a number small enough to make instructional planning feasible.
2. Identify those dimensions of common learner characteristics that carry different implications for instruction and that can lead to design differences that influence learning effectiveness.
3. Once common learner characteristics have been taken into account, provide a design appropriate for those learner variations that can be shown to make a difference in learning results.

We intend in this chapter to address these questions, and in so doing to convey the knowledge that will lead to good decisions in instructional design. Our discussion will be based upon considerations of what learners are like, and what different sets of learners are like. Instructional design needs to take account of the learners who are to participate in the activities that constitute instruction.

LEARNER CHARACTERISTICS

Learners possess certain qualities that relate to instruction—for example, they are able to hear orally delivered communications and to read communications printed on the page. Each of these common qualities varies in degree from learner to learner—one person may be able to read pages of printed text rapidly, whereas another reads slowly and haltingly. Regardless of variations in degree, the characteristics of concern to instructional design are those that affect the entire information-processing chain of learning. These are qualities that may pertain to sensory input, to the internal processing, storage, and retrieval of information, and finally to the organization of learner responses.

The Nature of Learner Qualities

Some qualities of the human individual that relate to learning are innately determined. For example, *visual acuity*, although it may be aided by artificial lenses, is a fundamental property of a person's sensory system that is "built in," and that cannot be changed by learning. Such a property needs to be considered in instruction, however, only under certain extreme conditions, leading to the avoidance of fine print or fuzzy projection images.

Other learner qualities, however, may affect learning at junctures of information processing that are more critical to instructional planning. For example, it has been proposed that the capacity of the working memory, where to-be-learned material is taken in and processed for memory storage, may have innate capacity limits. The number of items that can be "held in mind" at any one time is shown by the immediate memory span of seven plus or minus two. The speed with which previously learned concepts can be retrieved and identified may be measured by requiring individuals to respond as rapidly as possible to indicate whether pairs of letters match or not, when the letters are physically different (as A,a or B,b). Because of the physical difference, the letters must be retrieved as concepts in order to be matched (Hunt, 1978). The speed and efficiency of this process is another individual quality that may be innately determined.

For these and other learner characteristics that are genetically determined, instructional design cannot have the aim of altering these qualities by means of learning. Instead, instruction must be designed in such a way as to *avoid exceeding human capacities*. For example, in the early stages of learning to read, it is possible that some decoding tasks may exceed children's working memory capacity, when words of many letters are encountered. For readers in general, long sentences place demands upon working memory that may exceed the limits of capacity. The design technique to be employed in such instances is to use words, sentences, diagrams, or other varieties of communication that stay well within the capacity of the working memory, and thus avoid testing its limits.

Qualities That Are Learned

Besides qualities that may be innate, and therefore impervious to change by learning, there is a large set of characteristics that are learned. Many of these have critical effects upon learning. It is these that constitute the *internal conditions* described in previous chapters in connection with each of the learning varieties.

Intellectual Skills

The typical intellectual skill, a rule, is considered to be stored as a set of concepts that are syntactically organized. Specifically, it is generally believed that a rule has the functional form called by Newell and Simon (1972) a *production*. An example of production is:

> IF the goal is to convert x inches to centimeters,
> THEN multiply x by 2.54.

Concepts can similarly be represented as productions, as in the following example:

> IF the closed two-dimensional figure has all sides equal,
> THEN classify the figure as a polygon.

Obviously, we are describing procedures, and that is why the phrase *procedural knowledge* is customarily used to refer to stored sets of intellectual skills. It is also true that productions, considered as stored entities, have the syntactic and semantic properties of propositions. The complex rules that are typical intellectual skills are composed of simpler rules and concepts. The latter are usually learned as *prerequisites* to the skill that is the targeted instructional objective. When retrieved from memory, the complex skill readily activates these simpler prequisite skills since they are actual components. The example in Figure 6-1 shows the component skills that have entered into the learning of the target skill of "pronouncing multisyllable printed words."

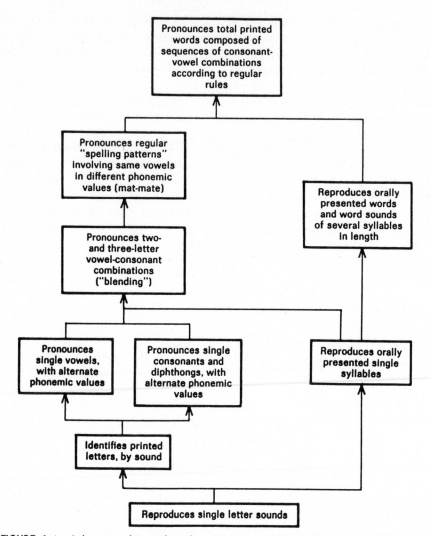

FIGURE 6-1 A learning hierarchy, showing prerequisite skills, for the skill of pronouncing multisyllable printed words.

Cognitive Strategies

Since strategies are mental procedures, they are a form of intellectual skill. Accordingly, they may be conceived as *productions* and represented in that manner. For example, youngsters may acquire a cognitive strategy that enables them to "self-edit" their writing of sentences, and by so doing, generate sentences that are more mature. Thus, an original sentence such as "John went to the store", when acted upon by a strategy of question asking, may be expanded to the sentence "In the morning, John walked to the hardware store in the center of town."

The production (strategy) in this case may be represented somewhat as follows:

IF the goal is to revise a sentence to achieve full communication,
THEN add component phrases that answer when, how, where, and why.

Two characteristics of cognitive strategies are particularly notable. First, they are procedures that govern the selection and use of intellectual skills. Thus, the strategy of "self-editing" can be used only when the intellectual skills of sentence making and phrase making are already known. And second, it may be noted that the structure of the strategy is not itself complex—it is simply a matter of asking four well-known questions. The typical cognitive strategy has broad generality; in this case it is applicable to virtually any sentence, of whatever subject. Yet it is *domain specific*—it is a strategy relevant to sentence editing and revision and not to something else.

Verbal Information

Knowledge in this declarative form may be stored as individual propositions (facts) or as networks of propositions, organized around central ideas or generic concepts. As indicated in Figure 6-2, the fact that "igneous rock is very hard because it has been subjected to high temperature and pressure" is connected with a very large network of other interrelated propositions (E. D. Gagné, 1985). Furthermore, elements of this complex network make connections at various points with productions (intellectual skills) (see Figure 6-2).

When a search of memory makes contact with a single proposition, other interconnected propositions are "brought to mind" as well. The process is known as *spread of activation* (Anderson, 1985) and is considered to be the basis for the retrieval of knowledge from the long-term memory store. When the learner attempts to recall a single idea, the initial search activates not only that idea but many related ones also. Thus, in searching for the name Helen one may be led by spreading activation through Troy and Poe and Greece and Rome and the Emperor Claudius to the Battle of Britain, and to many things in between. Spreading activation not only accounts for what we perceive as random thoughts, as in free association, but is also the basis for the great flexibility that is apparent when we engage in reflective thinking.

Attitudes

An attitude in memory appears to be somewhat complex, and therefore difficult to represent in a schematic fashion. For an attitude, what is stored in memory would appear to include: (1) the choice of personal action, as exhib-

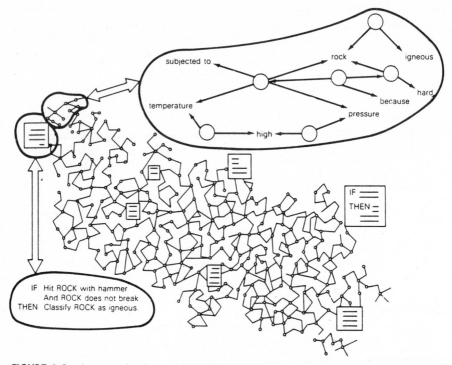

FIGURE 6-2 A network of interrelated propositions, including many items of verbal information and interspersed intellectual skills (productions), shown here as boxes. (From E. D. Gagné, *The cognitive psychology of school learning,* copyright 1985 by Little, Brown; Boston. Reprinted by permission.)

ited by a human model; (2) a representation of a standard of conduct for the self that reflects the standard of the human model; and (3) a feeling of satisfaction derived from reinforcement of the chosen action or from vicarious reinforcement, as described in the previous chapter. Overlayed on these stored memories may be an emotional disposition arising from the need to resist contrasting ideas or dissonant beliefs.

Attitudes, too, are likely to be embedded within a complex of interconnected propositions. It has often been observed that internal states affecting choices of personal action are strongly influenced by *situational factors*. It may be supposed, therefore, that attitudes occur within propositional networks that are organized by situation. A person may have an attitude of neatness, for example, that applies to the situation of storing kitchen tools, but that does not apply to the storage of papers on a working desk. When memory of the situation is revived, retrieval of the objects and events usually brings with it a revival of the attitude as well. Thus, for many older Americans, even the word *depression* can reactivate a fairly strong attitude along with memories of a set of situational factors pertaining to the 1930s.

Of particular importance to the representation of attitude in memory is a human model's report (or demonstration) of personal action choices. The human model is remembered as an admirable person and one who is both credible and powerful (Gagné, 1985). Memory storage also includes the model's action choices: The model may be remembered as rejecting harmful drugs, or as liking running as exercise, or as preferring to listen to classical music. When these memories are retrieved along with the situational factors appropriate to each, conditions become right for the choice of personal action that reveals an attitude.

Motor Skills

The core memory of a motor skill appears to consist of a highly organized and centrally located motor program (Keele, 1968). Such a program is established by practice, attains automaticity, and becomes only incidentally responsive to variations in external stimulation and in kinesthetic feedback. In addition, a motor skill has some prerequisites, as is the case with intellectual skills. One set of simpler components of a motor skill may be its part skills— sometimes easy to identify, and sometimes not. In threading a needle, for example, one can detect the part skills of (1) holding the needle steady, (2) inserting the thread in the hole, and (3) grasping the end of the thread after insertion.

Even more essential as an aspect of motor skill storage is its procedure, or *executive subroutine,* usually learned as the earliest component of a motor skill (Fitts and Posner, 1967). The basic sequence of movement, having no requisite smoothness and timing, has the character of an intellectual skill (a procedural rule). Such a procedure may have been acquired as a prerequisite to the motor skill or learned in an early stage of practice. When a motor skill has been unused for many years, the executive subroutine is likely to remain intact, even though the performance has become hesitant and rough. A person will remember how to finger a clarinet, even though the musical quality of the performance may show the effects of years of disuse.

MEMORY ORGANIZATION

The unitary things that are learned and stored in long-term memory may be conveniently thought of as propositions (both declarative and procedural) and images and motor programs. These unitary entities are organized into interconnected networks that may be searched out and retrieved by the learner to serve the requirements of some activity or of further learning.

The networks representing various kinds of learned capabilities often assume a form called a *schema,* in which ideas are organized in terms of a general topic or use function. As writers on the subject have pointed out (Rumelhart, 1980; Schank and Abelson, 1977), we carry around with us knowledge structures organized in terms of such general topics as "going to a restaurant" and "shopping in a supermarket."

An even more general conception of the stored capabilities of human learners is the notion of *ability*. Abilities are measured by psychological tests, which sample the quality of performance in a number of areas of activity. Some well-known ability areas are called verbal, numerical, visual, and spatial. These are usually differentiated into more specific abilities such as verbal fluency, numerical reasoning, memory for visual form, spatial orientation, and the like (Cronbach, 1970; Guilford, 1967). Still other general characteristics of human learners belong in the *affective* and *personality* domains. They are often referred to as *traits* and include such qualities as *anxiety* and *motivation to learn* (Tobias, 1986). The importance of abilities and traits as human qualities lies in the possibility that they may affect learning differently depending upon differences in the nature of instruction. Thus, learners with high verbal ability may respond well to instruction consisting of tersely written printed text. Learners who are very anxious may learn best from instruction that has a highly organized structure. These are simply examples and will be further discussed in a later section.

Schemas

A schema is an organization of memory elements (propositions, images, and attitudes) representing a large set of meaningful information pertaining to a general concept (Anderson, 1985). The general concept may be a category of objects like *house, office, tree, furniture.* Or the concept may be of an event, as in *going to a restaurant* or *watching a baseball game.* Regardless of subject, a schema contains certain features common to the set of objects or events contained in that category. Thus, the schema for *house* contains information on certain well-understood features, such as construction material, rooms, walls, roofs, windows, and the function of human dwelling. These features are called *slots,* to imply that the values they take are there to be "filled in" (how many windows, what kind of roof, etc.).

Event schemas have similar properties, including slots. Thus, the slots to be filled in in a restaurant schema (Schank and Abelson, 1977) include such actions as entering the restaurant, deciding where to sit, examining a menu, ordering from a waiter or waitress, eating food, receiving the bill, paying the bill, leaving a tip, and leaving the restaurant. Thus, everyone who has had sufficient experience in dining in a restaurant carries around in memory a schema of roughly this sort, containing slots into which new facts from new experiences are fitted. Obviously, depending on his experience, one person's "restaurant schema" will differ from that of another person. Schemas may differ in their comprehensiveness, as well as in the details stored in their slots.

These differences may have a particular importance to the planning of instruction whose content is related to the schemas.

In approaching learning from newly presented instruction, learners come to the task with various schemas already in their memories. Thus, in beginning a lesson on the period of U.S. history encompassed by the presidential terms of Franklin Delano Roosevelt, learners are likely to have available schemas that contain such interconnected concepts as the Great Depression, the New Deal, and World War II. These and other features are filled out with more elaborate historical information by the new lesson. In learning to solve arithmetic word problems, youngsters may solve certain problems by reference to schemas of *change, combine,* or *compare* (Riley, Greene, and Heller, 1983). For instance, a *combine* schema is applicable to the kind of problem that states: "Joe and Charles have 8 pennies together; Joe has 5 pennies; how many pennies does Charles have?" A learner schema of *air pollution,* besides containing information about the oxides of sulfur and nitrogen and their possible origins, is likely to include also an attitude that influences personal choices of action toward votes on legislation pertaining to this subject.

Abilities and Traits

In addition to the organization resulting from specific instances of learning and experience, it has long been apparent that human performance is affected in quality by even more broadly influential structures called *abilities*. Over a period of many years, these factors relating to *how well* new problems can be solved have been differentiated from the original purpose of assessing *general ability* (Cronbach, 1970). In general, the abilities that can be assessed by psychological testing tend to be stable characteristics of each human individual, persisting over long periods of time, and not readily changed by regimens of instruction or practice focused upon them.

Other qualities of the performing human individual reflect *personality* and are usually referred to as *traits*. These aspects of human performance, like abilities, are also persistent over relatively long periods and not readily influenced by instruction aimed at changing them. Examples of traits are *introversion, conscientiousness, impulsiveness, self-sufficiency*. So many traits have been assessed in so many ways, that it is difficult and perhaps pointless to keep track of them. Nevertheless, the possibility exists that differences in one or more traits will exhibit an influence on learning that makes desirable the adaptation of instructional approaches to these differences. For example, perhaps *anxious* learners will be better served by instruction that differs from the kind used with the *nonanxious* learner.

Differential Abilities

It is still an unanswered question as to whether it is most useful to measure individual differences in general ability (or *intelligence*) or to measure a number of differential abilities. Scores on the latter kinds of tests show positive correlations with each other, and low to moderate correlations with measures of intelligence obtained from tests of general ability (such as the Stanford-Binet or the Wechsler Scales). Such abilities, therefore, are not truly distinctive one from another. And those who favor a view of intelligence as a general ability take satisfaction in noting that the various measures of different abilities all contain a *g* factor (for general intelligence). Various factors of differential ability have been proposed and investigated. Among the best known systems for classifying abilities are those of Thurstone (1938) and Guilford (1967). Some of the best known differential abilities are contained in the following list, which includes an indication of the kind of system used to measure each:

Reasoning (completing nonsense syllogisms)
Verbal Comprehension (comprehension of printed prose)
Number Facility (speeded tests of addition, division)
Spatial Orientation (identifying rotated figures)
Associative Memory (recalling object or number pairs)
Memory Span (immediate recall of digit lists)

There are several varieties of commercially available tests designed to measure abilities of these sorts among others. A description of ability tests may be found in Anastasi (1976), Cronbach (1970), and Thorndike and Hagen (1985).

Traits

The tendencies of people to respond in characteristic ways to a broad variety of particular situations gives rise to the inference that they possess certain relatively stable *personality traits* (Cronbach, 1970; Corno and Snow, 1986). Many kinds of personality traits have been defined and studied in students of a variety of ages and types. In recent years, a great deal of research has been done on a few traits that appear to have strong conceptual relation to human abilities, as well as with academic achievement. Some of the most widely studied traits of this sort are *achievement motivation* (McClelland, 1965), *anxiety* (Tobias, 1979), *locus of control* (Weiner, 1980), and *self-efficacy* (Bandura, 1982).

Research on achievement-related traits often takes the form of a search for *aptitude treatment interactions* (ATI). The hypothesis being investigated is that some variety of instruction (called a *treatment*) will differ in its effectiveness for learners scoring high and learners scoring low on a trait. In line with this idea, studies have shown such relations with several of the traits mentioned previously (Cronbach and Snow, 1977; Snow, 1977). Thus, learners with high achievement motivation appear to achieve better than those with low motivation when the instruction permits a considerable degree of *learner control* (Corno and Snow, 1986). Another example of ATI comes from studies of the anxiety trait. Anxious students who were given the option of reviewing instruction on videotape were found to learn more than anxious students who viewed the same tapes in groups without the review option (Tobias, 1986). Some students feel that external factors (such as luck) are responsible for the results of their learning, whereas others attribute outcomes to their own efforts. This personality difference is called *locus of control*. Students with an internal attribution might be expected to work harder and to have an active orientation to learning; this hypothesis may account for findings of ATI with respect to this trait.

Summary of Memory Organization

The unitary things that are learned and stored in human memory may be conceived of as *learning outcomes* and are called *learned capabilities* (Gagné, 1985). These learning capabilities are intellectual skills, verbal information, cognitive strategies, attitudes, and motor skills. They can be acquired through learning in a reasonably short time as a result of suitably designed instruction. These unitary capabilities (such as particular concepts, rules, or verbal propositions) are stored in memory as part of larger complexes called *schemas*.

A schema is a network of memory entities associated with each other through propositions, relating to an organizing general concept. There are event schemas (shopping in a supermarket), object and place schemas (your living room), problem schemas (elapsed clock time), and many others in the repertory of the average person. Schemas are characterized by commonly existing features, sometimes called *slots,* into which newly learned information is fitted. Thus, a supermarket schema may contain such slots as grocery cart, produce area, bread counter, meat counter, refrigerator, check-out counter, and the like. Newly acquired information about any particular shopping experience tends to be stored in these slots.

In addition to the large complexes of learned information and skills represented by schemas, there are even more general dispositions that can be revealed by psychological tests—*abilities* and *traits*. These features of human performance, although influenced by learning over long periods of time, are usually considered relatively stable characteristics that are not readily affected by instruction. An ability like spatial orientation may, however, have an effect on the ease with which particular learners acquire the skills of map reading. Likewise, a personality trait such as anxiety may affect the readiness for learning of certain learners when faced with a task that has severe time constraints. Relations of this sort are studied in investigations of *aptitude treatment interaction* (ATI). In practical terms, this type of research seeks ways in which instruction can be adaptively designed to allow for individual differences in abilities and traits.

LEARNERS AS PARTICIPANTS IN INSTRUCTION

Learners come to learning situations and to new learning tasks with certain performance tendencies already present. In the simplest case, learners may approach instruction on some subject or topic they already know. More frequently, however, the new material may be only partially known, and gaps in knowledge may need to be filled. Frequently also, learners may have background knowledge, or knowledge that is *prerequisite* to the learning of new material. Besides these direct relationships between the stored effects of prior learning and new learning, there may be more general ability differences in learners, or in groups of learners, that can profitably be taken into account in the design of instruction.

Designing Instruction for Learner Diffferences

As might be expected from our previous discussion, each of the kinds of learner characteristics carries different implications for the design of instruction. The most direct effects of the memory storage of prior learning can be seen in the entities we call learned capabilities, which include intellectual skills, verbal information, cognitive strategies, attitudes, and motor skills. Retrieval of these previously acquired kinds of memories by the learner has a specific influence on the learning of new material. Similar effects result from the recall of organized information in the form of schemas, which may provide direct support to the accomplishment of a new learning task. Effects that are more indirect, however, are provided by learner abilities and traits. These dispositions do not enter directly into the new learning, but they may greatly influence the ease with which learning processes are carried out.

When a new learning task is undertaken, the learner begins with several varieties of memory structures already in place, which are available for retrieval as part of the processing of the new learning. The kinds of effects exerted by what has been learned previously depend primarily on the objective of the new learning (Gagné, 1980). It will be clearest, therefore, to consider how new learning is affected by prior learning in terms of the expected outcomes of the new learning. We shall deal with this question in the following paragraphs, in terms of the learning outcomes that may be the principal objectives of new learning.

New Learning of Intellectual Skills

The learning of intellectual skills is most clearly influenced by the retrieval of other intellectual skills that are *prerequisite*. Usually, these are simpler skills and concepts that, when analyzed, are revealed to be actual components of the skill to be newly learned (Gagné, 1985). Results of analysis of this sort may be expressed as a *learning hierarchy*, an example of which is shown in Figure 6-3. For the skill of calculating velocity from a position–time graph, the various skills in subordinate boxes are prerequisites. As the studies of White (1973) have shown, the retrieval of these prerequisite skills has a direct supporting effect on the learning of the targeted intellectual skill. In fact, the

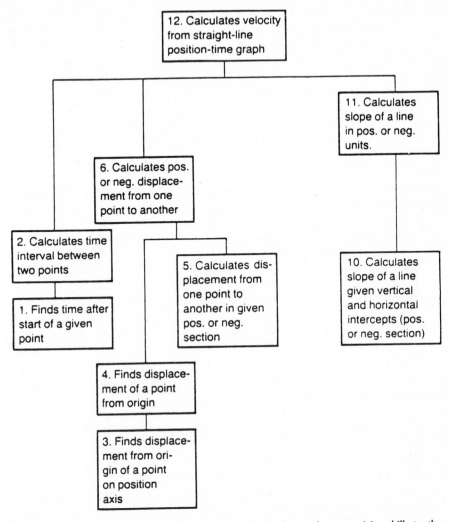

FIGURE 6-3   A learning hierarchy showing relationships of prerequisite skills to the task: Calculating velocity from a straight-line graph of *position* and *time*. Subordinate skills 7, 8, and 9 were found to be invalid prerequisites, and are not shown here.
(Based on a description in R. T. White & R. M. Gagné. Formative evaluation applied to a learning hierarchy, *Contemporary Educational Psychology*, 1978, 3, 87–94.)

absence of any of the skills in a subordinate box markedly decreases the ease of learning the superior skill to which it is connected. Clearly, for intellectual skills, the most direct effect of prior learning is through the retrieval of other intellectual skills that are prerequisite components.

To be most effective for new learning, prerequisite skills must be thoroughly learned, that is, learned to *mastery*. Presumably, this degree of learning makes the prerequisite skills easier to recall and, therefore, more readily accessible for new learning. Another condition that affects ease of retrieval is the number of cues available to the memory search process. Aids to memory search are provided by the cues of a schema. Accordingly, embedding the prerequisite skill (or skills) within the organized network of a schema may be expected to have a beneficial effect on instruction.

How will the presence of abilities and traits affect the learning of intellectual skills? This question deserves a separately headed section since its answer applies generally to other kinds of learning outcomes as well as to intellectual skills.

Effects of Abilities and Traits on New Learning

Human abilities are likely to affect new learning by contributing strategic techniques of processing the learning task and its material. For example, the ability called *numerical facility,* when possessed in a high degree, makes the processing of mathematics material easier and more rapid than it is for those whose numerical facility is low. The *verbal comprehension* ability has a similar facilitative effect on the processing of material presented in the form of connected prose. *Spatial orientation* as an ability aids in the processing of learning tasks that include information presented as figures and spatial arrays. In each case the ability contributes to learning in an *indirect* fashion by making the learning process easier. This effect contrasts with that of learned capabilities and of schemas, which enter directly into the new learning in a substantive manner.

Traits may also be expected to have an indirect effect upon the learning of intellectual skills (and other learning outcomes). Learners who are very anxious may be reassured by frequent feedback on their performances during practice of a newly learned skill, and they accordingly learn more readily than under conditions of infrequent feedback. Learners with high achievement motivation may learn rapidly when challenged by discovery learning, whereas learners with low achievement motivation may perform poorly.

An account of abilities and traits in instruction by Corno and Snow (1986) conceives of their effects in the following manner (p. 618). Abilities and traits in the individual learner give rise to an *aptitude complex* for performance in a particular situation. This complex engenders an attitude of *purposive striving.* Together with the learner's intellectual abilities, this prevailing attitude affects the *quality* of learning; along with personality traits, purposive striving influences the *quantity* of learning activity ("level of effort" and "persistence"). The combined effect of these factors of quality and quantity is to determine *learner engagement,* which in turn leads to learner *achievement.* Quite evidently, Corno and Snow see abilities and traits as having indirect effects upon learning. That is, these qualities of individuals influence how they go about learning, although they do not enter into the substance of the learning itself.

Bearing this point in mind, how can instructional design take abilities and traits into account? Corno and Snow suggest two alternatives: (1) by circumventing lack of aptitude, and (2) by developing aptitude. The first of these may be simply exemplified by instructional design that uses easily-read text and elaborate learning guidance for learners exhibiting a lack of aptitude in verbal comprehension. Developing aptitude is the other route, and it involves instruction and practice in cognitive strategies (O'Neil, 1978; Snow, 1982). Although much progress has been made in this area of investigation, a conservative view is that cognitive strategies applicable to specific task domains are readily learnable, but generalizable strategies may require many years to develop.

New Learning of Cognitive Strategies

As is true for intellectual skills, cognitive strategies in their initial learning call upon previously learned memories. One would expect that there might be readily identifiable prerequisite skills whose retrieval would aid the learning of new cognitive strategies. For example, could a strategy such as "remembering a list of names by associating each name in order with an object of furniture in its place around a familiar room," be aided by retrieval of some prerequisite skills? The answer is, of course, that prerequisite skills do in fact support this cognitive strategy. But note the fact that these prerequisite skills are often very simple and well-known. The prerequisite skills are simply: (1) identifying familiar furniture pieces; (2) imaging familiar furniture pieces; and (3) matching the sounds of names and objects (such as *Haire* and *chair*). Actually, it would seem that the more general the cognitive strategy, the simpler its prerequisites. For example, a general problem-solving strategy is to "work the problem in steps backward from the goal." To acquire this strategy, the learner must make use of previous skills of the following sort: (1) identify the problem goal; and (2) place a set of steps in backward order.

Do schemas come into play as recalled entities in the new learning of a cognitive strategy? Again as in the case of intellectual skills, a schema may provide cues for the retrieval of a strategy and its prerequisite skills. The familiar room in the list-learning strategy is itself a schema and is useful because it is easy to retrieve from memory. The "working backwards" strategy, or others more specifically oriented toward task domains, may be retrieved because it is analogous to the strategy employed in another, different problem. The entire schema representing the previously encountered problem may be retrieved to reveal the analogy.

As for abilities and traits, they may be assumed to operate in the manner described in the previous section. They have indirect effects on the learning of new cognitive strategies. Some investigators propose that the aim of instruction in cognitive strategies is to bring about *aptitude development*. In that case, instruction may begin with a task that uses an ability that has been only partially developed, and proceed to develop it further. For example, the ability to "mentally rotate figures" (spatial visualization) might be initially assessed as poorly developed in particular individuals. By instruction composed mainly of practice and feedback, the development of this ability could be attempted. A successful demonstration of training in spatial visualization has been conducted by Kyllonen, Lohman, and Snow (1981).

New Learning of Verbal Information

The new learning and storage of verbal information requires a number of intellectual skills relevant to the understanding and use of language. These are skills including the employment of synonyms and metaphors in word meanings, the rules of syntax in the formation of sentences, and the logical sequencing of ideas among related propositions. It is these basic skills of language *comprehension* and *usage* that strongly affect how readily learners will acquire new knowledge of the declarative variety and, ultimately, what quantity of such knowledge will be available in their long-term memories.

The learning of new verbal information is strongly affected by retrieval of previously learned information. The cues provided by the to-be-learned information are believed to activate concepts in the long-term memory that spread to other items in a propositional network—the process known as *spreading activation* (Anderson, 1985). Presumably, the more such ideas have been subject to *elaborative processing*, the more readily will the retrieval of knowledge take place. And the greater the elaboration of knowledge—that is, the larger and more intricate is the complex of ideas that can be retrieved from prior learning—the more readily will new verbal information be learned and remembered.

The complex of previously learned verbal information that is retrieved to incorporate new information occurs most often in the form of a schema. This form of verbal information carries the meaning of an organizing concept (such as "taking an airplane trip"). Schemas contain slots into which new information is fitted, and these help to ensure later recall. Designing instruction of the verbal information sort thus requires that you attempt to determine what schemas are already available to the prospective student. Learning new information about Queen Elizabeth I, for example, is best undertaken with a readily accessible schema that includes at least previously learned information about English royal succession during that era.

As you might expect, the most important set of abilities affecting the learning of new verbal information falls in the category of *verbal comprehension*. Measures of this ability appear to evaluate cognitive strategies relating to the facile understanding and use of language. Obviously, measures of verbal comprehension also assess, in part, the intellectual skills of word usage and syntactic and semantic language facility, as previously mentioned. It is little wonder, then, that this ability has been shown in many studies to predict the ease with which new verbal knowledge is acquired (Cronbach, 1970). If it were possible to obtain only a single ability test score for a group of students about to learn new verbal information, verbal comprehension would be the one to choose.

New Learning of Attitudes

When learners acquire new attitudes, the retrieval of certain relevant intellectual skills and verbal information may be essential. For example, an attitude regarding safety in handling certain chemical substances may require the kinds of intellectual skills that make possible the estimation of concentrations of those substances. The attitude involved in following a dietary prescription may require the use of previously learned intellectual skills that make possible the calculation of caloric intake. For a number of reasons, verbal information may also be of importance to the learning of modification of attitudes. If a human model is employed to communicate the choice of personal action (see Chapter 5), previously learned information must be available that identifies the model as a familiar, respected person and attests to his or her credibility.

The most typical form for essential verbal information in the learning of attitudes is the schema. In this case, schemas have the function of representing the *situation* or *situations* in which the attitude will be displayed. For example, an attitude toward association with people of different ethnic origins is evidenced by choices of action made in a number of situations (Gagné, 1985). Is the association likely to be in a large crowd, in an intimate family group, or on the job? Each of these possibilities is a different situation in which the attitude might be displayed. Schemas representing each of these situations must be accessible in memory for the new (or reactivated) attitude to be learned. Another example of the necessity of situational schemas may be illustrated by the attitude "declining to drive after drinking." The social situations that make this attitude a desirable one are what need to be represented as schemas—parties with friends, extended after-dinner meetings, and the like. The attitude of refusing to drive will be most effective if the situational schemas can be readily recalled at the time of learning.

Are there abilities and traits that make attitude learning easier or more rapid? Abilities have perhaps no different effects on attitude learning than they do for other types of learned capabilities. An ability like verbal comprehension naturally facilitates the understanding and learning of verbal communications used in instruction. As for traits, it is possible that traits such as *sociability* and *external locus of control* may affect the ease with which learners acquire an attitude communicated by a human model. Little evidence is available on these relationships, however, and they are judged not to be of great significance. In any case, it is evident that these effects are of the indirect sort previously described.

New Learning of Motor Skills

Two kinds of prior learning are likely to be of importance in the new learning of motor skills. One kind consists of the part skills that are components of the total skill being acquired. The kicking part of swimming the crawl may have been learned separately for retrieval and use in combination with other part skills in the learning of the total skill. When children learn to print letters, the drawing of curved parts and straight-line parts may be previously learned as part skills. If already present as a result of prior learning, they may be retrieved and integrated with the total letter-printing skill.

The other kind of prior learning essential to learning a new motor skill is actually an intellectual skill—a *procedural rule* (see Chapter 5). This is the aspect of motor skill learning that Fitts and Posner (1967) identified as an early cognitive learning stage. The skill of throwing darts at a target, for example, requires retrieval of the procedure of holding the dart, balancing for the throw, aiming, and releasing the dart. Whatever degree of skill may come from practice, the procedure must always be followed for skill improvement to occur. Although itself an intellectual skill, the *procedure* may occur and be retrieved as part of a schema. It is reasonable to suppose, for example, that the tennis backhand and the golf swing may be conceived as schemas.

Abilities have their usual function in the learning of motor skills. Abilities such as *speed of movement* and *motor coordination* may be found to aid the learning of some motor skills. Also, motor skill learning can often be shown to be affected by spatial abilities such as *spatial visualization* and *space relations*. Correlations between these abilities and motor skill learning are usually found to be low to moderate in value.

Applications to Instructional Design

This review of relations between characteristics of the learner and the ease and effectiveness of learning has a number of implications for the practical task of instructional design. The designer needs to take account of the outcomes of learning, as indicated in Table 6-1.

SUMMARY

Learner characteristics that affect the learning of new instructional material assume several kinds of organization in human memory. The learned capabilities of intellectual skills, cognitive strategies, verbal information, attitudes, and motor skills have direct effects on the learning of new instances of these same kinds of capabilities. Another kind of memory organization is represented by the notion of *abilities,* which are measured by psychological tests (such as those of Reasoning and Number Facility). These are measures of human qualities that predict how well certain general types of performances will be accomplished by different individuals. Still other characteristics of human learners fall in the domain of *traits* (such as anxiety, locus of control). Abilities and traits affect new learning in indirect ways.

Relations between characteristics of the learner and the ease and effectiveness of learning have a number of implications for the practical task of instructional design. The designer needs to take account of the outcomes of learning, as described in the preceding chapter, and be cognizant of how these different outcomes may be brought about in different learners. After all, various types of learners may be addressed by instruction. They may be children or adults and may, therefore, differ in the amount of prior learning they have experienced. They may have different learned capabilities, different schemas, and different abilities and traits. The main implications of these differences are summarized in Table 6-1.

As can be seen from the table, *intellectual skills* and *cognitive strategies* are usually of help to new learning, and their retrieval needs to be provided for

TABLE 6-1. Instructional Design for Different Learner Characteristics

Learner Characteristics	Design Procedure for New Learning
Intellectual Skills	Stimulate retrieval of: (1) prerequisite skills as components of new skill; (2) subordinate skills essential to cognitive strategies; and (3) basic skills involved in verbal information learning, attitude learning, and motor skill learning.
Cognitive Strategies	Provide for retrieval when available.
Verbal Information	Stimulate recall of propositions that may cue retrieval of newly learned intellectual skills. Provide retrieval of a meaningful context (schema) for new learning of verbal information. Stimulate retrieval of situational context for attitude learning.
Attitudes	Activate for learning motivation.
Motor Skills	Recall essential part skills.
Schemas	Activate retrieval of schemas consisting of complex networks of propositions to aid new learning of intellectual skills, cognitive strategies, verbal information, attitudes, motor skills.
Abilities	Adapt instruction to differences in ability whenever possible. Example: Use easily read printed text for learners low in verbal comprehension.
Traits	Adapt instruction to learner differences in traits when possible. Example: Provide detailed learning guidance and frequent feedback for learners of high anxiety.

in design. Stimulating the recall of *verbal information* makes provision for cue retrieval and the activation of a meaningful context within which new information can be subsumed. Previously acquired positive *attitudes* contribute to motivation for learning. Motor skills that are *part skills* need to be retrieved as components of the learning of new skills.

Many of these previously learned capabilities are incorporated into meaningful complexes called *schemas*. These networks of meaningful propositions and concepts are of considerable importance to new learning. Instructional design procedures include provisions for detecting the presence of relevant schemas and activating them by means of questions, advance organizers, or other devices.

Instruction for new learning can be adapted for learner differences in *abilities* and *traits* to the extent that feasibility considerations permit. When instructions are verbal in nature, ease of *verbal comprehension* is of particular importance in the instructional design.

REFERENCES

Anastasi, A. (1976). *Psychological testing* (4th ed.). New York: Macmillan.

Anderson, J. R. (1985). *Cognitive psychology and its implications* (2nd ed.). New York: Freeman.

Bandura, A. (1982). Self-efficacy mechanism in human agency. *American Psychologist, 37,* 122–148.

Corno, L., & Snow, R. E. (1986). Adapting teaching to individual differences among learners. In M. C. Wittrock (Ed.), *Handbook of research on teaching* (3rd ed.). New York: Macmillan.

Cronbach, L. J. (1970). *Essentials of psychological testing* (3rd ed.). New York: Harper & Row.

Cronbach, L. J., & Snow, R. E. (Ed.), (1977). *Aptitudes and instructional methods.* New York: Irvington.

Fitts, P. M., & Posner, M. I. (1967). *Human performance.* Monterey, CA: Brooks/ Cole.

Gagné, E. D. (1985). *The cognitive psychology of school learning.* Boston: Little, Brown.

Gagné, R. M. (1980). Preparing the learner for new learning. *Theory into Practice, 19*(1), 6–9.

Gagné, R. M. (1985). *The conditions of learning* (4th ed.). New York: Holt, Rinehart and Winston.

Guilford, J. P. (1967). *The nature of human intelligence.* New York: McGraw-Hill.

Hunt, E. B. (1978). Mechanics of verbal ability. *Psychological Review, 85,* 271–283.

Keele, S. W. (1968). Movement control in skilled motor performance. *Psychological Bulletin, 70,* 387–403.

Kyllonen, P. C., Lohman, D. F., & Snow, R. E. (1981). *Effects of task facets and strategy training on spatial task performance* (Tech. Rep. No. 14). Stanford, CA: Standford University, School of Education.

McClelland, D. C. (1965). Toward a theory of motive acquisition. *American Psychologist, 20,* 321–333.

Newell, A., & Simon, H. A. (1972). *Human problem solving*. Englewood Cliffs, NJ: Prentice-Hall.

O'Neil, H. F., Jr. (1978). *Learning strategies*. New York: Academic Press.

Riley, M. S., Greeno, J. G. & Heller, J. I. (1983). Development of children's problem-solving ability in arithmetic. In H. P. Ginsburg (Ed.), *The development of mathematical thinking*. New York: Academic Press.

Rumelhart, D. E. (1980). Schemata: The building blocks of cognition. In R. J. Spiro, B. C. Bruce, & W. F. Brewer (Eds.), *The theoretical issues in reading comprehension*. Hillsdale, NJ: Erlbaum.

Schank, R. C. & Abelson, R. P. (1977). *Scripts, plans, goals, and understanding*. Hillsdale, NJ: Erlbaum.

Snow, R. E. (1977). Individual differences and instructional theory. *Educational Researcher, 6(10)*, 11–15.

Snow, R. E. (1982). The training of intellectual apptitude. In D. K. Detterman & R. J. Steinberg (Eds.), *How and how much can intelligence be increased*. Norwood, NJ: Ablex.

Thorndike, R. L., & Hagen, E. (1985). *Measurement and evaluation in psychology and education* (5th ed.). New York: Wiley.

Thurstone, L. L. (1938). Primary mental abilities. *Psychometric Monographs*, No. 1.

Tobias, S. (1979). Anxiety research in educational psychology. *Journal of Educational Psychology, 71*, 573–582.

Tobias, S. (1986). Learner characteristics. In R. M. Gagné (Ed.), *Instructional technology: Foundations*. Hillsdale, NJ: Erlbaum.

Weiner, B. (1980). *Human motivation*. New York: Holt, Rinehart and Winston.

White, R. T. (1973). Research into learning hierachies. *Review of Educational Research, 43*, 361–375.

PART THREE

DESIGNING INSTRUCTION

7 DEFINING PERFORMANCE OBJECTIVES

Instructional design technology involves correction and revision of instruction based upon the results of empirical testing. Therefore, it is essential that the desired outcomes of the designed instruction be clearly and unambiguously stated. These outcomes are variously referred to as *behavioral objectives, learning objectives,* or *performance objectives.*

We define a performance objective as a precise statement of a capability that, if possessed by the learner, can be observed as a performance. The question the designer must be able to answer before starting the development of any instruction is, "What will these learners be able to do after the instruction, that they couldn't (didn't) do before?" or, "How will the learner be different after the instruction?"

Precision in the definition of objectives meets the need for communication of the purposes of instruction and the need for evaluation of instruction. Objectives that are precisely defined provide a common technical basis for meeting both of these needs. The instructor wants to communicate the intended outcomes of instruction to students, teachers, and parents (when appropriate). Although these communications usually differ from one another in the ways they are stated, the instructor nevertheless wants all of them to express the same idea. This goal can best be achieved by having available a technically complete definition of the objective. Likewise, the instructor wants the observations (or test) used in evaluating the outcome of learning to reflect the commonly understood idea of this outcome. Again, this can best be achieved by reference to a technically adequate definition of the outcome. Precise definitions of objectives become the specification for "domains" from which items for achievement testing are drawn (Popham, 1975). Objective definitions thus ensure that what is evaluated and what is communicated as an intended learning outcome have a common meaning.

As has often been suggested, the procedure for overcoming the ambiguity of course purpose statements, and thereby achieving greater precision, runs somewhat as follows: "All right, I will accept this statement as reflecting upon one of the purposes of the course. The question now is, how will I know when this purpose has been achieved?"

How will one know that the student "understands the principle of commutativity?"

How will it be known that the student "appreciates allegory in *A Midsummer Night's Dream*"?

How can it be told that the student "comprehends spoken French"?

How will one tell that the student "reads short stories with enjoyment"?

Statements of course purposes may be quite successful in communicating general goals to fellow teachers, yet they are often not sufficiently precise for unambiguous communication of the content and outcomes of instruction. The key to their ambiguity is simply that they do not tell how a person could observe what has been accomplished without being present during the lesson itself. Another teacher, who accepts the general purpose of the course, may wish to know how to tell when it has been accomplished. It may be of interest to a parent who may not know exactly what "commutative" means but wishes to assure himself that his son or daughter can in fact use this principle in performing arithmetic operations. It is likely to be of interest to students who want to be able to tell when their own performance reaches the goal that the teacher or textbook had in mind.

ACHIEVING PRECISION IN OBJECTIVES

An objective is precisely described when it communicates to another person what would have to be done to observe that a stated lesson purpose has in fact been accomplished. The statement is imprecise if it does not enable the other person to think of how to carry out such an observation. Consider the following instances:

1. "Realizes that most plant growth requires sunshine." Such a statement doesn't say or imply how such an outcome would be observed. Does it mean that the teacher would be satisfied with the answer to the question "Is sunshine necessary for the growth of most plants?" Evidently not. How, then, would such an objective be observed?
2. "Demonstrates that sunshine affects plant growth." This statement implies that the teacher must observe instances in which the student shows that he

knows the relation between sunshine and plant growth. The observation assumes the relation between sunshine and plant growth. The observation might be made in various ways (by using actual plants, pictures, or verbal statements). The main point is, it tells in a general way what sort of observation is required.

A number of types of capabilities (learning outcomes) have been described in previous chapters. In writing objectives to specify these outcomes we use a convention that we call the "five-component objective". The five-component objective specifies the situation in which the performance is performed; the type of learned capability; the object of the performance; the specific action the learner takes in employing the capability; and the tools, constraints, or special conditions associated with the performance. This chapter discusses the five-component convention and the classification of objectives by learning type.

Components of Objectives

One purpose for unambiguous objectives is to enable the designer to determine which *conditions of learning* should be included in the instructional materials. The five-component objective is more specific than the definitions suggested by other authors (Mager, 1975; Popham and Baker, 1970). The reason for greater specificity is to communicate more information about the type of learning outcome that is desired. We cannot directly observe that someone has acquired a new capability. We can only infer that the capability has been attained through the observation of satisfactory performance by the learner on a task that employs that capability. Often the particular performance (action) exhibited by the learner is confused with the capability. The five-component method of writing objectives seeks to avoid this confusion by specifying two verbs: one to define the capability, and a second to define the observable action. Each component of a five-component objective serves an express purpose as described in the following paragraphs.

Situation

What is the stimulus situation faced by the student? For example, when asked to "type a letter," is the student given parts of the letter in longhand copy? Is the letter to be produced from an auditory message or from notes? Obviously, what the student actually does is highly dependent on the situation. An objective must specify the features of this situation.

Sometimes, it may be desirable for the objective to include a description of the environmental conditions under which the behavior is to be performed on the job. For example, in the case of typing a letter, is this to be done in a quiet room with no other disturbances, or is it more likely that it will be in a busy office with phone interruptions, people walking by, or other tasks coming in? For many types of learned behavior the environment in which the behavior is performed may not be terribly important. However, for other performances, for example, donning a gas mask, it may be critical.

Learned Capability Verb

Some of the problems with the use of behavioral objectives arise from ambiguity about what type of learning outcome the demonstrated behavior actually represents. For example the statement, "Given an IBM typewriter, types a business letter in 15 minutes of less," tells us very little about the type of learned capability intended. It might mean "types a copy of a letter from handwritten draft," or it might mean a quite different capability, "composes a business letter." This ambiguity can be reduced by including within the objective an indicator of the type of learned capability being demonstrated.

There are nine different learned *capability verbs* as shown in Table 7-1. These pertain to four of the learned capabilities described in previous chap-

TABLE 7-1.   Standard Verbs to Describe Human Capabilities, with Examples of Phrases Incorporating Action Verbs

Capability	Capability Verb	Example (Action Verb in Italics)
Intellectual Skill		
Discrimination	discriminates	discriminates by *matching* French sounds of *u* and *ou*
Concrete Concept	identifies	identifies, by *naming* the root, leaf, and stem of representative plants
Defined Concept	classifies	classifies, by *using a definition*, the concept family
Rule	demonstrates	demonstrates, by *solving* verbally stated examples, the addition of positive and negative numbers
Higher-Order Rule (Problem solving)	generates	generates, by *synthesizing* applicable rules, a paragraph describing a person's actions in a situation of fear
Cognitive Strategy	adopts	adopts a strategy of imagining a U.S. map, to recall the states, in *writing* a list
Verbal Information	states	states *orally* the major issues in the presidential campaign of 1932
Motor Skill	executes	executes *backing* a car into a driveway
Attitude	chooses	chooses *playing golf* as a leisure activity

ters, and to the five subordinate types of intellectual skills. These verbs may be used to *classify* each of the nine types of learning outcomes. By including one of these verbs in the objective, the intended behavior is more clearly communicated, and the conditions of learning appropriate to that type of learning outcome are more readily applied.

Object

The object component indicates the content of the learner's performance. For example, if the learned capability is the procedure for calculating the sum of two three-digit numbers (a rule), the learned capability and its object might be stated as: Demonstrates (the learned capability verb) *the calculation of the sum of two three-digit numbers* (the object).

The letter example given previously could be stated as: "generates a business letter" (problem solving) or "executes the typing of a business letter" (motor skill). However, there is still some ambiguity in the objective: How is generation of the business letter to be observed? An indicator of the observable performance is needed, and this is the function of the *action verb*.

Action Verb

The action verb describes how the performance is to be completed, "executes a copy of a business letter by typing" describes the action one would observe (the typing). For the problem-solving objective, the observable behavior is also typing: "generates a business letter by typing a reply to a job inquiry." There are innumerable action verbs: matching, writing, speaking, discussing, pointing, selecting, drawing, etc. Table 7-1 demonstrates how the learned capability verbs and action verbs work together to describe a task. We will describe the process of writing performance objectives shortly, but keep one rule in mind: *never use one of the nine learned capability verbs as an action verb.* This will avoid confusion when it comes to sequencing the objective later on.

Tools, Constraints, or Special Conditions

In some situations the performance will require the use of special tools, certain constraints, or other special conditions. For example, the letter may have to be typed using a Savotti Model 11 Teletypewriter. Notice that the objective is not aimed at the acquisition of skill with the Savotti; instead, it is a special condition placed on the performance of typing the letter. An example of a constraint could be a criterion of performance; a letter might have to be completed within a specified time, with fewer than three errors. As is true of the situation, the indication of any special conditions or tools may imply other prerequisite skills that must be learned before the target skill can be adequately evaluated.

Generating Five-Component Objectives

As implied by the title of this section, writing a five-component objective is a problem-solving task. There are many rules that must be applied. The first problem to be solved is deciding what type of learning outcome the instruction aims to produce. The next section of chapter will discuss how five-component objectives would be written for each of the nine types of learning, beginning with the five subordinate kinds of intellectual skills.

Discrimination

Discrimination performance always involves being able to see, hear, or feel sameness or differences between stimuli. There are many reasons persons cannot discriminate; for example, a person who is color blind cannot discriminate red from blue. However, even persons who are physically able must learn important discriminations. For example, sighted people must learn to discriminate a *b* from a *d* if they are to learn to read. Objectives for this type of skill would read like the following. (The acronym LCV is used for *learned capability verb*.)

> [Situation] Given an illustration of three plane figures, two the same and one different [LCV] discriminates [object] the figure that is different [action] by pointing to it.

Another discrimination objective might be:

> [Situation] Given an illustration of the letter *b* and instructions to select other illustrations that look the same from a set containing *d, p, b,* and *q* [LCV] discriminates [object] *b* [action] by circling.

Not all discriminations are visual; they may also be aural, tactile, or olfactory. For example, a discrimination objective that might be appropriate for someone becoming a chef might be:

> [Situation] Given a piece of fresh beef as a reference [LCV] discriminates [object] the smell of fresh from the smell of beef that is on the verge of spoiling [action] by indicating that the smells are the same or different.

Notice that in the previous objective the situation did not include a description of the environment in which the action is to take place. Is it important that this be specified? Maybe in an operating kitchen there are likely to be many smells of food in preparation that would make the discrimination more difficult than it would be in an isolated situation.

Concrete Concept

Concrete concepts require that the student be able to identify one or more instances of a class of items. For example, how would you know if a student understood the concept *bodkin*? You could ask him to describe a bodkin. If he could describe it you might infer that he indeed knew the concept bodkin. However, he may know what a bodkin is without being able to describe it. You could ask him to point to, touch, or select a bodkin. You could ask for the recognition of pictures of bodkins. All these are acceptable, although different, ways to demonstrate that the learner possesses the concept bodkin.

The learned capability verb [LCV] used in conjunction with concrete concepts is *identify*. Of course, in order to identify anything, the student must be able to discriminate critical physical attributes. How could a blind person acquire the concept bodkin? Obviously, he must distinguish the essential characteristic physical attributes through a sense of touch.

Defined Concept

A defined concept is a rule that classifies objects or events (Gagné, 1985). By rule we mean a definition that expresses the relationships among the concept's attributes and its function. For example, a dish is a class of things including saucers, cups, plates, etc. A saucer might be described as a dish designed to hold a cup. In this case a dinner plate is not a saucer (even though it may hold a cup) because that is not its express purpose.

The LCV associated with defined concepts is *classify,* since what the learner has to do is classify some instance into one or more categories, based on a verbal description of its attributes and function, or to be able to appropriately apply the concept when classifying an instance.

Boundary is a defined concept. If we want to express the objective of determining if the student possesses the concept *boundary,* we might say:

> [Situation] When asked to tell what a boundary is [LCV] classifies [object] *boundary* [action] by describing or illustrating a boundary, by reference to its definition (a line indicating a limit or extent).

Another way to approach the objective would be to see if the student could recognize the attributes of the *boundary* concept, as in the following:

> [Situation] Given descriptions of instances of lines which do and lines which do not indicate the extent of an area [LCV] classifies [object] *boundaries* [action] by selecting those which conform to the definition.

Still another way to observe if students possess a defined concept is to have them apply it (use it) properly in a sentence; for example, "The boundary of a lake is marked by the line of its shore when the lake is full."

Rule

A rule is an internal capability that governs one's behavior, and enables one to demonstrate a relationship among concepts in a class of situations (Gagné,

1985). The inferred capability is that of demonstrating an appropriate response to a class of stimulus situations possessing established relationships. For example, a student applies a rule when confronted with the stimulus ($236/4 = n$). The same rule applies when the stimulus is different ($515/5 = n$). Rules are often given names. In this case the rule might be called "short division," and the student possessing the capability of applying this rule would be able to show mastery by solving any number of short division problems.

The learned capability verb associated with rules is *demonstrate*. A typical example of an objective for rule outcomes is:

[Situation] Given a set of 10 numerical expressions indicating short division (abc/d) [LCV] demonstrates [object] division [action] by writing the answers [tools and constraints] with 90% accuracy, using no special aids.

Is the objective representative of the examples of short division given above? It is not specific enough if the problems are to be limited to the type shown. A more accurate object statement would be, "short division of three-digit numbers by a one-digit number, with no remainder." Are the rules different for division by a number that leaves a remainder? To the extent they are, it is desirable to be specific, so that the components of the task can be accurately described, and appropriate instruction accordingly designed.

To learn this rule the student must be familiar with rules of multiplication and subtraction. To multiply, the student must be familiar with rules of addition. One reason that most rules are complex is that they require previously learned concepts and rules.

Problem Solving

Gagné (1985) defines a problem as one in which the learner selects and uses rules to find a solution in a novel situation. What the learner acquires during the process of problem solving is a new higher-order rule. The new rule is a synthesis of other rules and concepts. This higher-order rule may then be used by the learner to solve other problems of the same type.

One problem in writing learning outcomes associated with problem-solving skills is separating the process of problem solving from the statement of a desired learning outcome. The desired learning outcome is an expression of relationships among other rules and concepts. The student must generate this higher-order rule and also apply it, in order to achieve a solution to a novel problem. We suggest that when writing statements of learning outcomes, the instructor focus on what is being constructed by the learner.

The verb we associate with problem solving is *generate*. An example of a problem solving objective is:

[Situation] Given the description of a construction project in a playground [LCV] generates [object] an expression regarding the relationship between the time it takes to do a task and the number of workers performing the task [action] explaining the rule orally to the teacher.

Problem-solving objectives are not always easy to write, mainly because problem-solving skills are usually not formally taught. Instead, most teachers present problem-type situations, and then verify whether or not the student has the problem-solving skill. A source of confusion among novice designers may occur in differentiating between rule-using and problem-solving objectives. The distinction can be kept clear by asking the questions "demonstrate what rule?" and "generate what rule?" If the learners are given the rule and asked to apply it, the skill is rule learning; if they must generate the new rule, or develop a synthesis of already existing rules, it is problem solving.

Cognitive Strategy

By cognitive strategies we mean internally directed control processes that regulate and moderate other learning processes. Gagné (1985) describes a number of types of cognitive strategies, including those which control attending, encoding, retrieval, and problem solving. Bruner (1971) distinguishes between the skills of problem solving and problem finding, the latter of which is a cognitive strategy requiring the location of "incompleteness, anomaly, trouble, inequity, and contradiction" (p. 111). Again it is important to differentiate between the product of a cognitive strategy and the statement of the strategy itself as an outcome of instruction. For example, most persons can memorize a list of 10 items if given enough time. However, some can memorize the list much faster and remember it longer. Perhaps this is the result of their having a more efficient and effective encoding strategy. Research has demonstrated (Rohwer, 1975) that encoding strategies that facilitate learning can readily be suggested to the learner.

When learners acquire new ways of focusing their attention, encoding material to be learned, or retrieving previously learned knowledge, they may be using novel cognitive strategies that they have discovered themselves. Otherwise, strategies may be acquired by being directly described to learners and subsequently practiced. Generally, students apply existing strategies that have worked in the past. A verb to describe such a process is *adopt*. That is, learners may adopt as control processes the kinds of strategies that aid various other processes relevant to particular learning tasks.

Using the verb *adopt*, a cognitive strategy related to encoding may be stated as:

[Situation] Given a list of 10 items to be memorized, the student [LCV] adopts [object] the key-word mnemonic technique [action] for memorizing the list, using no mechanical aids, within 30 seconds, and with a retention for at least 49 hours.

Notice that this objective does not give the student the mnemonic. Instead it implies that the student will adopt a mnemonic using an already known technique. Can a student actually "originate" a cognitive strategy? The answer to this question is probably yes, but we suggest that this would be an instance of problem solving—the student would be "generating a strategy" which, if it is effective, could be implemented for similar learning tasks. For instructional purposes we are usually more interested in the adoption of strategies rather than in their generation.

Verbal Information

Verbal information refers to information (names, facts, propositions), that can be *verbalized*. It is also called *declarative knowledge* (Anderson, 1985). According to Gagné (1985), a distinction should be made between the learning of verbal chains and the learning of verbal information. The former is a type of association learning, where each element of the chain must be previously learned before the entire chain can be reconstructed. Individuals can learn extremely long verbal chains, and recall them verbatim, without having any comprehension of what the words mean. An essential characteristic of verbal information learning, on the other hand, is that it consists of propositions that are semantically meaningful.

The verb we associate with verbal information is *states*. We differentiate the learned capability of stating from the actions of stating, either "in writing" or "orally." An example of a verbal information objective is:

[Situation] Given a verbal question, [LCV] states [object] three causes of the Civil War [action] orally or in writing [constraints] without references.

Could students memorize three causes of the Civil War as a verbal chain, and lead the teacher to conclude that they had mastered the above objective? Yes, and many students probably do this because the teacher doesn't require that they present the information in any other way. A modification of this objective would be to add the condition "in your own words" to the objective. However, the fact that the student may recall something verbatim does not necessarily mean that it is not stored in memory as a set of meaningful propositions. The proof of this would come when the student is required to use the same information in some meaningful way. For example, the student might learn that hamsters eat grain. If asked, "What do hamsters eat?", it would be possible to receive the answer "grain" as a verbal chain. However, the same answer to a different question, "What will you feed the hamster?", shows that the student has acquired the information in a meaningful way.

Motor Skills

Some behaviors require their expression in coordinated, precise muscular movements. Examples of these kinds of performances would be gymnastic skills such as a back-flip or a jack-knife dive. Less obvious are common skills like walking or riding a bicycle. Forming a letter on paper, using a pen, requires the coordinated use of muscles, and most teachers recognize that some students have greater proficiency at this skill than do others.

The verb we associate with motor skills is *execute*. A typical objective might be:

> [Situation] At a swimming pool on a 3-meter board [LCV] executes [object] a jack-knife dive [action] by diving [constraints] with smooth and continuous movement, entering the water in a vertical position. Notice that the statement of the action, "by diving", is redundant here, as there is only one way to demonstrate execution of the dive.

Study the following objective: "Given a blood pressure apparatus, executes the procedure of taking someone's blood pressure." Is this a motor skill? Probably not, as the student already has the requisite motor skills and is only applying a procedure (rule learning). However, "Given a hypodermic syringe, executes an intravenous injection, missing the vein no more than one time in twenty," probably is a motor skill, in that precision, timing, and coordination of muscular performance are involved.

Attitude

When an attitude is specified as the desired learning outcome, it is a choice of personal action that the learner is expected to exhibit. For example, a description of behavior might be: "the student will choose to vote when the opportunity presents itself." Obviously the concept of choice means that people are free to "not vote" if they so desire. There are many determinants of attitudes, including situational factors. For instance, one person may choose to vote if it is physically convenient, but choose not to vote if it is inconvenient. It might be said that a person has a positive attitude toward voting as long as it is not inconvenient. We might wish, as an alternative, to specify an attitude change that would involve learner inconvenience, that is, "will choose to vote even if it is inconvenient."

The learned capability verb used to classify attitude objectives is *choose*. A typical attitude objective that might be the focus of an educational program might be:

> [Situation] When harmful drugs are being used by peers [LCV] chooses [action] to refuse [object] drugs when offered.

Martin and Briggs (1986) point out that many cognitive behaviors have affective components. For example, mathematical operations are taught as cognitive skills, but it is hoped that the learner will *choose* to feel that mathematics knowledge is important to learn (for reasons other than a course grade). If designers attempt to specify the attitudinal components, they will likely pay attention to the context in which the cognitive skills are presented, attempting to make the newly learned skill meaningful (relevant) for the learner, and building as much reinforcement into the learning situation as possible.

Statements of Objectives and Criteria of Performance

Some systems of writing performance objectives (Mager, 1975) require the inclusion of a criterion of performance in the objective. The system we suggest does not. That is, the objective statements themselves do not necessarily describe "how many times the student is to demonstrate the addition of mixed numbers," or how many "errors" will be permitted. There are two reasons why a designer might wish to avoid including the criterion statement in the objective. First, the necessary criteria are likely to be different for each type of human capability, and it is desirable to avoid the error of thinking they can be the same. For example, if a skill is prerequisite to learning another skill the level of performance specified to denote mastery might have to be higher than if the skill is not prerequisite to learning another skill. Second, the question of criteria of performance is a measurement question, and should be considered with assessment procedures. At the point in instructional planning when objectives are being described, it is potentially confusing to be concerned with assessment procedures. (Such procedures are described in Chapter 13.)

However, if the criteria of performance for a given task are already known at the time of writing the objective, the criterion statement can become a component of the objective, falling under the category of "tools, constraints, and special conditions." It would be perfectly acceptable to have an objective like the following:

> Given a 3-page handwritten manuscript, executes typing of the manuscript within 20 minutes with fewer than six errors.

EXAMPLES OF OBJECTIVES

One of the first questions from new learners of the five-component format is, "Is it really practical to require that all five components be specified?" Our answer is that an objective statement is written for the purpose of unambiguous communication of intent. If you can communicate unambiguously without all five components, then do so. For example, take the objective, "States the names of the 50 states in the United States of America." This seems like a pretty unambiguous statement, and all that is included is the learned capability verb and the object. But notice that even in this objective there are many assumptions; a classroom situation, a verbal question, 100% recall, written or oral response. There is nothing wrong with the objective as stated, but it leaves some of the components open for interpretation.

The following examples show how the five-component format can be used in a number of different subject areas to make vaguely stated objectives more specific.

Examples in Science Instruction

Suppose that the instructional designer formulates in a written statement the purposes to be accomplished by a course of instruction. If the lesson is one of science, the following purposes might be considered. These have been abstracted from a list of objectives for junior high school science instruction prepared by the Intermediate Science Curriculum Study (1973).

1. Understanding the concept of an electric circuit
2. Knowing that a major advantage of the metric system in science is that its units are related by factors of ten
3. Taking personal responsibility for returning equipment to its storage places

Objective No. 1—The Concept of an Electric Circuit

This is a fairly straightforward purpose for instruction. The first question to be asked by an instructional designer is: "What kind of capability am I looking for here?" Do I mean by "understanding" something like "stating what an electric circuit is?" No, that would not be convincing, since it might merely indicate that the student had acquired some verbal information which he could repeat, perhaps in his own words. Do I mean "distinguishing an electric circuit from a noncircuit when shown two or more instances?" No, I cannot be sure that the student has the understanding I wish in this case, because he may simply be able to pick up the cue of an open wire in the instances shown him and respond on that basis. What I actually want the student to do is to *show me that he can use a rule for making an electric circuit* in one or more specific situations. The rule to be learned has to do with the flow of electric current from a source through a connected set of conductors and back to the source. The student could be asked to exhibit this performance in one or more situations.

The result of this line of reasoning is an objective statement that puts together the necessary components as follows:

[Situation] Given a battery, light bulb and socket, and pieces of wire [LCV] demonstrates [object] the making of an electric circuit [action] by connecting wires to battery and socket and [constraint] testing the lighting of the bulb.

Objective No. 2—Knowing Something about the Metric System

The statement of purpose in this instance implies that some verbal information is to be learned. Again, the first question to be asked by the instructional designer is, "What do I mean by 'knowing' this fact about the metric system?

What will convince me that the student 'knows'?" In this instance, the designer may readily come to the conclusion that "knowing" means being able to state the particular fact about the metric system. Accordingly, the identification of the required capability as verbal information is fairly straightforward. The resulting objective can then be constructed as follows:

> [Situation] Given the question: "What major advantage for scientific work do the units of the metric system have?" [LCV] states [object] the "tens" relationship among units [action] by writing [constraints] in his own words.

Objective No. 3—Taking Responsibility for Equipment

In thinking over this instructional purpose, the designer will immediately realize that it is not concerned with whether the students are able to put equipment back in its place, but rather with whether they tend to do so on all appropriate occasions. The word "responsibility" implies that the actions of a student may occur at any time, and are not expected to result from any specific direction or questions. The designer must ask, "What would convince me that the student is 'taking responsibility' of this sort?" The answer to this question implies that the objective in this case deals with choices of personal action, in other words, with an attitude.

The standard method of constructing the objective would therefore take this form:

> [Situation] Given occasions when laboratory activities are completed or terminated [LCV] chooses [object] courses of action [action] returning equipment to its storage places.

An Example from English

A second example of the procedure for constructing statements of objectives comes from a hypothetical course in English literature. Suppose that a set of lessons in such a course had the following purposes:

1. Identifying the major characters in *Hamlet*.
2. Understanding Hamlet's soliloquy.
3. Being able to recognize a metaphor.

Objective No. 1—Identifying the Major Characters in Hamlet

This objective, according to our model, involves using definitions to classify. In this case the student is being asked to classify characters in *Hamlet* in accordance with their functions within the plot of the play. Under most circumstances, it would be assumed that doing this by way of verbal statements would be convincing. That is, the student answers a question like: "Who was Claudius?" by defining Claudius as the king of Denmark, Hamlet's uncle, who is suspected by Hamlet of having killed his father. The objective can be constructed as follows:

[Situation] Given oral questions about the characters of *Hamlet,* as "Who was Claudius?" [LCV] classifies [object] the characters [action] by defining their relationship to the plot.

Objective No. 2—Understanding Hamlet's Soliloquy

Here is a much more interesting and presumably more important instructional purpose. The instructional designer needs to ask, "How will I know if the student 'understands' this passage?" In all likelihood, the answer to this question is to "ask the student to express the thoughts of the passage in words that simplify or explain their meaning." (An example would be explaining that "to be, or not to be" means "to remain living or not".) To accomplish such a task, the student must solve a series of problems, bringing a number of intellectual skills to bear upon them such as rules for using synonyms, rules for defining, and the concepts of figures of speech. In sum, the student will be asked to generate a paraphrase of the soliloquy. It is, then, a problem-solving task, or more precisely a whole set of problems, in which subordinate rules must be applied to the generation of higher-order rules. The latter cannot be exactly specified, of course, since one does not know exactly how the student will solve each problem.

As a result of this analysis, the following objective might be composed:

[Situation] Given instructions to interpret the meaning of Hamlet's soliloquy in simple terms [LCV] generates [object] an alternative communication of the soliloquy [action] by writing appropriate sentences of simple content.

Objective No. 3—Recognizing Metaphors

Even in its expression, this objective has the appearance of representing a somewhat less complex purpose than No. 2. It may be evident, also, that if students are able to generate a paraphrased soliloquy, they must be able to detect the metaphoric meaning of such phrases as "to take arms against a sea of troubles." In this simpler example of a purpose, then, the question for the instructional designer is, "What will convince me that the student can 'recognize' a metaphor?" Obviously, a metaphor is a concept, and since it is not something that can be denoted by pointing, it must be a *defined* concept. The performance to be expected of the student, then, will be one of *classifying a metaphor in accordance with a definition.*

The resulting objective might be stated as follows:

[Situation] Given a list of phrases, some of which are metaphors and some not [LCV], classifies [object] the metaphors [actions] by picking out those that conform to the definition, rejecting those that do not.

An alternative objective (and possibly a better one) for this instructional purpose would be:

[Situation] Given a phrase containing a verb participle and an object (as, "resisting corruption") [LCV] classifies [object] a metaphor [action] by selecting an example which accords with the definition (as, "erecting a bulwark against corruption").

An Example from Social Studies

A course in social studies in junior high school might have the following purposes:

1. Knowing terms of office for members of the two houses of Congress
2. Interpreting bar charts showing growth in agricultural production
3. Applying knowledge of the "judicial review" process of the Supreme Court.

Objective No. 1—Terms of Office for Congress

The intended outcome in this case is information. It is, of course, rather simple information, and therefore something that might be learned in grades before high school. As an objective this purpose may be stated as follows:

[Situation] Given the question, "In what terms of office do members of both houses of Congress serve?" [LCV] states [object] the terms for House and Senate members [action] orally.

Objective No. 2—Interpreting Bar Charts

Often an important kind of objective in social studies is an intellectual skill. Interpreting charts is a rule-using skill. There may be several such skills of increasing complexity to be learned. Consequently, particular attention has to be paid to the description of the situation. More complex charts may require several intellectual skills, or a combination of them. This objective may be illustrated by the following example:

[Situation] Given a bar chart showing production of cotton bales by year during the period 1950–1960 [LCV] demonstrates [object] the finding of years of maximal and minimal growth [action] by checking appropriate bars.

Objective No. 3—Applying Knowledge of Judicial Review

The statement of this goal is somewhat ambiguous. It might best be interpreted as one of solving problems pertaining to the Supreme Court's judicial review function, and exhibiting knowledge by so doing. Such an objective might be stated in the following way:

[Situation] Given the statement of an issue of constitutionality contained in a fictitious Act of Congress, and reference to the constitutional principle to be invoked [LCV] generates [object] a proposed judicial opinion [action] in written form.

USING OBJECTIVES IN INSTRUCTIONAL PLANNING

When instructional objectives are defined in the manner described here, they reveal the fine-grained nature of the instructional process. This in turn reflects the fine-grained characteristics of what is learned. There may be scores of objectives for the single topic of a course, and several for each individual lesson.

How does the instructional designer employ these objectives in his development of topics, courses, or curricula? And how does the teacher use objectives? Can the teacher, as the designer of an individual lesson, make use of lengthy lists of objectives? Many such lists are available, it may be noted, for a variety of subjects in all school grades.

Objectives and Instruction

The instructional designer, or design team, faces the need to describe objectives as part of each individual lesson. Typically, there will be several distinct objectives for a lesson. Each may then be used to answer the question, "What kind of a learning outcome does this objective represent?" The categories to be determined are those corresponding to the major verb indicating capability. That is, the objective may represent verbal information, an intellectual skill in one of its sub-varieties, a cognitive strategy, an attitude, or a motor skill. Having determined the categories of a lesson's objectives, the designer will be able to make decisions about the following matters:

1. whether an original intention about the lesson's purpose has been overlooked, or inadequately represented;
2. whether the lesson has a suitable "balance" of expected outcomes; and
3. whether the approach to instruction is matched to the type of objective in each case.

The Balance of Objectives

The objectives identified for each lesson are likely to represent several different categories of learning outcome. First of all, it may be possible to identify a primary objective—one without which the lesson would seem hardly worthwhile. In addition, however, there are necessarily bound to be other objectives that must be learned prior to the desired objective. Thus, the lesson that has an intellectual skill as its primary objective is likely to be supported by other objectives classifiable as cognitive strategies, information, or attitudes. As an example, one might expect a lesson having as its primary objective the intellectual skills of "demonstrating chemical equations for the oxidation of metals" to also include objectives pertaining to information about common metallic oxides, and to favorable attitudes toward possessing knowledge about metallic oxides. How to reflect these several objectives in lesson design is a subject for later chapters. The first step, however, is to see that a reasonable balance of the expected outcomes is attained.

Designing Instruction

Clearly, then, the systematic design of lessons making up a topic or course will result in the development of a sizeable collection of statements of objectives. This collection will grow as lessons are developed and assembled into topics. Decisions about the correspondence of these objectives with original intentions for the topic and course, and judgments about the balance of objectives, can also be made with reference to these larger instructional units. As in the case of the individual lesson, these decisions are made possible by the categorization of objectives into types of capabilities to be learned.

The teacher's design of the single lesson also makes use of individual statements of objectives and the classes of capabilities they represent. The instructional materials available to the teacher (textbook, manual, or whatever) may identify the objectives of the lesson directly. More frequently, the teacher may need to (1) infer what the objectives are; and (2) design the lesson so that the objectives represented in the textbook are supplemented by others. For purposes of planning effective instruction, the determination of categories of expected learning outcomes is as important to the teacher as it is to the design team. The teacher, for tomorrow's lesson, needs to make decisions about the adequacy with which the lesson's purpose is accomplished, and about the relative balance of the lesson's several expected outcomes.

Objectives and Assessment

Fortunately, the lists of individual objectives developed in a systematic design effort have a second use. Descriptions of objectives, as we have said, are descriptions of what must be observed to verify that the desired learning has taken place. Consequently, statements of objectives have direct implications for assessing student learning (see Chapter 13).

The teacher may use objective statements to design situations within which student performance can be observed. This is done to verify that particular outcomes of learning have in fact occurred. Consider the objective: "Given a terrain map of the United States and information about prevailing winds, demonstrates the location of regions of heavy rainfall by shading the map (applying a rule)." This description more or less directly describes the situation a teacher can use to verify that the desired learning has taken place. A student or group of students could be supplied with terrain maps, prevailing wind information, and asked to perform this task. The resulting records of their performances would serve as an assessment of their learning the appropriate rule.

With comparable adequacy, statements of objectives can serve as bases for the development of teacher-made tests. These in turn may be employed for formal kinds of assessment of student performance, when considered desirable by the teacher. Alternately, they can be used as "self-tests" that students employ when engaging in individual study or self-instruction.

The classes of objectives described in this chapter constitute a taxonomy that is applicable to the design of many kinds of assessment instruments and tests. A somewhat different, although not incompatible, taxonomy of objectives is described in the work of Bloom (1956), and in that of Krathwohl, Bloom, and Masia (1964). The application of this latter taxonomy to the design of tests and other assessment techniques is illustrated in many subject matter fields in the volume edited by Bloom, Hastings, and Madaus (1971). This work describes in detail methods of planning assessment procedures for most areas of the school curriculum. Further discussion of methods for developing tests and test items based on the categories of learning outcomes described in this chapter is contained in Chapter 13.

SUMMARY

The identification and definition of performance objectives is an important step in the design of instruction. Objectives serve as guidelines for developing the instruction and for designing measures of student performance to determine whether the course objectives have been reached.

Initially, the aims of instruction are frequently formulated as a set of purposes for a course. These purposes are further refined and converted to operational terms by the process of defining the performance objectives. These describe the planned outcomes of instruction, and they are the basis for evaluating the success of the instruction in terms of its intended outcomes. It is recognized, of course, that there are often unintended or unexpected outcomes, judged, when later observed, to be either desirable or undesirable.

This chapter has presented a five-component guide to the writing of performance objectives. The five elements are:

1. situation
2. learned capability
3. object
3. action
5. tools and constraints

Examples are given, showing how these components can be used to make unambiguous statements of objectives for different school subjects. The examples chosen also illustrate objectives for various categories of learned capabilities.

Special attention is called to the need for care in choosing action verbs suitable for describing both the learned capability inferred from the observed performance and for describing the nature of the performance itself. Table 7-1 presents a convenient summary of major verbs and action verbs.

The kinds of performance objectives described for the various categories of learned capabilities play an essential role in the method of instructional design presented in this book. Precisely formulated definitions of objectives within each category serve as a technical base from which unambiguous communications of learning outcomes can be derived. Different communications of objectives, conveying approximately a common meaning, may be needed for teachers, students, and parents. At the same time, precisely defined objectives relate the same common meanings to the construction of tests for evaluation of student performance, as will be indicated later in Chapter 13.

REFERENCES

Anderson, J. R. (1985). *Cognitive psychology and its implications* (2nd ed.). New York: Freeman.

Bloom, B. S. (Ed.). (1956). *Taxonomy of educational objectives. Handbook I: Cognitive domain*. New York: McKay.

Bloom, B. S. (1971). Learning for mastery. In B. S. Bloom, J. T. Hastings, & G. F. Madaus (Eds.), *Handbook on formative and summative evaluation of student learning*. New York: McGraw-Hill.

Bloom, B. S., Hastings, J. T., & Madaus, G. F. (Eds.). (1971). *Handbook on formative and summative evaluation of student learning*. New York: McGraw-Hill.

Briggs, L. J., & Wager, W. (1981). *Handbook of procedures for the design of instruction*. Englewood Cliffs, NJ: Educational Technology Publications.

Bruner, J. S. (1971). *The relevance of education*. New York: Norton.

Gagné, R. M. (1985). *The conditions of learning* (4th ed.). New York: Holt, Rinehart and Winston.

Intermediate Science Curriculum Study (1973). *Individualizing objective testing*. Tallahassee, FL: ISCS, Florida State University.

Krathwohl, D. R., Bloom, B. S., & Masia, B. B. (1964). *Taxonomy of educational objectives. Handbook II: Affective domain*. New York: McKay.

Mager, R.F. (1975). *Preparing objectives for instruction* (2nd ed.). Belmont, CA: Fearon.

Martin, B., and Briggs, L. J. (1986). *The cognitive and affective domain*. Englewood Cliffs, NJ. Educational Technology Publications.

Popham, W. J. (1975). *Educational evaluation*. Englewood Cliffs, NJ: Prentice-Hall.

Popham, W. J., & Baker, E. L. (1970). *Establishing instructional goals*. Englewood Cliffs, NJ: Prentice-Hall.

Rohwer, W. D., Jr. (1975). Elaboration and learning in childhood and adolescence. In H. W. Reese (Ed.), *Advances in child development and behavior* (Vol. 8). New York: Academic Press.

8 ANALYSIS OF THE LEARNING TASK

Designing instruction for a course or topic must surely begin with an idea of the purpose of what is being designed. The greatest clarity in conception of the outcomes of instruction is achieved when human performances are described in the form of objectives. The question initially asked by the designer is not, "What will the students be studying?" but rather, "What will students be doing after they have learned?" This means that design begins with a consideration of the instructional objectives.

In describing a procedure for instructional design, it is difficult to decide whether *all* the objectives (of a course or topic) should be specified as a first step, or only *some* of them. This difficulty arises because there are at least two kinds of objectives involved in instructional design: (1) those to be attained at the end of a course of study; and (2) those that must be attained during a course of study because they are prerequisites to the former type. The first kind may be called *target objectives* and the second *enabling objectives*. In a course in reading, "classifying the main idea of a paragraph" is a target objective, whereas "classifying the meaning of unfamiliar words from context," might be considered an enabling objective.

The procedure adopted for description here is one that begins with the target objectives. A "top-down" procedure is then employed to determine what enabling objectives are prerequisite to the attainment of the target objective.

SCOPE OF THE ANALYSIS

When doing an instructional task analysis, the scope of the task must be taken into consideration. Is the analysis related to a course (generally containing many skills), or is it related to a lesson (generally a particular skill)? Actually the process of task analysis is the same for both, but the scope of the analysis and the number of steps in the analysis are different.

Starting with a course, it is necessary to identify the *purposes* of the course. At this stage of course development it isn't necessary to attempt to formulate specific five-component objectives; however, many of the same guidelines apply to defining purposes.

1. The statement of purpose of a course should be concerned with what the student will be like *after* the instruction, not what he is doing *during* the course. For example, the statement, "To provide the learner experience with chemical apparatus," describes what the student will be doing in the course, not what he will be learning. What is the *purpose* of experience with chemical apparatus? One might be, "To be able to state the names of different kinds of chemical apparatus." Another might be, "To be able to assemble apparatus for chemical experiments."

2. In stating course purposes, avoid the tendency to identify those that are too far removed, too far in the future. Purposes should be stated in *expected current outcomes* of instruction. For example, rather than a purpose such as, "acquires a lifelong respect for chemistry," it would be more realistic to say, "states how chemistry is important for an understanding of the world around us." There is nothing wrong with the lifelong goal, but it is probably not going to occur as the result of a single course, which is but a part of all instruction leading to lifetime respect.

In summary, a good way to start the task analysis process for a course is to define the course purpose. Examples of acceptable course purposes include:

1. Understands commutative properties in multiplication
2. Can recognize tonal differences among different instruments of the same family
3. Reads with enjoyment short stories with a simple plot

As discussed in Chapter 7, these *course purposes* can be translated into more specific learning outcome statements, such as:

1. Demonstrates that $A \times B$ is equivalent to $B \times A$, by example, citing the commutative property
2. Presented with a recording of a violin, viola, or cello, identifies each instrument by writing its name.
3. During free reading, chooses to read short stories with a simple plot as evidenced by a reading log.

These examples show how course purposes, even though originally broadly stated, can be classified by type of outcome and stated so that the outcomes of instruction are clear.

TYPES OF TASK ANALYSIS

There are two major classes of task analysis. The first is what we will call an information-processing or procedural analysis, and the second is a learning-task analysis.

A *procedural task analysis* describes the steps in the process of performing a task or skill. For example, the task of making sentences with indefinite pronouns as subjects may be seen to comprise a series of steps as shown in Figure 8-1. Such a sequence involves *choice* and the *alternative actions* the choice implies. Information-processing analysis goes beyond the observed overt behavior. It includes recalling a verb and making a decision whether to use a singular or plural form. The distinction between choice and action implies that a diagram of the resulting analysis needs to be more than a series of steps. This is accomplished by using a flow chart for the diagramming of decisions. The convention in flow charts is to represent an input with a trapezoid, an action with a rectangle, and a decision with a diamond. The flow chart for writing sentences with the pronoun *everyone* as the subject would look like Figure 8-2.

Two kinds of information useful to instructional designers result from procedural analysis. First, such an analysis and the diagram that results from it provide a clear description of the target objective, including steps involved in the procedure. A flow chart makes it possible for the instructional designer to specify the sequence of the target performance for presentation to the learner.

A second use of procedural analysis arises from the revelation of individual steps that might not otherwise be obvious. This is particularly true of the decision steps since they are, after all, instances of internal processing rather than of overt behavior. For example, the decisions indicated by the diagram of Figure 8-2 imply that the learner must be able to distinguish the singular

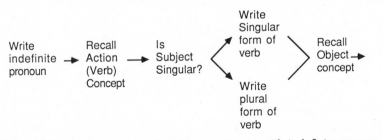

FIGURE 8-1   Steps in the process of making sentences with indefinite pronouns as subjects.

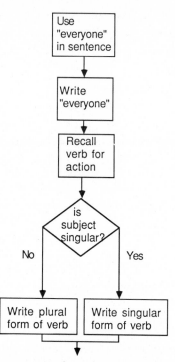

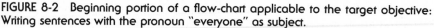

FIGURE 8-2 Beginning portion of a flow-chart applicable to the target objective: Writing sentences with the pronoun "everyone" as subject.

from plural subjects to complete the task. This then becomes a skill that must be acquired if it is not known already. It becomes one of the enabling objectives that make up the total task of writing sentences with the subject *everyone*.

A flow chart resulting from the procedural analysis of another task is illustrated in Figure 8-3. The chart includes several decisions and a number of actions, leading to the performance of finding the difference between two two-digit numbers.

A *learning-task analysis* identifies the skills it would take to be able to learn to perform each step of the procedural task. For example, at step 1 the learner must know the concept of smaller than. We might presume that if students were at a point where they would be learning this task, they would have acquired this concept. However, at step 3, learners have to decrease the tens digit by 1. This operation could be taught without any understanding of why 1 is subtracted. However, if the learners are expected to transfer the newly learned skill to other problems, it would probably be helpful if they understood the operation of subtracting the 1 from the 10s column as the concept of "borrowing." This same concept is used when subtracting numbers containing more than two digits. The concept of borrowing (classifying

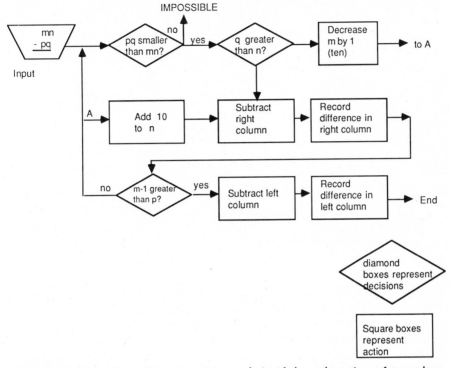

FIGURE 8-3   An information-processing analysis of the subtraction of two-place numbers.
(From R. M. Gagné, Analysis of objectives. In L. J. Briggs (Ed.), *Instructional design.* Copyright 1977 by Educational Technology Publications, Englewood Cliffs, NJ. Reprinted by permission.)

an instance appropriate for borrowing) and the rule for borrowing, "when the top digit being subtracted from is less than the bottom digit being subtracted, subtract 1 from the top digit to the left and add 10 to the digit being subtracted from," must both be learned. It will be evident that this is a complex rule, involving knowing the place values of digits and a number of other concepts. Furthermore, there are many special cases of this rule; for example, if the problem has more than two digits and the second digit from the right is a 0. The relevant skill in this case would presumably be taught at a later time after the simple case has been learned. This example illustrates the point that a learning task analysis may uncover objectives that are not specified in the procedural analysis.

Prerequisite Skills

In its most general sense a *prerequisite* is a skill that is learned prior to the learning of a target objective, and that aids or enables that learning. Any given task may be a target objective for a lesson and at the same time an enabling objective for a subsequent lesson because it is a prerequisite of the task to be learned in the later lesson. For example, the target objective of calculating the diagonal distance across a rectangular piece of land, has as prerequisites the skills of (1) measuring the distance along the sides of a rectangle, and (2) applying the rule for computation of the hypotenuse of a right triangle. These two enabling objectives may have been learned some years prior to the lesson designed to teach the target objective (finding diagonal distance), or they may have been learned at an immediately prior time, as during the same lesson.

Types of Prerequisites

It may help to classify prerequisite objectives into one of two types: essential prerequisites and supportive prerequisites.

Examples of essential prerequisites may be found by analyzing the task of "supplying the definite article" for a noun in writing a sentence in the German language. In acquiring such a capability, a student must learn the prior tasks of (1) identifying gender, (2) identifying number (singular or plural), and (3) applying grammatical rules of case. Such capabilities may have been learned as a result of formal instruction or, in an incidental fashion, by experience with the language. However, the latter possibility is not a relevant fact if one is concerned with the systematic design of instruction. What is relevant is that these "subordinate" capabilities are actually part of the total skill of supplying the definite article. This means that they are essential prerequisites, not merely helpful or supportive. These component skills must be learned if the total task of supplying the definite article is to be learned and performed correctly. Early or late in the course of instruction, they must be acquired prior to the learning of the objective.

A prerequisite may, however, be simply supportive. This means that the prerequisite may aid the new learning by making it easier or faster. For example, a positive attitude toward learning to compose proper German sentences may have been acquired by the learner because of an anticipated visit to Germany. Such an attitude is likely to be helpful to the learner of the language. In other words, it is supportive of the learning, although not essential. Another example is provided by the possibility that the learner has previously acquired a cognitive strategy for remembering the gender of German nouns. Such a strategy might involve associating each newly encountered noun with a visual image representing, respectively, *masculine, feminine,* or *neuter.* Again, such a strategy could be a supportive prerequisite because it makes learning easier and faster.

For each category of task (as identified by task classification), one can identify both essential and supportive prerequisites. However, these prerequisites are considerably different in character for each task category. Keeping these differences straight is important for instructional design. This is one of the major reasons for conducting task classification and for determining such task categories before one attempts to identify prerequisites. In the following sections, our discussion of prerequisites begins with those that are essential and proceeds to describe some that are supportive.

PREREQUISITES IN LEARNING INTELLECTUAL SKILLS

Intellectual skills, like other kinds of learning, are affected by both essential and supportive prerequisites. For this class of learning task, however, essential prerequisites are particularly evident and also likely to be directly involved in the planning of individual lessons.

Essential Prerequisites for Intellectual Skills

A target objective representing an intellectual skill is typically composed of two or more subordinate and simpler skills. These latter skills are prerequisites to the learning of the target skill, in the sense that they must be learned first, before the target skill is "put together." The prior learning of prerequisites may have occurred some considerable time previously. Often, though, it occurs just prior to the learning of the target skill, within the same lesson.

An example of the meaning of essential prerequisites is provided by Gagné (1977) for the task of subtracting whole numbers. Such a task may be represented by problems such as the following:

(a) $\begin{array}{r} 473 \\ -342 \end{array}$ (b) $\begin{array}{r} 2132 \\ -1715 \end{array}$ (c) $\begin{array}{r} 953 \\ -676 \end{array}$ (d) $\begin{array}{r} 7204 \\ -5168 \end{array}$

A commonly taught method of performing subtraction is by borrowing, and we assume that is the method to be learned as part of the target skill. The four examples illustrate four prerequisite skills (rules) that are involved in the skill of subtracting whole numbers. Example (a), the simplest, is "subtracting one-place numbers in successive columns, without borrowing." Example (b) is "subtracting when several borrowings are required, in nonadjacent columns." Example (c) may be described as "successive borrowing in adjacent columns"; borrowing must be done in the first column on the right, so that 6 can be subtracted from 13, and again in the next column, so that 7 can be subtracted from 14. Example (d) is the prerequisite skill "borrowing across 0."

Each of these prerequisite skills represent a rule that is involved in the total skill of subtracting whole numbers. The latter task cannot be learned in any complete sense without the prior learning of these subordinate skills. They therefore deserve to be called *essential* prerequisites.

Other examples of prerequisites for intellectual skills may be found by examination of the results of the information-processing analyses described earlier in this chapter. The analysis of subtraction, of course, includes the borrowing skill, which is comparable to Example (a). When the objective is writing sentences with the subject *everyone,* the essential prerequisites indicated by the diagram of Figure 8-1 are (1) identifying verb names for actions and (2) using rules to make verbs singular or plural.

Hierarchies of Prerequisite Skills

Although learning-task analysis is often concerned with the prerequisite of a target skill, the analysis may be applied to the enabling skills as well, since these skills themselves have prerequisites. Accordingly, it is possible to continue the learning-task analysis until a point is reached at which the skills identified are quite simple (and perhaps assumed to be known by all students).

When a learning-task analysis is carried out on successively simpler components of a target skill, the result is a learning hierarchy (Gagné, 1985). This outcome may be displayed as a diagram containing boxes describing the successively identified subordinate skills (i.e., essential prerequisite skills). Figure 8-4 is an example of a learning hierarchy for the target objective of "subtracting whole numbers." At a first level of analysis, this learning hierarchy incorporates the four prerequisites skills involved in subtracting, as described in the previous section (numbered VII, VIII, IX, and X). Proceeding from that point, analysis makes possible the identification of the simpler skill VI, "subtracting when a single borrowing is required, in any column." This can readily be perceived as an essential prerequisite to the more complex skills of borrowing (VIII, IX, and X). Skill VI, however, may also be analyzed to reveal the prerequisites described in the boxes labeled IV and V. The process may be continued to the level of simple subtraction "facts" (I).

The learning hierarchy produced by a learning task analysis displays a pattern of progressively simpler intellectual skills. These skills are enabling objectives for a given target objective (which is also an intellectual skill). Furthermore, any particular skill, which may itself be an enabling objective for a target skill, is composed of other subordinate skills.

When instructional design interest centers upon an intellectual skill as an objective, the enabling skills of primary relevance are the immediate prerequisites. Thus, the most important questions to be answered by a learning task analysis come from any two adjacent levels of the learning hierarchy. Is there, then, any usefulness to the display of the entire pattern of enabling skills, involving several levels? The main uses for a fully worked-out learning hierarchy are (1) as a guide in the design of a sequence of instruction (Cook and Walbesser, 1973), and (2) as a guide for student and teacher in the planning of instructional assignments.

Conducting a Learning-Task Analysis for Intellectual Skills

Intellectual skill analysis is carried out by "working backwards" from a target skill. Obviously, the purpose of the analysis is to reveal the simpler compo-

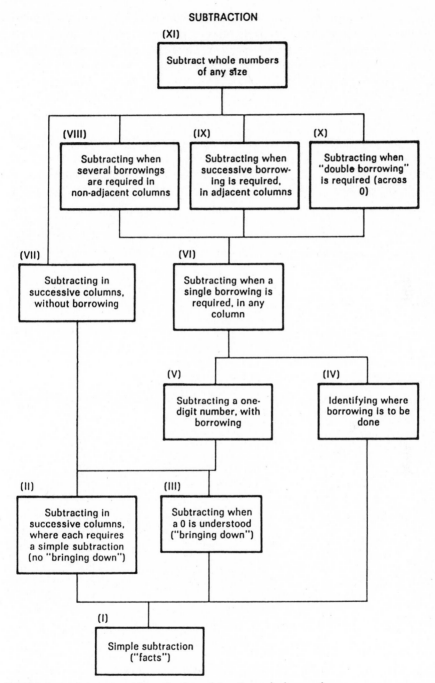

FIGURE 8-4   A learning hierarchy for subtracting whole numbers.

nent skills that constitute the target skill. Often there is a correspondence at the first level of analysis between the component steps resulting from an information-processing analysis and the component skills revealed by a learning task analysis. But the distinction between sequential steps and subordinate skills should be carefully maintained. The steps are what an individual does in sequence when exhibiting a target performance, the capability of which has been learned. In contrast, the subordinate skills are what the individual must learn, in a sequence beginning with the simplest.

Subordinate skills are derived by asking a question of any given intellectual skill: "What simpler skill(s) would a learner have to possess to learn this skill?" (Gagné, 1985). Once the first set of subordinate (enabling) skills is identified, the process may be repeated by addressing the same question to each of these. Of course, the subordinate skills so generated become simpler and simpler. Normally, the process stops at a point that is determined pragmatically. One decides, on the basis of knowledge about student characteristics, that the skills at the lowest level of the hierarchy are already known and do not have to be learned. Naturally, this stopping point varies with the educational background of the students. A learning hierarchy on grammatical rules in a foreign language, for example, would have many more levels for students who had learned few grammatical rules of their native language, than for student who had learned many of these rules.

Supportive Prerequisites for Intellectual Skills

A number of kinds of learning, when undertaken on a prior basis, may be supportive in the acquisition of an intellectual skill. This means that the previously acquired capability may be helpful, although not essential, for the learning of a target skill.

Verbal information, for example, is often useful to the learner in acquiring intellectual skills, presumably because it facilitates the verbal communications that are a part of instruction. Some good examples of the relation of verbal information to the learning of an intellectual skill occur in a study by White (1974). White developed and validated a learning hierarchy for the intellectual skill "finding velocity at a prescribed point on a curved graph relating position (of an object) and time." In its initial form, the trial hierarchy included capabilities of the information sort, such as "states that the slope of a position–time graph is velocity," and "states units of slope in units labeling axes." The results of White's investigation showed that although these capabilities may have been helpful for intellectual skill learning, they did not reveal themselves as being essential. In this, they contrasted markedly with other subordinate components of the intellectual skill category. Further verification of the supportive nature of information, contrasting with prerequisites that are essential, is provided in a study by White and Gagné (1978).

The instructional designer may find it desirable to take into account several kinds of supportive prerequisites. Depending on other circumstances, these may be introduced into a single lesson, or they may be given a place in the sequence of a topic or course. The major possibilities of supportive prerequisites for intellectual skill learning would appear to be as follows:

Information as Supportive Prerequisites

As our previous example illustrates, verbal information may support the learning of intellectual skills by aiding the communication of instruction. Often, examples of this function take the form of labels for the concepts involved in rule learning. Another possible function of verbal information is as a context providing cues for retrieval of the intellectual skill (Gagné, 1985). Normally, when intellectual skills are being learned, a fair amount of information is included at the same time (see, for example, any typical science textbook). Presumably, a supportive function is being performed by the latter sort of material. The circumstance to be avoided in design is not the "mixing" of intellectual skills and information during instruction but the potential confusion of these two types of learning as target objectives.

Cognitive Strategies as Supportive Prerequisites

The learning of intellectual skills may be aided in a supportive sense by the use of cognitive strategies. For example, if students learning to "add positive and negative integers" have available the cognitive strategy of imagining a "number line," their learning of the requisite rules may be facilitated. Generally speaking, cognitive strategies may speed the learning of intellectual skills, make them easier to recall, or aid their transfer to novel problems. Although these actions of cognitive strategies have broad acceptance, it must be said that empirical evidence of the effectiveness of their support is, as yet, scanty and still sorely needed.

Attitudes as Supportive Prerequisites

The supportive effect of positive attitudes on the learning of intellectual skills is widely recognized. Presumably, the attitudes a learner has toward a subject strongly influences the ease with which the subject is learned, retained, and put to use (Mager, 1968). The relation between positive attitude and learning may often be readily observed in students' conduct toward a subject like mathematics. Evidence concerning the effects of "affective entry characteristics" on achievement in school subjects is reviewed by Bloom (1976). A recent review of the literature by Martin and Briggs (1986) shows attitude learning strongly related to the learning of cognitive skills.

LEARNING-TASK ANALYSIS AND OTHER LEARNING TYPES

The rationale of learning-task analysis can also be brought to bear on learning tasks other than intellectual skills—on the learning of cognitive strategies, information, attitudes, and motor skills. The purpose of analysis remains unchanged—that is, the identification of both essential and supportive prerequisites. However, the picture that emerges in considering these applications is surely different and not as clear. Capabilities like information and attitude are not learned by putting together subordinate parts, as is the case with intellectual skills. Consequently, prerequisites for these other types of learning tend to be of a supportive rather than an essential nature.

Table 8-1 summarizes the essential and supportive prerequisites resulting from analysis of the five types of learning outcomes.

Prerequisites: Cognitive Strategies

Presumably, the prerequisites for cognitive strategies of learning, remembering, and thinking are some very basic (and perhaps very simple) mental abilities. For example, an effective cognitive strategy for remembering a list of items may be to generate a different mental image for each item. The essential prerequisite in such a case must be the *ability to have visual images,* which surely must be considered a very basic ability. Again, an often effective cognitive strategy in solving complex mathematical problems is breaking the problem into parts and seeking the solution to each part. What prerequisite ability

TABLE 8-1.   Essential and Supportive Prerequisites for Five Kinds of Learning Outcome[a]

Type of Learning Outcome	Essential Prerequisites	Supportive Prerequisites
Intellectual Skill	Simpler component intellectual skills (rules, concepts, discriminations)	Attitudes, cognitive strategies, verbal information
Cognitive Strategies	Specific intellectual skills (?)	Intellectual skills, verbal information, attitudes
Verbal Information	Meaningfully organized sets of information	Language skills, cognitive strategies, attitudes
Attitudes	Intellectual skills (sometimes) Verbal information (sometimes)	Other attitudes, verbal information
Motor Skills	Part skills (sometimes) Procedural rules (sometimes)	Attitudes

[a]From Analysis of objectives (p. 141) by R. M. Gagné, 1977. In L. J. Briggs, (Ed.), *Instructional design.* Englewood Cliffs, NJ: Educational Technology Publications. Copyright 1977 by Educational Technology Publications. Reprinted by permission.

is involved in such a strategy? Evidently, an *ability to divide a verbally described situation into parts.* This, too, seems to be a fundamental kind of ability of a rather simple sort.

Whatever the essential prerequisites of cognitive strategies may be, there is disagreement concerning how much they depend upon innate factors (which develop through maturation) and how much they are learned. A discussion of these issues has been presented by Case (1978), and more briefly by Gagné (1977). The factor of maturation plays a prominent role in the developmental theory of Piagét (1970). In contrast, Gagné (1985) proposes that cognitive strategies of the executive type are generalizations from learned intellectual skills. It is interesting to note, however, that either of these processes (maturation or learning) takes its effect on such strategies over a considerable period of time, as viewed from the standpoint of intellectual development.

Supportive prerequisites for the learning of cognitive strategies include the intellectual skills that may be useful in learning the particular material or solving the particular problems presented to the learner. Relevant verbal information may also play this supportive role. Just as in learning other kinds of capabilities, favorable attitudes toward learning are likely to be helpful.

Prerequisites: Verbal Information

To learn and store verbal information, the learner must have some basic language skills. A number of learning theories propose that information is stored and retrieved in the form of propositions. If this is the case, then the learner must already possess the essential prerequisite skills of forming propositions (sentences) in accordance with certain rules of syntax. These skills, of course, are likely to have been learned fairly early in life.

Verbal information, whether of single items or longer passages, appears to be most readily learned and retained when it occurs within a larger context of meaningful information. This context may be learned immediately before the new information to be acquired, or it may have been learned a long time previously. The provision of this meaningful context has been described as a learning condition in Chapter 5, and deserves to be classified as a supportive prerequisite of information learning.

Attitudes support the learning of verbal information in much the same way as they do other kinds of learning tasks. A number of different cognitive strategies have been found to be supportive of the learning of word lists (cf. Gagné, 1985; Rohwer, 1973). Presumably, it should be possible to identify particular strategies that aid the retention of prose passages, as in remembering the gist of a textbook chapter (Palincsar and Brown, 1984).

Prerequisites: Attitudes

The acquiring of particular attitudes may require the prior learning of particular intellectual skills or particular sets of information. In this sense, then, these learned capabilities may be essential prerequisites to attitude learning. For example, if a positive attitude toward "truth in labeling" of packaged foods is to be acquired, the learner may need to have (1) the intellectual skills involved in comprehending the printed statement on the label, and (2) a variety of verbal information about food ingredients.

As Table 8-1 indicates, attitudes may have a mutually supportive relation to each other; related attitudes may be supportive of the acquisition of another given attitude. For example, preference for a political candidate makes it easier for a person to prefer also the political views of that candidate's party. In a more general sense, the degree to which a human model is respected affects the readiness with which the model's attitude is adopted.

Besides its essential role in a specific sense, verbal information also has a supportive function in establishing attitudes. Knowledge of the situations in which the choice of personal action will be made contributes to the ease of attitude acquisition. For example, an attitude such as "don't drive after drinking" will likely be more readily acquired if the individual has knowledge about the various social situations in which the temptation will occur to drive after drinking.

Prerequisites: Motor Skills

As described in Chapter 5, motor skills are often composed of several part skills. It is sometimes the case that the most efficient learning comes about when part skills are practiced by themselves and later combined in practice of the total skill. In such instances, the part skills may be said to function as essential prerequisites for the learning of the total skill.

Another component of a motor skill that has this role is the *executive subroutine* (Fitts and Posner, 1967), which is sometimes learned as an initial step in the acquiring of a motor skill. Swimming the crawl, for example, involves an executive subroutine that selects a sequence of movements for arms, legs, body, and head. Even before the total skill is practiced very much, the swimmer may receive instruction in the correct execution of this sequence. In Table 8-1, these subroutines are referred to as *procedural rules*. When learned separately and prior to the skill itself, they may be classed as essential prerequisites.

Positive attitudes toward the learning of a motor skill, and toward the performance it makes possible, are often supportive prerequisites of some significance.

Instructional Curriculum Maps

We have described the diagramming of skills in the intellectual skills domain, involving hierarchial relationships among intellectual skills, with discriminations as prerequisites for concepts, concepts as prerequisites for rules, and rules as prerequisites for problem solving. These relationships can be diagrammed as shown in Figure 8-5.

It is somewhat more difficult to visualize what a diagram of the relationships among objectives from different domains might be; for example, the relationship between an intellectual skill and an attitude. Briggs and Wager (1981) have described a system called *instructional curriculum mapping* for illustrating these relationships. Instructional curriculum mapping simply tries to represent the functional relationships among instructional objectives. It starts by identifying the target objective and asking the question what other objectives are related to the attainment of this objective (either essential skills or supportive prerequisites). Hierarchical relationships of essential prerequisites are drawn in much the same fashion as the illustration in Figure 8-5. However, supportive objectives are shown connected to the target objective, with an indication in each case that they are *not* from the same domain.

For example, this target objective is an attitude: "Given access to microcomputers, the student chooses to use a computer as a word processor to type assignments rather than writing them by hand.

Obviously the student will have to have the intellectual skills necessary to apply the skills, but they alone may be insufficient to formulate the attitude. Supporting objectives might be:

1.   States functions of a word processor
2.   States the advantages of word processors over type writers
3.   States advantages of typed over hand-written material

All three of the above objectives are verbal information that is not needed (essential prerequisites) by the learner to learn the intellectual skills associated with employing word processors. However, the student probably already knows a fair amount about the advantages of neatly presented work as a result of having come through the school system. The information is supportive

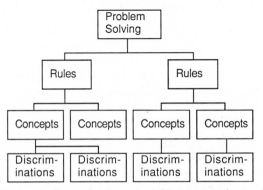

FIGURE 8-5   Hierarchical relationships among objectives in the intellectual skills domain.

of formulating an attitude toward use of the computer as diagrammed in Figure 8-6. The triangle between the verbal information and the attitude objective illustrates the fact that this is an intersection of two different domains and alerts the designer that perhaps different conditions of learning will probably be needed to accomplish the target objective. The three verbal information objectives are not essential prerequisites to each other, but they may be taught in contextual (supportive) sequence. The relationships among objectives from the same domain may be shown with solid lines.

Figure 8-7 shows the highest level intellectual skill associated with this target objective—rule using: "Given hand-written copy, demonstrates the use of a word processor by typing, editing, and printing the text."

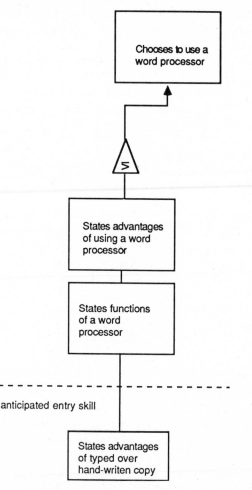

FIGURE 8-6 An instructional curriculum map (ICM) showing supportive relationships of verbal information (V.I.) objectives to an attitudinal target objective.

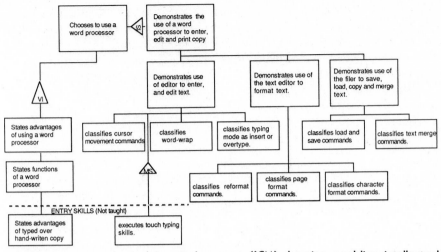

FIGURE 8-7   An instructional curriculum map (ICM) showing enabling intellectual skills objectives for the task of using a word processor.

This objective is functionally related to the target attitude objective as an essential prerequisite. The relationship is again diagrammed to show that there is a domain change by the inclusion of the ⚠IS symbol in the connecting line, meaning that the intellectual skill is functionally related to attainment of the attitude.

Of course, ". . . demonstrates the use of word processors . . " is an intellectual skill with other prerequisite intellectual skills, including the concepts of *editor, word wrap, filer, blocks, formatting,* and others. Objectives related to these concepts are also diagrammed in a hierarchial manner as shown in Figure 8-7.

Notice that the motor skill of touch-typing may be related to the entry of text and knowledge of advantages of typing. However, touch-typing skills are not essential to learning the use of the word processor.

The technique of instructional curriculum mapping facilitates the designer's task of relating objectives to one another when they are from different domains, and makes it possible to see when there are holes in the instruction and when there are extraneous "dead-wood" objectives that don't seem to relate to any target objectives.

Subordinate and Entry Skills

The terms entry skill and subordinate objectives generally apply to the description of objectives related to a particular lesson. For example, a single lesson may have one or more target objectives. Subordinate to these objectives are the enabling objectives that will be *taught* as part of the lesson. Entry skills, which the learner is expected to have acquired before attempting this lesson are also subordinate to the target objective. They are essential or supportive prerequisites, but they will not be taught during the lesson. Entry skills are identified in both hierarchy diagrams and instructional maps as those skills listed below the dotted line. This is illustrated in Figure 8-8. Identifying the entry skills in this manner makes it possible to construct a pretest to see if the student has the skills necessary to continue with this lesson.

SUMMARY

Task analysis refers to several different, though interrelated, procedures that are carried out to yield the systematic information needed to plan and specify the conditions for instruction. The two procedures described in this chapter are: (1) information-processing analysis; and (2) learning-task analysis. Both types of analysis begin with target objectives for lessons or courses.

Information-processing analysis describes the steps taken by the human learner in performing the task he has learned. Included in these steps, in the typical case, are (1) input information, (2) actions, and (3) decisions. Of particular importance is the fact that this type of analysis usually reveals mental operations that are involved in the performance but are not directly observable as overt behavior. Together, the various steps in the performance may be shown in a flow chart. The results of the analysis exhibit (or imply) capabilities that must be learned as components of the performance described as the target objective.

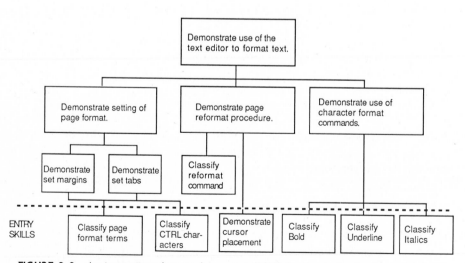

FIGURE 8-8   An instructional curriculum map (ICM) for the task of using a text editor, with entry skills indicated.

These components are themselves instructional objectives, called enabling objectives, that support the learning of the target objective. In addition, they may need to be analyzed further (in the manner of learning task analysis) to reveal additional enabling objectives.

Task classification has the purpose of providing a basis for designing the conditions necessary for effective instruction. The objectives of instruction are categorized as intellectual skills, cognitive strategies, information, attitudes, or motor skills. As indicated in previous chapters, each of these categories carries different implications regarding the conditions necessary for learning, which can be incorporated into the design of instruction.

Learning-task analysis has the purpose of identifying the prerequisites of both target and enabling objectives. Two kinds of prerequisites are distinguished—essential and supportive. Essential prerequisites are so called because they are components of the capability being learned, and therefore their learning must occur as a prior event. Other prerequisites may be supportive in the sense that they make the learning of a capability easier or faster.

Target objectives of the intellectual skill variety may be analyzed into successive levels of prerequisites, in the sense that complex skills are progressively broken down into simpler ones. The result of this type of analysis is a learning hierarchy, which provides a basis for the planning of instructional sequences. Prerequisites for other categories of learning objectives do not form learning hierarchies since their prerequisites do not relate to each other in the manner of intellectual skills.

A number of kinds of supportive prerequisites may be identified for particular types of target objectives. For example, task-relevant information is often supportive of intellectual skill learning. Positive attitudes toward lesson and course objectives are also an important source of learning support. Cognitive strategies of attending, learning, and remembering may be brought to bear by the learner in supporting these processes. These supportive relationships may be diagrammed through the use of instructional curriculum mapping.

REFERENCES

Bloom, B. S. (1976). *Human characteristics and school learning.* New York: McGraw-Hill.

Briggs, L. J., & Wager, W. W. (1981). *Handbook of procedures for the design of instruction.* Englewood Cliffs, NJ: Educational Technology Publications.

Case, R. (1978). Piaget and beyond: Toward a developmentally based theory and technology of instruction. In R. Glaser (Ed.), *Advances in instructional psychology* (Vol. 1.). Hillsdale, NJ: Erlbaum.

Cook, J. M. & Walbesser, H. H. (1973). *How to meet accountability.* College Park, MD: University of Maryland, Bureau of Educational Research and Field Services.

Fitts, P. M., & Posner, M. I. (1967). *Human performance.* Monterey, CA: Brooks/Cole.

Gagné, R. M. (1977). Analysis of objectives. In L. J. Briggs (Ed.), *Instructional design*. Englewood Cliffs, NJ: Educational Technology Publications.

Gagné, R. M. (1985). *The conditions of learning* (4th ed.). New York: Holt, Rinehart and Winston.

Mager, R. F. (1968). *Developing attitude toward learning*. Belmont, CA: Fearon.

Martin, B. L., & Briggs, L. J. (1986). *The affective and cognitive domains: Integration for instruction and research*. Englewood Cliffs, NJ: Educational Technology Publications.

Palincsar, A. S., & Brown, A. L. (1984). Reciprocal teaching of comprehension-fostering and comprehension-monitoring activities. *Cognition and Instruction, 1,* 117–175.

Piaget, J. (1970). Piaget's theory. In P. H. Mussen (Ed.), *Carmichael's manual of child psychology*. New York: Wiley.

Rohwer, W. D., Jr. (1970). Images and pictures in children's learning. *Psychological Bulletin, 73,* 393–403.

White, R. T. (1974). The validation of a learning hierarchy. *American Educational Research Journal, 11,* 121–136.

White, R. T., & Gagné, R. M. (1978). Formative evaluation applied to a learning hierarchy. *Contemporary Educational Psychology, 3,* 87–94.

9

DESIGNING INSTRUCTIONAL SEQUENCES

Learning directed toward achieving the goals of school education takes place on a number of occasions over a period of time. The learning of any particular capability is preceded by the learning of prerequisite capabilities and is followed, on other occasions, by learning more complex capabilities. We generally refer to the specification of a set of capabilities as a *curriculum* or *course of study*.

A curriculum or course requires decisions about the sequencing of objectives. The goal of an educational institution is to establish sequences within courses that promote effective learning. The most obvious sequence follows the order of simple (prerequisite skills) to complex (target) skills, the latter of which take longer to accomplish. Another sequencing principle is one of sequencing objectives in increasing order by the degree of meaning in what is being learned. We know from cognitive learning theory (Anderson, 1985), that one very important determinant of what students will learn, and how fast they will learn, is what they already know. Reigeluth and Stein (1983) address the sequencing problem at a macrolevel in their "elaboration theory of instruction." The elaboration theory proposes that sequencing should be structured so that the student is presented first with an epitome (a generalization) of a concept, procedure, or principle to be learned. The student is then presented with an elaboration or extension of the epitome. The concepts, procedures, and rules are organized from simple to complex; general to specific.

As will be seen, the question of sequencing is encountered at several levels of curriculum and course design, and the issues are somewhat different between levels. But basically, the matter of effective sequences of instruction is closely related to the matter of course organization. This chapter will describe a procedure for organizing a course from the top downward, going from general to more specific objectives, and utilizing the functional relationships among the types of learning described in previous chapters.

In Chapter 8 we indicated that once the objectives or goals are specified for a given curriculum or course, it is desirable next to identify major *course units,* each of which may require several weeks of study. Under each such unit, one may next identify specific objectives to be reached by the end of the unit, or by the end of the course. These specific objectives are then grouped together into lessons, which in turn may require the identification of several enabling objectives.

Intellectual skills objectives are usually the starting point for consideration of the sequencing of instruction. This is primarily due to the importance we place on intellectual skills as components of the curriculum, but also because more is known about the sequencing relationships among intellectual skills than about other types of learned capabilities. Objectives from other domains of learned capabilities are then woven into the intellectual skills structure insofar as they support the learning of the intellectual skills. This procedure assumes that intellectual skill objectives are the principal target objectives. However, if the terminal objective is not an intellectual skill, but instead an attitude, then the intellectual skills that are supportive of it must be identified. The integration of objectives from different domains may be expressed in the form of instructional maps, as described in Chapter 8. Ultimately, individual lessons are planned to integrate related skills into an overall curriculum of lessons to accomplish the purposes of the course.

One problem with terminology is that the word *course* has many different meanings. For example, a course on cardio-pulmonary resuscitation (CPR) is quite different from a course in computer literacy. The first has very definite criteria for judging mastery of the required skills. There would be substantial consensus among persons who teach CPR about the objectives, the criteria for performance, and the amount of time needed to teach the course. Also, the number of objectives in the CPR course is relatively small. In contrast, the curriculum for a computer literacy course is much broader, there is less agreement on terminal objectives, and the total number of objectives for such a course is quite large.

Another problem in defining a course is the constraint imposed by the designation of a course as a specific number of hours of instruction. For example, a one-semester-hour course in a university typically represents 16 contact hours of formal instruction. In public school a course represents about 180 hours of formal instruction. Despite these variations, time is a very important part of planning instruction, and the amount of time available must be carefully considered in planning course and unit objectives.

There are no standard levels to be employed in organizing a course (except for the possible assumption that it consists of two or more lessons). However, even the largest course can be described in five different levels of performance outcomes as follows:

1. *Lifelong objectives* which imply the continued future use of what is learned, after the course is over
2. *End-of-course objectives* which state the performance expected immediately after instruction is completed for the course
3. *Unit objectives* which define the performance expected on clusters of objectives (topics) having a common purpose in the organization of the total course
4. *Specific performance objectives* which are the specific outcomes attained during a segment of instruction and which are likely to be at the appropriate level for task analysis
5. *Enabling objectives* which are either essential or supportive prerequisites for specific target objectives

AN EXAMPLE OF COURSE ORGANIZATION

Considering the nature of the content of this book, it is perhaps not inappropriate to illustrate the levels of course organization for a graduate course in the design of instruction. Such a course would be part of a doctoral program in instructional systems design. Related courses in the curriculum would pertain to learning theory, research methods, statistics, varieties of instructional design, design theories, and models of instructional delivery. Students entering the course would typically have master's degrees, often in an area of teaching such as science education, or in fields such as educational media or educational administration. Most of them would have completed an introductory course in the theory base for systems models of the design of instruction.

Students are taught to design their own courses based on some identified instructional need or goal. Following the "general-to-specific" basis for course design, students are asked to state their course objectives at several levels.

The levels of objectives for the graduate course in the design of instruction may be illustrated as follows:

1. Lifelong objective. After completion of this course, the students will continue to add to their course design skills by (1) enrolling for other design courses, and (2) seeking a variety of opportunities to apply design skills in circumstances that require them to modify learned models or to originate new models. Students will choose to employ or originate systematic course design procedures based on theory, research, and consistent rationales; they will choose to use empirical data to improve and evaluate their designs.

2. End-of-course objective. By the end of the course, the students will have demonstrated the ability either to perform or to plan each step in a systems model of instructional design, from needs analysis to summative evaluation. (In our hypothetical course, efforts are concentrated on stages 4 through 9, as listed in Table 2-1).

3. Unit objectives. Students will complete four successive assignments by completing stages of design representing the following course units:

Unit A. The student will generate a course organization map showing lifelong objectives, end-of-course objectives, and unit objectives, with accompanying measures of learner performance for those levels of objectives at which the learners' work is to be evaluated.

Unit B. The student will generate, in writing, a learning hierarchy for an intellectual skill objective, and will also devise an instructional map to show how the prerequisite skills in the hierarchy are to be sequenced in relation to each other and to objectives in other domains of outcomes.

Unit C. The student will generate, in writing, either a lesson plan or a module of instruction, showing a rationale for media selection and prescriptions for instruction to be prepared for each medium selected. (See Chapter 11).

Unit D. The student will generate a written script that implements the media prescriptions made in unit C.

4. Specific performance objectives (for unit C, above).

1. State the objective(s) or enabling objective(s) for the lesson being planned.
2. Classify the objective(s) by domain (and subdomain, if appropriate).
3. List the instructional events to be employed and give a rationale for use or nonuse of each of the nine events.
4. List the type of stimuli for each event.
5. List the media choices appropriate for each event (see Chapter 11).
6. Identify the theoretically best medium for each event.
7. Make the final medium selection for each event.
8. Give a rationale for decisions 4 through 7.
9. Write the prescriptions (to the media producers) for each event.

Specific performance objectives for all four units of such a course, along with criteria appropriate for evaluating students' work, are described in detail by Briggs (1977, pp. 464–468). Many of the prerequisite objectives supporting the specific performance objectives for this course are represented in the form of practice tests and exercises that students take to test their readiness to write each of the four unit assignments (Briggs and Wager, 1981).

The levels of objectives just described may be seen as one way to organize a course. Note that this organization progresses downward to the level of objectives for individual lessons. However, the materials themselves must also be organized and sequenced; that is, the sequence of the instructional events that constitute the lesson must also be planned. This part of the planning depends a great deal on the sophistication of the learners since the completeness of instructional events designed into the lesson depends upon which events the learners are expected to provide for themselves.

The sequencing of the four course units follows the design sequence described in Chapter 2. Although learning could presumably take place in other sequencing arrangements, in this case it appeared reasonable to have the learning sequence follow that which an experienced designer might employ in practice. However, there is often reason to have a learning sequence that is different from the performance sequence. This might happen when a complex skill needed early in the procedure is more easily learned after acquiring skills needed later. For example, teaching a student how to write good test items may be easier after teaching him how to write five-component objectives. However, on the job he may write the test items first, and then write the objective.

An example of four levels of sequence planning is illustrated by a curriculum in English writing at the level of junior high school (Table 9-1). The sequence problem obviously arises here at the course level. Also, there may be a problem to be solved for the single-course topic, such as "writing the paragraph." A third and critically important level of the sequence question concerns the sequence of skills within the individual lesson, such as "constructing sentences with dependent clauses." And finally, there is the matter of the sequence of events that occur or is planned to occur to bring about the acquisition of an individual lesson component objective, such as "making subject and verb agree in number."

TABLE 9-1.   Four Different Levels of the Problem of Instructional Sequence

	Unit	*Example*	*Sequence Question*
Level 1	Course or course sequence	Essay composition	How will the topics of "achieving unity," "paragraph arrangment," "writing the paragraph," "summarizing," and so on be arranged in sequence?
Level 2	Topic or unit	Writing the paragraph	How will the subtopics of "topic sentence," "arranging ideas for emphasis," "expressing a single idea," and so on be arranged in sequence?
Level 3	Lesson	Composing a topic sentence	How will the subordinate skills in composing a topic sentence be presented for learning in sequence?
Level 4	Lesson component	Constructing a complex sentence	In what sequence will concepts of parts of speech and grammatical rules be presented?

It is important to distinguish among these four levels since quite different considerations apply to each one. As will be apparent from the contents of this chapter, we are mainly concerned here with levels 1 and 2, and we will be dealing later on with questions posed by levels 3 and 4.

Sequence of the Course and Curriculum

Course sequence decisions deal mainly with answering the question, "in what sequence should the units be presented?" Presumably, one wants to assure that the prerequisite information and intellectual skills necessary for any given topic have been previously learned. For example, the topic of adding fractions is introduced in arithmetic after the student has learned to multiply and divide whole numbers, because the operations required in adding fractions require these simpler operations. In a science course, one is concerned that a topic like "graphically representing relations between variables" be preceded by attainment of the skill of "measuring variables." And obviously, one would expect the student to have an understanding of the concept "culture" before teaching a social studies topic on "comparing family structures across cultures."

A model for sequencing instruction in a course is referred to as *macrolevel* sequencing by Reigeluth and Stein (1983) in their account of the elaboration theory of instruction. The content of ideas with which this theory deals includes concepts, procedures, and principles. The theory proposes that instructional content be structured so that the student is first presented with a special kind of overview called an *epitome,* which includes a few general, simple, and fundamental ideas. Instruction then proceeds by presenting more detailed ideas that *elaborate* on the earlier ones. Following this phase is a *review* of the overview and a delineation of the *relationships* between the most recent ideas and those presented earlier. This pattern of overview, elaboration, summary, and synthesis is continued until the desired level of coverage of all aspects of the subject has been reached.

Course and curriculum sequences are typically represented in *scope and sequence* charts, which name the topics to be studied in a total course or set of courses and lay them out in matrices. This approach was utilized by Tyler (1949), and it makes a good first step in defining different levels of skills across content topics. For example, an introductory course on computers might be represented by Table 9-2.

This scope and sequence matrix is by no means complete and represents only four types of learning outcomes, yet it demonstrates how the designer can structure topics and skills. It is especially useful to specify the affective outcomes that may be desired. In the computer course, it is evident that most of the outcomes are directed toward achieving intellectual skills. However, in the social issues unit the attitude outcomes are the most important if the learner is to respond to using computers in a positive manner.

TABLE 9-2.   Scope and Sequence Matrix of Topics and Types of Learning Objectives for a Course on Computer Usage

Topical Content	Verbal Information State:	Defined Concepts Classify:	Rule Using Demonstrate:	Attitudes Choose:
Components of computers	Parts names Definitions	Storage devices RAM memory Output devices Input devices CPU Hardware Software	Booting machine Turning machine off	Caring for computer
Operating systems	Definitions Purpose	Program commands I/O file names file types	Listing files Formatting disk Copying files Renaming files	Learning the functions associated with operating systems
Language	Definition	Commands Statements Lines Editor Interpreter	Input a program Edit a program Run a program Save a program	Structuring operations in logical order
Social issues	Five social issues	Computer theft Computer fraud Copyright Equity Health		Using computers for social good

(Column group header: *Types of Learning Outcomes*)

The target objectives of a unit can be related to the course objective or goal in a course level instructional curriculum map (ICM). Figure 9-1 shows such an ICM for an introductory computer course. In this example the sequence in which Units 1 and 4 are taught is not critically important, since the intellectual skills objectives are fairly independent. However, the skills in Unit 1 are prerequisite to the skills in Unit 2, and those in Unit 2 are prerequisite to those in Unit 3. Also, basic terminology and use of the computer (Unit 1) is prerequisite to the rule-using skills in Unit 5.

Sequence of Skills Within the Topic

Specifying a teaching sequence within a topic is a problem upon which some systematic techniques can be brought to bear. Initially, it may be recognized that a topic can, and often does, have several components. A topic on computer hardware, for instance, is likely to include objectives such as

Introduction to Computers in Education

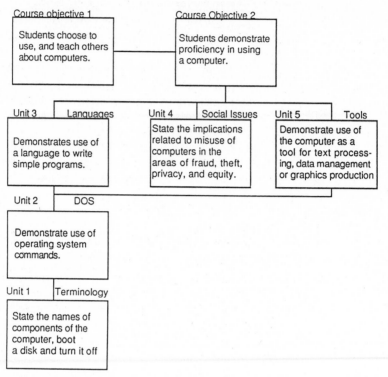

FIGURE 9-1   Instructional curriculum map (ICM) for a course on computers and their uses in education.

(1) identify the components of a microcomputer; (2) demonstrate the booting procedure to boot the microcomputer into the operating system; and (3) choose to handle the equipment and software so it won't be damaged. Notice that all three of these objectives are stated in performance terms. It would not be helpful to have objectives like "understand the booting procedure" or "appreciate computer equipment." These kinds of statements are too ambiguous and, accordingly, may imply different performance objectives to different people.

Analyzing Topic Objectives to Determine Learning Outcomes

The use of performance objectives is particularly important at the unit level, since the designer's objective is to determine what lessons are needed. This can get a bit out of hand, however, because each unit objective may have a good many subordinate essential and supportive prerequisites. At this point in design planning, we suggest that the outline be kept rather broad, and that only the *major* objectives of the unit be specified. These objectives may include any or all of the types of learning outcomes. The specific unit objectives can be represented in an ICM, just as the course and unit objectives were represented. The unit map for the second unit in the computer course is shown in Figure 9-2.

As you can see, this map has more detail, and it shows the relationships among the objectives within the topical unit. The relationship between the course map and the unit map may be compared to the relationship between a world globe and a series of flat maps of each country. The flat maps are going to show less scope than the globe, but more detail.

The unit maps are also going to show the relationships of the objectives from the different domains of learning outcomes. Some will be prerequisite to others and, therefore, they must be taught in the earlier lessons.

Identifying Lessons

The next question is, how does one identify lessons? A lesson is generally considered to occur in a given period of time; that is, the learner expects to spend a given amount of time on a lesson. Obviously lesson times vary. A lesson for a small child may be shorter than one for an adult because the attention span of a child is shorter. Sometimes, the designer tries to have a single lesson deal

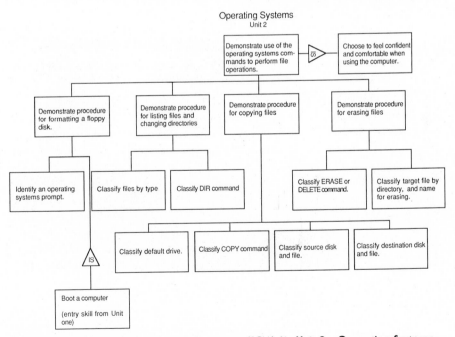

FIGURE 9-2 An instructional curriculum map (ICM) for Unit 2—Operating Systems—of a course on Introduction to Computers.

with a single learning outcome. The reason for this procedure is that each type of learning outcome requires a different set of learning conditions, as described in Chapters 4 and 5. However, since the time it takes to teach a single objective may be quite short, it is not feasible to think of every objective as having its own lesson. For this reason, specific objectives are often grouped into lessons.

It is probably more important to plan lessons around the order in which the performance objectives are best learned rather than worry about having different kinds of learning outcomes in the same lesson. In fact, once the decision is made to group different types of learning outcomes together based on their functional relationships and the amount of time spent at a single sitting, the process of integrating the necessary conditions of learning becomes quite straightforward.

The unit map shown in Figure 9-3 shows how specific objectives within the previous map may be grouped into lessons. In this case the unit has two lessons, each about one hour in length. If the instruction period were two hours long, the whole unit could possibly be taught in a single lesson.

The sequencing of the lessons within the units then should be based on the prerequisite relationships among the objectives. Although these are very loose guidelines, they have the following requirements: (1) that new learning is supported by previous learning, (2) that it is important to do a learning analysis

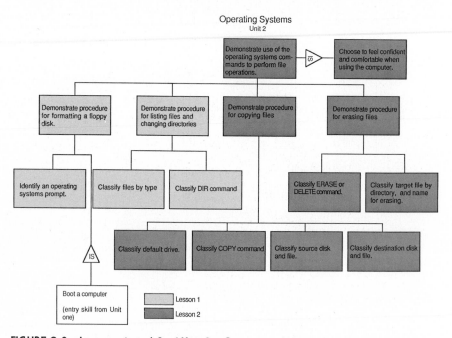

FIGURE 9-3   Lessons 1 and 2 of Unit 2—Operating Systems—showing subordinate skills for each target objective.

to determine that the skills are being taught in sequential order, (3) that the sequences are complete, and (4) that objectives irrelevant to the learning task at hand are eliminated or taught at a different time.

Table 9-3 summarizes the major considerations regarding the sequential arrangement within a topic for each of the types of learned capability. The middle column of the table indicates sequencing principles applicable to the particular type of capability that represents the central focus of the learning. The right-hand column lists sequence considerations relevant to this learning but arising in other domains.

Sequencing of Skills Within Lessons

The next level for mapping is the lesson map as shown in Figure 9-4. A lesson map is to a unit map what a state road map is to a map of the United States. It contains less scope than the unit map and even greater levels of detail. Al-

Table 9-3.   Desirable Sequence Characteristics Associated with Five Types of Learning Outcome

Type of Learning Outcome	Major Principles of Sequencing	Related Sequence Factors
Intellectual Skills	Presentation of learning situation for each new skill should be preceded by prior mastery of subordinate skills.	Verbal information may be recalled or newly presented to provide elaboration of each skill and conditions of its use.
Cognitive Strategies	Learning and problem-solving situations should involve recall of previously acquired relevant intellectual skills.	Verbal information relevant to the new learning should be previously learned or presented in instructions.
Verbal Information	For major subtopics, order of presentation is not important. New facts should be preceded by meaningful context.	Prior learning of necessary intellectual skills involved in reading, listening, etc. is usually assumed.
Attitudes	Establishment of respect for source is an initial step. Choice situations should be preceded by mastery of any intellectual skills involved.	Verbal information relevant to choices should be previously learned or presented in instructions.
Motor Skills	Provide intensive practice on part skills of critical importance and practice on total skill.	First of all, learn the executive subroutine (rule).

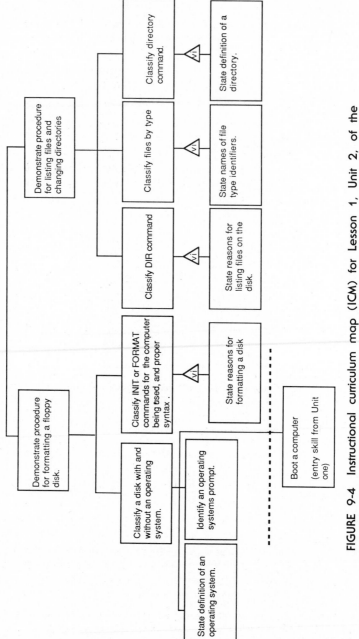

FIGURE 9-4 Instructional curriculum map (ICM) for Lesson 1, Unit 2, of the computer course.

though Chapter 12 is concerned with the designing of individual lessons, we introduce the lesson map here so that you may see how it relates to the course and unit map.

The lesson map shown in Figure 9-4 has as its target objectives one or more objectives from the unit map. In addition, it has subordinate objectives related to the attainment of these target objectives. These subordinate objectives were obtained by asking the question, "What must the learner know to learn these new skills?" It is also necessary to ask, "What do these learners already know of importance in the learning of these new skills?" The learner's present general knowledge and that portion of his knowledge that may be necessary for learning the new skills become the entry skills for the lesson. That is, the designer must begin with some information about the audience and must make some assumptions about the skills they bring to a particular learning task. Usually this means that the designer will have to make a somewhat detailed analysis of the intellectual skills involved in a lesson. This is the organizing factor, and the device for revealing the requirements for sequencing needs to be described in greater detail.

During the process of constructing a lesson map it may become obvious that the skills that need to be taught cannot be accomplished in a single instructional period. In this case the map may be divided into two lesson maps, each representing a single period of instruction. This will be discussed in more detail in Chapter 11.

In some cases the unit may center on a specific domain of learning such as motor skills, information, intellectual skills, attitudes, or cognitive strategies.

LEARNING HIERARCHIES AND INSTRUCTIONAL SEQUENCE

The nature of intellectual skills makes it possible to design effective conditions for their learning with considerable precision. When a proper sequence of prerequisite skills is established, the learning of intellectual skills becomes easy for a teacher to manage. In addition, the process of learning becomes highly reinforcing for the learners because they frequently realize that with apparent and satisfying suddenness they know how to do some things that they didn't know how to do before. Thus the activity of learning takes on an excitement that is at the opposite pole from "drill" and "rote recitation."

As described in Chapter 5, the learning hierarchy that results from a learning-task analysis is the arrangement of intellectual skill objectives into a pattern that shows the prerequisite relationships among them. An additional example of a learning hierarchy, this time for a skill in solving a type of physics problem, is show in Figure 9-5.

Here the lesson objective is one of finding the horizontal and vertical components of forces as vectors of a system in equilibrium. To learn to perform such a task correctly, students must have some prerequisite skills; these are

VECTOR RESOLUTION OF FORCES

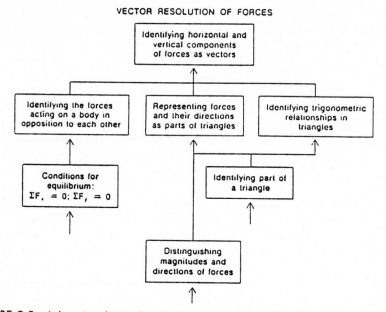

FIGURE 9-5 A learning hierarchy for the target skill of identifying horizontal and vertical components of forces as vectors.
(From R. M. Gagné, *The conditions of learning*, 4th ed., copyright 1985 by Holt, Rinehart & Winston, New York, Reprinted by permission.

indicated on the second level of the hierarchy. Specifically, they must be able to (1) identify the forces in the situation that are acting in opposition to each other when the body being acted upon is in equilibrium, (2) represent these opposing forces as sides of triangles that include vertical and horizontal sides, and (3) identify trigonometric relationships of the sides and angles of right triangles (sine, cosine, and so forth). Each of these subordinate skills (1), (2), and (3) also has prerequisites, which are shown below them in the hierarchy.

What is meant by a prerequisite? Evidently, a prerequisite is a simpler intellectual skill, but such a characterization is inadequate to identify it properly since one could name a number of intellectual skills that are simpler than the lesson objective described in the figure. A prerequisite skill is integrally related to the skill to which it is subordinate, in the sense that the latter skill cannot be done if the prerequisite skill is not available to the learner. Consider what students are doing when they identify (more correctly, demonstrate) "horizontal and vertical components of forces as vectors." They must show the values and directions of the horizontal and vertical vectors forces. The directions must be such as to "resolve" the forces that oppose each other to produce a state of equilibrium (prerequisite 1). And the values of these vectors must be obtained by using trigonometric relationships of right triangles (prerequisites 2 and 3). If the students do not know these prerequisites, they will not be able to perform the target (lesson) objective. Now turn this idea around: if students already know how to do each of these subordinate things, learning to do the lesson objectives will be easy and straightforward. The likelihood is that such students will learn to do it rapidly, perhaps even with the kind of immediacy implied by the word "discovery."

To identify a skill's prerequisites, one must ask, "What skill must the learner have to learn this (new) skill, the absence of which would make that learning impossible?" (cf. Gagné, 1985). In other words, prerequisite intellectual skills are those that are critical to the rapid, smooth learning of the new skill. There is a way of checking whether one's first attempt to answer the preceding question has been successful. This is to examine the demands the new skill makes on the learner and identify where he could go wrong. Applying this to the lesson objective of Figure 9-5, one can see that students who are attempting to "identify horizontal and vertical components of forces as vectors" might fail if they weren't able to (1) identify the forces acting in opposition (at equilibrium), (2) represent the forces as parts of triangles, or (3) identify trigonometric relationships in right triangles. Thus the specification of prerequisite skills should provide a complete description of those previously learned skills necessary for acquiring the new skill most readily.

Incidentally, the fact that prerequisite skills may be checked by considering ways in which the learner can fail serves to emphasize the direct relevance of learning hierarchies to the teacher's task of diagnosis. If one finds a learner who is having trouble acquiring a new intellectual skill, the first diagnostic question should probably be, "What prerequisite skills has this person failed to learn?" The contrast between the preceding question and these—"What genetic deficiency does this person have?" or, "What is the person's general intelligence?" will be apparent. The latter questions may suggest solutions that merely serve to remove the learner from the learning environment by putting him or her in a social group or class. Responsible diagnosis, in contrast, attempts to discover what the learner needs to learn. The chances are high that this will be a prerequisite intellectual skill, as indicated by a learning hierarchy. If it is, suitable instruction can readily be designed to get the learner "back on track" in a learning sequence that continues to be positively reinforcing.

SUMMARY

This chapter opens with an account on how the organization of a total course relates to questions about the sequencing of instruction. Sequencing decisions are identified at the four levels of course, topic, lesson, and lesson component. We suggested ways for deciding upon instructional sequences at the levels of the course and the topic. Course planning for a sequence of topics is typically done by a kind of common-sense logic. One topic may precede another because it describes earlier events, because it is a component part, or because it provides a meaningful context for what is to follow.

In proceeding from course purposes to performance objectives, it may not always be necessary to describe all the intermediate levels of planning in terms of complete lists of performance objectives for the topic. The method suggested here involves choosing representative samples of objectives within each domain of learning outcomes. It may be noted, however, that the more complete procedure can be followed and may sometimes be desirable, as is illustrated in Briggs (1977), and in Briggs and Wager (1981).

The designing of sequences for intellectual skills is based upon learning hierarchies. These hierarchies are derived by working backwards from target objectives, and by so doing, we can analyze the sequences of skills to be learned (see Chapter 8). The learning of a new skill will be accomplished most readily when the learner is able to recall the subordinate skills that compose it. When an instructional sequence has been designed for an intellectual skill, related learning of other capabilities may be interjected at appropriate points, as when the learning of information is required or the modification of an attitude is desired. In other instances, instruction aimed at other capabilities may come before or after the intellectual skill represented in the learning hierarchy. Designing sequences for other types of learned capabilities also requires an analysis of prerequisite learnings and the identification of supportive and enabling objectives.

The next three chapters describe how the plans for instructional sequence are carried into the design of a single lesson or lesson component. It is in the latter context that the events of instruction are introduced. These events pertain to the external supports for learning provided by the teacher, the course materials, or the learner himself. They depend upon previous learning that has been accomplished in accordance with a planned sequence.

REFERENCES

Anderson, J. R. (1985). *Cognitive psychology and its implications* (2nd ed.). New York: Freeman.

Briggs, L. J. (Ed.) (1977). *Instructional design: Principles and applications.* Englewood Cliffs, NJ: Educational Technology Publications.

Briggs, L. J., & Wager, W. (1981). *Handbook of procedures for the design of instruction.* Englewood Cliffs, NJ: Educational Technology Publications.

Gagné, R. M. (1985). *The conditions of learning* (4th ed.). New York: Holt, Rinehart and Winston.

Reigeluth, C. M. & Stein, F. S. (1983). The elaboration theory of instruction. In C. M. Reigeluth (Ed.), *Instructional-design theories and models.* Hillsdale, NJ: Erlbaum.

Tyler, R. W. (1949). *Basic principles of curriculum and instruction.* Chicago: University of Chicago Press.

10 THE EVENTS OF INSTRUCTION

Planning a course of instruction makes use of the principles described in the preceding chapters: determining what the outcomes of instruction are to be, defining performance objectives, and deciding upon a sequence for the topics and lessons that make up the course. When these things have been done, the fundamental "architecture" of the course is ready for more detailed planning in terms of both teacher and student activities. It is time to give consideration to the bricks and mortar of the individual lesson.

Supposing, then, that the course of instruction has been planned so that the student may reasonably progress from one lesson to the next. How does one ensure that he or she takes each learning step and does not falter along the way? How is the student coaxed along during the lesson itself? How does one, in fact, *instruct* the student?

THE NATURE OF INSTRUCTION

In designing the architecture for the course, we have said virtually nothing about how instruction itself may be done. During a lesson there is progress from one moment to the next as a set of events acts upon and involves the student. This set of events is what is specifically meant by *instruction*.

The instructional events of a lesson may take a variety of forms. They may require the teacher's participation to a greater or lesser degree; and they may be determined by the student to a greater or lesser degree. In a basic sense, these events constitute a set of *communications to the student.* Their most typical form is as verbal statements, whether oral or printed. Of course, communications for young children may not be verbal but instead use other media of communication such as gestures or pictures. But whatever the medium, the essential nature of instruction is most clearly characterized as a set of communications.

The communications that make up instruction have the sole aim of aiding the process of learning—that is, of getting the student from one state of mind to another. It would be wrong to suppose that their function is simply "to communicate" in the sense of "informing." Sometimes it appears that teachers are inclined to make this mistake—they "like to hear themselves talk," as has sometimes been said. There is perhaps no better way to avoid the error of talking too much than to keep firmly in mind that communications during a lesson are to facilitate learning, and that anything beyond this is mere chatter. Much of the communicating done by a teacher is essential for learning. Sometimes a fairly large amount of teacher communication is needed; on other occasions, however, none may be needed at all.

Self-Instruction and the Self-Learner

Any or all of the events of instruction may be put into effect by the learner himself when he is "self-instructing." Students engage in a good deal of self-instruction, not solely when they are working on programmed materials, but also when they are studying textbooks, performing laboratory exercises, or completing projects of various sorts. Skill at self-instruction may be expected to increase with the age of the learners, as they gain in experience with learning tasks. Events of the lesson, designed to aid and support learning, require teacher activities to a much greater extent in the first grade than they do in the tenth. As learners gain experience and continue to pursue learning activities, they acquire more and more of the characteristics of "self-learners." That is, they are able to use skills and strategies by which they manage their own learning.

The events of instruction to be described in this chapter, therefore, should not be viewed as being invariably required for every lesson and every learner. In practice, a judgment must be made concerning the extent of self-instruction the learner is able to undertake. A more extensive consideration of self-instruction in systems of individualized instruction is contained in Chapter 14.

Instruction and Learning

The purpose of instruction, however it may be done, is to provide support to the processes of learning. It may therefore be expected that the kinds of events that constitute instruction should have a fairly precise relation to what is going on within the learner whenever learning is taking place. To undertake instructional design at the level of the individual learning episode, it appears necessary to derive the desirable characteristics of *instructional events* from what is known about *learning processes.*

Although we are unable within the confines of this book to describe modern theoretical notions of learning processes in detail, it appears worthwhile to provide a brief account of learning theory, which the reader may wish to supplement by reference to other works. In particular, we are concerned with establishing a sound basis in learning theory for the derivation of instructional events. Each of the particular events that make up instruction functions to aid or otherwise support the acquisition and the retention of whatever is being learned. These functions of external events may be derived by consideration of the internal processing that makes up any single act of learning. The kinds of internal processing to which we are referring are those involved in modern cognitive learning theories (Anderson, 1985; Estes, 1975; Klatzky, 1980).

The sequence of processing envisaged by cognitive theories of learning is approximately as follows (Gagné, 1977, 1985). The stimulation that affects the learner's receptors produces patterns of neural activity that are briefly "registered" by *sensory registers.* This information is then changed into a form that is recorded in the *short-term memory,* where prominent features of the original stimulation are stored. The short-term memory has a limited capacity in terms of the number of items that can be held in mind. The items that are so held, however, may be internally rehearsed, and thus maintained. In a following stage, an important transformation called *semantic encoding* takes place when the information enters the *long-term memory* for storage. As its name implies, in this kind of transformation, information is stored according to its meaning. (Note that in the context of learning theory, *information* has a general definition that includes the five kinds of learned capabilities distinguished in this book).

When learner performance is called for, the stored information or skill must be searched for and *retrieved.* It may then be transformed directly into action, by way of a *response generator.* Frequently, the retrieved information is recalled to the *working memory* (another name for the short-term memory), where it may be combined with other incoming information to form new learned capabilities. Learner performance itself sets in motion a process that depends upon external *feedback,* involving the familiar process of *reinforcement.*

Figure 10-1 shows the relation between the structures involved in cognitive theories of learning and memory, and the processes associated with them.

In addition to the learning sequence itself, cognitive theories of learning and memory propose the existence of *executive control* processes (not shown in Figure 10-1). These are processes that select and set in motion cognitive strategies relevant to learning and remembering. Control processes of this sort modify the other information flow processes of the learner. A control process

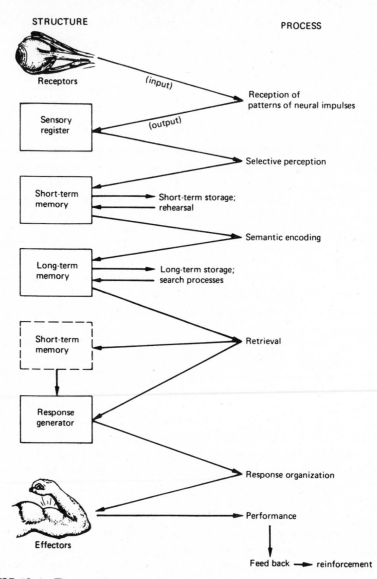

STRUCTURE   PROCESS

Receptors

(input)

Sensory
register

(output)

Reception of
patterns of neural impulses

Selective perception

Short-term
memory

Short-term storage;
rehearsal

Semantic encoding

Long-term
memory

Long-term storage;
search processes

Short-term
memory

Retrieval

Response
generator

Response organization

Effectors

Performance

Feed back ➡ reinforcement

FIGURE 10-1   The postulated structures of cognitive learning theories, and the pro-
cesses associated with them.
(From R. M. Gagné, *The conditions of learning,* 3rd ed, copyright 1977 by Holt, Rinehart and Winston, New
York. Reprinted by permission).

may select a strategy of continued rehearsal of the contents of short-term
memory, for example, or a cognitive strategy of imaging sentences to be
learned. They may exercise control over attention, over the encoding of incoming information, and over the retrieval of what has been stored.

The *kinds of processing* presumed to occur during any single act of learning
may be summarized as follows:

1. *Attention* determines the extent and nature of *reception* of incoming stimulation.
2. *Selective perception* (sometimes called *pattern recognition*) transforms this stimulation into the form of object-features, for storage in short-term memory.
3. *Rehearsal* maintains and renews the items stored in short-term memory.
4. *Semantic encoding* prepares information for long-term storage.
5. *Retrieval,* including *search,* returns stored information to the working memory or to a response generator.
6. *Response organization* selects and organizes performance.
7. *Feedback* provides the learner with information about performances and sets in motion the process of *reinforcement.*
8. *Executive control processes* select and activate cognitive strategies; these modify any or all of the previously listed internal processes.

Instructional Events

The processes involved in an act of learning are, to a large extent, activated internally. That is to say, the output of any one structure (or the result of any one kind of processing) becomes an input for the next, as Figure 10-1 indicates. However, these processes may also be influenced by *external* events, and this is what makes instruction possible. Selective perception, for example, may obviously be affected by particular arrangements of external stimuli. The features of a picture or text organized by perception may be influenced by highlighting, underlining, bold printing, and other measures of this general sort. Similarly, the particular kind of semantic encoding that is done in learning may be specified or suggested by meaningful information provided externally.

From these reflections on the implications of learning theory, one can derive a definition. Typically, *instruction consists of a set of events external to the learner designed to support the internal processes of learning* (Gagné, 1977, 1985). The events of instruction are designed to make it possible for learners to proceed from "where they are" to the achievement of the capability identified as the target objective. In some instances, these events occur as a natural result of the learner's interaction with the particular materials of the lesson; as, for example, when the beginning reader comes to recognize an unfamiliar printed word as something familiar in her oral vocabulary, and thus receives feedback. Mostly, however, the events of instruction must be deliberately arranged by an instructional designer or teacher. The exact form of these events (usually communications to the learner) is not something that can be specified in general for all lessons, but rather must be decided for each learning objective. The particular communications chosen to fit each set of circumstances, however, should be designed to have the desired effect in supporting learning processes.

The functions served by the various events of instruction in an act of learning are listed in Table 10-1, in the approximate order in which they are

TABLE 10-1.   Events of Instruction and Their Relation to Processes of Learning

Instructional Event	Relation to Learning Process
1. Gaining attention	*Reception* of patterns of neural impulses
2. Informing learner of the objective	Activating a process of *executive control*
3. Stimulating recall of prerequisite learning	*Retrieval* of prior learning to working memory
4. Presenting the stimulus material	Emphasizing features for *selective perception*
5. Providing learning guidance	*Semantic encoding;* cues for *retrieval*
6. Eliciting the performance	Activating *response organization*
7. Providing feedback about performance correctness	Establishing *reinforcement*
8. Assessing the performance	Activating *retrieval;* making *reinforcement* possible
9. Enhancing retention and transfer	Providing cues and strategies for *retrieval*

typically employed (Gagné, 1968, 1985). The initial event of *gaining attention* is one that supports the learning event of reception of the stimuli and the patterns of neural impulses they produce. Before proceeding further, another instructional event is designed to prepare the learner for the remaining sequence. This is Event number 2, *informing the learner of the objective,* which is presumed to set in motion a process of executive control by means of which the learner selects particular strategies appropriate to the learning task and its expected outcome. Event number 3 is also preparatory to learning and refers to the *retrieval of items of prior learning* that may need to be incorporated in the capability being newly learned. Events 4 through 9 of Table 10-1 are each related to the learning processes shown in Figure 10-1.

It should be realized that these events of instruction do not invariably occur in this exact order, although this is their most probable order. Even more important, by no means are all of these events provided for every lesson. Their role is to stimulate internal information processes, not to replace them. Sometimes, one or more of the events may already be obvious to the learner and, therefore, may not be needed. Also, one or more of these events are frequently provided by learners themselves, particularly when they are experienced self-learners. In designing instruction, the list of instructional events usually becomes a checklist. In using the checklist the designer asks, "Do these learners need support at this stage for learning this task?"

Gaining Attention

Various kinds of events are employed to gain the learner's attention. Basic ways of commanding attention involve the use of stimulus change, as is often done in moving display signs or in the rapid "cutting" of scenes on a television screen. Beyond this, a fundamental and frequently used method of gaining attention is to appeal to the learner's interests. A teacher may appeal to some particular student's interests by means of a verbal question such as "Wouldn't you like to know what makes a leaf fall from a tree?" in introducing a lesson dealing with leaves. One student's interest may be captured by such a question as "How do you figure a baseball player's batting average?" in connection with a lesson on percents. Naturally, one cannot provide a standard content for such questions—quite to the contrary since every student's interests are different. Skill at gaining attention is a part of the teacher's art, involving insightful knowledge of the particular students involved.

Communications that are partially or even wholly nonverbal are often employed to gain attention for school lessons. For example, the teacher may present a demonstration, perhaps exhibiting some physical event (a puff of smoke, an unexpected collision, a change in the color of a liquid), which is novel and appeals to the student's interest or curiosity. Or, a motion picture or television scene may depict an unusual event and thus command attention.

A good preplanned lesson provides the teacher with one or more options for communications designed to gain attention. When instruction is individualized, the teacher is able to vary the content and form of the communication whenever necessary to appeal to individual student interests.

Informing the Learner of the Objective

In some manner or other, the learner should know the kind of performance that will be used as an indication that learning has, in fact, been accomplished. Sometimes this aim of learning is quite obvious, and no special communication is required. For example, it would be somewhat ridiculous to make a special effort to communicate the objective to a novice golfer who undertakes to practice putting. However, there are many performance objectives that may not be initially obvious to students in school. For example, if the subject under study is the Preamble to the United States Constitution, being able to recite it verbatim is not at all the same objective as being able to state its major ideas. If decimals are being studied, is it obvious to the student during any given lesson whether he or she is expected to learn to (1) read decimals, (2) write decimals, or perhaps, (3) add decimals? The student should not be required to guess what is in the instructor's mind. The student needs to be told (unless, of course, he or she already knows).

On the whole, it is probably best not to take the chance of assuming that the student knows what the objective of the lesson is. Such a communication takes little time, and may at least serve the purpose of preventing the student from "getting entirely off the track." Communicating the objective also appears to be an act consistent with the frankness and honesty of a good teacher. In addition, the act of verbalizing the objective may help the teacher to stay "on target."

Of course, if objectives are to be communicated effectively, they must be put into words (or pictures, if appropriate) that the student can readily understand. For a six-year-old, an objective like "given a noun subject and object, and an active verb, formulate a correct sentence" must be translated into a communication that runs somewhat as follows: "Suppose I have the words 'boy,' 'dog,' and 'caught.' You could make them into a sentence, like 'The boy caught the dog.' This is called 'making a sentence,' and that's what I want you to do with the words I point to." Performance objectives, when used to describe a course of study, are typically stated in a form designed to communicate unambiguously to teachers or to instructional designers. The planning of instruction for a lesson, however, includes making the kind of communication of the lesson's objective that will be readily understood by students.

It is sometimes speculated that communicating an objective to students may tend to keep them from trying to meet still other objectives they may formulate themselves. No one has ever seen this happen, and the chances are it is a highly unlikely possibility. When one communicates a lesson's objective to students, they are hardly inclined to think that such a statement forbids them from giving further thought to the subject at hand. Working with an objective of "reading decimals," for example, it is not uncommon for a teacher to ask the question, "What do you suppose the sum of these decimals might be?" Thus still another objective is communicated, about which the students are perfectly free to think about, while making sure that they have achieved the first objective. Naturally, one also wants the students to develop in such a way that they will think of objectives themselves and learn how to teach them to themselves. Nothing in the communication of a lesson's objectives carries the slightest hint that such activities are to be discouraged. The basic purpose of such communication is simply to answer the student's question, "How will I know when I have learned?"

Stimulating Recall of Prerequisite Learned Capabilities

This kind of communication may be critical for the essential event of learning. Much of new learning (some might say all) is, after all, the combining of ideas. Learning a rule about *mass* (Newton's second law of motion) involves a combination of the ideas of *acceleration* and *force,* as well as the idea of *multiplying.* In terms of modern mathematics, learning the idea of *eight* involves the idea of the *set seven,* the *set one,* and *joining.* Component ideas (concepts, rules) must be previously learned if the new learning is to be successful. At the moment of learning, these previously acquired capabilities must be highly accessible to be part of the learning event. Their accessibility is ensured by having them recalled just before the new learning takes place.

The recall of previously learned capabilities may be stimulated by asking a recognition or, better, a recall question. For example, when children are being taught about rainfall in relation to mountains, the question may be asked, "Do you remember what the air is like in a cloud that has travelled over land in the summer?" (The air is warm). The further question may then be asked, "What is the temperature of the land on a high mountain likely to be?" (Cold). This line of questioning recalls previously learned rules and obviously leads to a strand of learning that will culminate with the acquistion of a new rule concerning the effects of cooling on a warm, moisture-laden cloud.

Presenting the Stimulus Material

The nature of this particular event is relatively obvious. The stimuli to be displayed (or communicated) to the learner are those involved in the performance that reflects the learning. If the learner must learn a sequence of facts, such as events from history, then these are facts that must be communicated, whether in oral or printed form. If the learner is engaged in the task of pronouncing aloud printed words, as in elementary reading, then the printed words must be displayed. If the student must learn to respond to oral questions in French, then these oral questions must be presented, since they are the stimuli of the task to be learned.

Although seemingly obvious, it is nevertheless of some importance that the proper stimuli be presented as a part of the instructional events. For example, if the learner is acquiring the capability of answering questions delivered orally in French, then the proper stimuli are *not* English questions, nor printed French questions. (This is not to deny, however, that such tasks may represent subordinate skills that have previously been used as learning tasks). If the learner is to acquire the capability of using positive and negative numbers to solve verbally stated problems, then the proper stimuli are verbally stated problems and not something else. If one neglects to use the proper stimuli for learning, the end result may be that the learner acquires the "wrong" skill.

Stimulus presentation often emphasizes *features* that determine selective perception. Thus, information presented in a text may contain italics, bold print, underlining, or other kinds of physical arrangements designed to facilitate perception of essential features. When pictures or diagrams are employed, important features of the concepts they display may be heavily outlined, circled, or pointed to with arrows. In establishing discriminations, distinctive features may be emphasized by enlarging the differences between the objects to be distinguished. For example, in programs of reading readiness, large differences in shapes (such as those of a circle and triangle) may be introduced first, followed by figures exhibiting smaller differences. Distorted features of a *b* and *d* may be initially presented, in order that the smaller differences of these letters will eventually be discriminated.

Stimulus presentation for the learning of concepts and rules requires the use of a *variety of examples.* When the objective is the learning of a concept such as *circle,* it is desirable to present not only large and small circles on the chalkboard or in a book, but also green circles, red ones, and ones made of rope or string. One might even have the children stand and join hands to form a circle. For young children, the importance of this event can hardly be over-emphasized. The failure to provide such a variety of examples accounts for the classic instance related by William James in which a boy could recognize a *vertical* position when a pencil was used as the test object, but not when a table knife was held in that position.

Comparable degrees of usefulness can be seen in the use of variety of examples as an event for rule learning. To apply the formula for area of a rectangle, $A = x \cdot y$, the student must not only be able to recall the statement that represents the rule, but he must know that A means area; he must understand what area means; he must know the x and y are the dimensions of two non-parallel sides of the rectangle, and he must know that the dot between x and y means to multiply. But even when all these subordinated concepts and rules are known, the learner must do a variety of examples to ensure that he understands and can use the rule. Retention and transfer are also likely to be enhanced by presenting problems stated in words, in diagrams, and in combinations of the two over a period of time.

Once such rules are learned, groups of them need to be selectively recalled, combined, and used to solve problems. Employing a variety of examples in problem solving might entail teaching the learner to break down odd-shaped figures into known shapes, like circles, triangles, and rectangles, and then to apply rules for finding the area of these figures as a way to arrive at the total area of the entire shape.

In the learning of both concepts and rules, one may proceed either inductively or deductively. In learning concrete concepts, like circle or rectangle, it is best to introduce a variety of examples before introducing the definition of the concept. (Imagine teaching a four-year-old the formal definition of a circle before exposure to a variety of circles!) But for older learners who are learning defined concepts, a simple definition might best come first, such as: "A root is the part of a plant below the ground." Assuming the learner understands the component concepts that are contained in the statement, this should be a good start, perhaps followed at once by a picture.

Providing Learning Guidance

Suppose one wishes a learner to acquire a rule (or it might be called a defined concept) about the characteristics of prime numbers. He might begin by displaying a list of successive numbers, say, 1 through 25. He then might ask the learner to recall that the numbers may be expressed as products of various factors: $8 = 2 \cdot 4 = 2 \cdot 2 \cdot 2 = 8 \cdot 1$, and so on. The learner could then be asked to write out all the factors for the set of whole numbers through 30. What is wanted now, as a learning outcome, is for the learner to discover the rule that there is a certain class of numbers whose only factor (or divisor) other than the number itself is 1.

The learner may be able to "see" this rule immediately. If not, he may be led to its discovery by a series of communications in the form of hints or questions. For example, such a series might run somewhat as follows: "Do you see any regularities in this set of numbers?" "Do the original numbers differ with respect to the number of different factors they contain?" "In what way are the numbers 3, 5, and 7 different from 4, 8, and 10?" "In what way is the number 7 like the number 23?" "Can you pick out *all* the numbers that are like 7 and 23?"

These communications and others like them may be said to have the function of *learning guidance*. Notice that they do not "tell the learner the answer"; rather, they suggest the line of thought which will presumably lead to the desired "combining" of subordinate concepts and rules to form the new to-be-learned rule. Again, it is apparent that the specific form and content of such questions and hints cannot be spelled out in precise terms. Exactly what the teacher or textbook says is not the important point. It is rather that such communications are performing a particular function. They are stimulating a direction of thought and are thus helping to keep the learner on the track. In performing this function, they contribute to the efficiency of learning.

The amount of learning guidance, that is, the number of questions and the degree to which they provide "direct or indirect prompts," will obviously vary with the kind of capability being learned (Wittrock, 1955). If what is to be learned is an arbitrary matter such as the name for an object new to the learner (say, a pomegranate), there is obviously no sense in wasting time with indirect hinting or questioning in the hope that somehow such a name will be "discovered." In this case, just telling the student the answer is the correct form of guidance for learning. At the other end of the spectrum, however, are cases where less direct prompting is appropriate, because this is a logical way to discover the answer, and such discovery may lead to learning that is more permanent than that which results from being told the answer.

Guidance for learning is an event that may readily be adapted to learner differences, as described in Chapter 6. Instruction that is highly didactic and that makes use of "low-level" questions is likely to find greatest appeal and effectiveness among learners of high anxiety, whereas low-anxious learners may be positively affected by the challenge of difficult questions. As previously noted, guidance taking the form of frequent pictures and oral communications may aid learners of low ability in reading comprehension, whereas these measures may be quite inefficient with skillful readers.

The amount of hinting or prompting involved in the learning guidance event will also vary with the kind of learners. Some learners require less learning guidance than do others; they simply "catch on" more quickly. Too much guidance may seem condescending to the quick learner, whereas too little can simply lead to frustration on the part of the slow learner. The best practical solution may sometimes be to apply learning guidance a little at a time and allow the student to use as much as he needs. Only one hint may be necessary for a fast learner, wheras three or four may work better with a slower learner. Providing adaptive learning of this sort can readily be made part of a total system of computer-based instruction (Tennyson, 1984).

In the learning of attitudes, a *human model* may be employed, as indicated in Chapter 5. Models themselves, as well as the communications they deliver, may be considered to constitute the learning guidance in attitude learning. Thus the total instructional event in this case takes a somewhat more complex form than is the case with the learning of verbal information or intellectual skills. The same function of semantic encoding is being served, however.

Eliciting the Performance

Presumably, having had sufficient learning guidance, the learners will now be carried to the point where the actual internal combining event of learning takes place. Perhaps they look less confused, or some indication of pleasure has crossed their faces. They have seen how to do it! We must now ask them to show that they know how to do it. We want them not only to convince us, but to convince themselves as well.

Accordingly, the next event is a communication that in effect says "show me," or "do it." Usually, this first performance following learning will use the same example (that is, the same stimulus material) with which the learners have been interacting all along. For example, if they have been learning to make plurals of words ending in *ix,* and have been presented with the word *matrix,* the first performance is likely to be production of the plural *matrices.* In most instances, the instructor will follow this with a second example, such as *appendix,* to make sure the rule can be applied in a new instance.

Providing Feedback

Although in many situations it may be assumed that the essential learning event is concluded once the correct performance has been exhibited by the learner, this is not universally the case. One must be highly aware of the aftereffects of the learning event and their important influence on determining exactly what is learned. In other words, as a minimum, there should be feedback concerning the correctness or degree of correctness of the learner's performance. In many instances, such feedback is automatically provided— for example, an individual learning to throw darts can see almost immediately how far away from the bull's eye the dart lands. Similarly, a child who has managed to match a printed word with one in her oral vocabulary, which at the same time conveys an expected meaning, receives a kind of immediate feedback that has a fair degree of certainty. But, of course, there are many tasks of school learning that do not provide this kind of "automatic" feedback. For example, in practicing using the pronouns *I* and *me* in a variety of situations, is the student able to tell which are right and which are not? In such instances, feedback from an outside source, usually a teacher, may be an essential event.

There are no standard ways of phrasing or delivering feedback. In an instructional program, the confirmation of correctness is often printed on the side of the page or on the following page. Even standard textbooks for such subjects as mathematics and science customarily have answers printed in the back of the book. When the teacher is observing the learner's performance, the feedback communication may be delivered in many different ways—a nod, a smile, or a spoken word. Again in this instance, the important characterisitic of the communication is not its content but its function: providing information to the learners about the correctness of their performance.

Assessing Performance

The immediate indication that the desired learning has occurred is provided when the appropriate performance is elicited. This is, in effect, an assessment of learning outcome. Accepting it as such, however, raises the larger questions of *reliability* and *validity* that relate to all systematic attempts to assess outcomes or to evaluate the effectiveness of instruction. These are discussed in a later chapter, and we shall simply state here their relevance to the single learning event.

When one sees the learner exhibit a single performance appropriate to the lesson objective, how does the observer or teacher tell that he or she has made a *reliable* observation? How does that person know the student didn't do the required performance by chance or by guessing? Obviously, many of the doubts raised by this question can be dispelled by asking the learner to "do it again," using a different example. A first-grader shows the ability to distinguish the sounds of *mat* and *mate*. Has he been lucky, or can the child exhibit the same rule-governed performance with *pal* and *pale*? Ordinarily, one expects the second instance of the performance to raise the reliability of the inference (concerning the student's capability) far beyond the chance level. Employing yet a third example should lead to a higher probability so far as the observer is concerned.

How is the teacher to be convinced that the performance exhibited by the learner is *valid*? This is a matter that requires two different decisions. The first is, does the performance in fact accurately reflect the objective? For example, if the objective is to "recount the main idea of the passage in your own words," the judgment must be made as to whether what the student says is indeed the *main* idea. The second judgment, which is no easier to make, is whether the performance has occurred under conditions that make the observation *free of distortion*. As an example, the conditions must be such that the student could not have "memorized the answer," or remembered it from a previous occasion. The teacher must be convinced, in other words, that the observation of performance reveals the learned capability in a genuine manner.

Obviously, the single, double, or triple observations of performance that are made immediately after learning may be conducted in quite an informal manner. Yet they are of the same sort, and part of the same piece of cloth, as the more formally planned assessments described in a later chapter. There need be no conflict between them and no discrepancies.

Enhancing Retention and Transfer

When information or knowledge is to be recalled, the existence of the meaningful context in which the material has been learned appears to offer the best assurance that the information can be reinstated. The network of relationships in which the newly learned material has been embedded provides a number of different possibilities as cues for its retrieval.

Provisions made for the recall of intellectual skills often include arrangements for "practicing" their retrieval. Thus, if defined concepts, rules, and higher order rules are to be well retained, course planning must make provision for systematic *reviews* spaced at intervals throughout weeks and months. The effectiveness of these spaced repetitions, each of which requires that the skill be retrieved and used, contrasts with the relative ineffectiveness of repeated examples given directly following the initial learning (Reynolds and Glaser, 1964).

As for the assurance of transfer of learning, it appears that this can best be done by setting some *variety* of new tasks for the learner—tasks that require the application of what has been learned in situations that differ substantially from those used for the learning itself. For example, suppose that what has been learned is the set of rules pertaining to "making the verb agree with the pronoun subject." Additional tasks that vary the pronoun and the verb may have been used to assess performance. Arranging conditions for transfer, however, means varying the entire situation more broadly still. This might be accomplished, in this instance, by asking the child to compose several sentences in which he himself supplies the verb and pronoun (rather than having them supplied by the teacher). In another variation of the situation, the student may be asked to compose sentences using pronouns and verbs to describe some actions shown in pictures. Ingenuity on the part of the teacher is called for in designing a variety of novel "application" situations for the purpose of ensuring the transfer of learning.

Variety and novelty in problem-solving tasks are of particular relevance to the continued development of cognitive strategies. As has previously been mentioned, the strategies used in problem solving need to be developed by the systematic introduction of occasions for problem solving, interspersed with other instruction. An additional event to be especially noted in the presentation of novel problems to the student is the need to make clear the general nature of the solution expected. For example, "practical " solutions may be quite different from "original" solutions, and the student's performance can easily be affected by such differences in the communication of the objective (cf. Johnson, 1972).

Instructional Events and Learning Outcomes

The events of instruction may be appropriately used in connection with each of the five kinds of learned capabilities described in Chapters 4 and 5. In the case of some instructional events, such as gaining attention, the particular means employed to bring about the event does not have to be different for intellectual skill objectives, say, and for attitude objectives. However, for learning guidance the specific nature of the event is likely to be very different indeed. As we have seen in the previous section, the encoding of an intellectual skill may be guided by verbal instructions, such as communicating to the learner a verbal statement of a rule to be learned. In contrast, effective encoding of an attitude often requires a complex event that includes observation of a human model. The requirement of differing treatments of instructional events extends also to step 3, stimulating recall of prior learning, and to step 4, presenting the stimulus.

A summary of events 3, 4, and 5 for each type of learned capability is contained in Table 10-2, along with examples of the function served by these events. For each kind of learning outcome, appropriate conditions of learning are listed under each of the three events. These descriptions are not intended to be all-inclusive; additional suggestions for the design of instructional events are given in the preceding paragraphs of this section.

As inspection of the table will show that the particular form taken by each of these three instructional events depends upon the capability to be learned. For example, when an intellectual skill is to be learned, the stimulation of recall pertains to the retrieval of prerequisite concepts or rules; whereas if verbal information is to be learned, the recall of a context of organized information is required. Similar differences in the specific form of Event 3, as well as Events 4 and 5, may be noted throughout the table.

THE EVENTS OF INSTRUCTION IN A LESSON

In using the events of instruction for lesson planning, it is apparent that they must be organized in a flexible manner, with primary attention to the lesson's objectives. What is implied by our description of these events is obviously not a standard, routine set of communications and action. The invariant features

TABLE 10-2. Functions of Instructional Events 3, 4, and 5 with Examples for Each of Five Kinds of Learning Outcomes

Intellectual Skill

Event 3: Stimulate Recall of Prior Learning Essential for learner to retrieve to working memory prerequisite rules and concepts.

Event 4: Present the Stimulus Display a statement of the rule or concept, with example giving emphasis to features of component concepts.

Event 5: Provide Learning Guidance Present varied examples in varied contexts; also, give elaborations to furnish cues for retrieval.

Cognitive Strategy

Event 3: Stimulate Recall of Prior Learning Recall task strategies and relevant intellectual skills.

Event 4: Present the Stimulus Describe the task and the strategy, and show what the strategy accomplishes.

Event 5: Provide Learning Guidance Describe the strategy and give one or more application examples.

Verbal Information

Event 3: Stimulate Recall of Prior Learning Recall familiar well-organized bodies of knowledge related to the new learning.

Event 4: Present the Stimulus Display printed or verbal statements, emphasizing distinctive features.

Event 5: Provide Learning Guidance Elaborate content by relating to larger bodies of knowledge; use mnemonics, images.

Attitude

Event 3: Stimulate Recall of Prior Learning Recall the situation and the actions involved in personal choice. Remind learner of human model.

Event 4: Present the Stimulus Human model describes the general nature of the choice of personal action to be presented.

Motor Skill

Event 3: Stimulate Recall of Prior Learning Recall the executive subroutine and relevant part skills.

Event 4: Present the Stimulus Display the situation existing at the beginning of the skill performance. Demonstrate executive subroutine.

Event 5: Provide Learning Guidance Continue practice with informative feedback.

of the single lesson are the functions that need to be carried out in instruction. Even these functions are adapted to the specific situation, the task to be accomplished, the type of learning represented in the task, and the students' prior learning. But each one of these functions should be specifically considered in lesson planning.

It is now possible to consider how these events are exemplified within an actual lesson. We have chosen, as an example, a set of instructions to the designer of a computer-based lesson, showing the implications of each instructional event for frame-by-frame design (Gagné, Wager, and Rojas, 1981). The lesson is about a *defined concept* in use of the English language, namely, the part of a sentence called the *object*. Instructions to the designer are outlined in Table 10-3.

It will be evident that this lesson in English grammar may best be conceived as part of a longer sequence in which such prior concepts as *sentence, subject,* and *predicate* have already been learned. For learners without such previous experience, instruction in the concept *object* would need to begin with simpler prerequisite concepts. It may be noted that the lesson is carefully planned in the sense that it reflects each of the instructional events described in this chapter. Obviously, it is an exercise in which the designer's art has considerable opportunity to flourish within the framework of events that support the desired learning.

Comparison with Lessons for Older Students

As instruction is planned for middle and higher grades, one can expect the events of instruction to be increasingly controlled by the materials of a lesson or by the learners themselves. Thus, when the units of instruction that make up a course of study are structurally similar, as may be the case in mathematics or beginning foreign language, for example, the objectives for each succeeding unit may be evident to students and, therefore, may not need to be communicated. For reasonably well-motivated students, it is often unnecessary to make any special provisions for controlling attention, since this event, too, may be appropriately managed by the learners themselves. This circumstance clearly contrasts with that prevailing in, say, a classroom of seventh-graders where the teacher may need to make special provisions for getting their attention.

Homework assignments, such as those that call for learning from a text, depend upon the learners to employ cognitive strategies available to them in managing instructional events. The text may aid selective perception by its inclusion of bold printing, topic headings, and other features of this general sort. The text may, and often does, include a context of meaningful material that provides for semantic encoding by relating new information to organized knowledge already in learners' memories. An important part of "studying," however, is the necessity for practicing appropriate performance, whether this is a matter of stating information in the learner's own words, applying a

TABLE 10-3. Events of Instruction in Design of a Computer-Based Lesson

Instructional Event	Procedure
1. Gaining attention	Present initial operating instructions on screen, including some displays that change second by second. Call attention to screen presentation, using words like "Look!", "Watch!", etc.
2. Informing learner of lesson objective	State in simple terms what the student will have accomplished once he or she has learned. *Example:* Two sentences, such as: "Joe chased the ball." "The sun shines brightly." One of these sentences contains a word that is an *object,* the other does not. Can you pick out the object? In the first sentence, *ball* is the *object* of the verb *chased.* In the second sentence, none of the words is an object. You are about to learn how to identify the *object* in a sentence.
3. Stimulating recall of prior learning	Recall concepts previously learned. *Example:* Any sentence has a *subject* and a *predicate.* The subject is usually a noun, or a noun phrase. The predicate begins with a verb. What is the *subject* of this sentence? "The play began at eight o'clock." What verb begins the predicate of this sentence? "The child upset the cart."
4. Presenting stimuli with distinctive features	Present a definition of the concept. *Example:* An *object* is a noun in the predicate to which the action (of the verb) is directed. For example, in the sentence, "The rain pelted the roof" *roof* is the *object* of the verb "pelted."
5. Guiding learning	Take a sentence like this: "Peter milked the cow." The answer is *the cow,* and that is the *object* of the verb. Notice, though, that some sentences do not have objects. "The rain fell slowly down." In this sentence, the action of the verb *fell* is not stated to be directed at something. So, in this sentence, there is no *object.*
6. Eliciting performance	Present three to five examples of sentences, one by one. Ask, "Type O if this sentence has an object, then type the word that is the object." *Examples:* "Sally closed the book." "The kite rose steadily."

7. Providing informative feedback

Give information about correct and incorrect responses.
Example: Book is the object of the verb *closed* in the first sentence. The second sentence does not have an object.

8. Assessing performance

Present a new set of concept instances and noninstances in three to five additional pairs of sentences. Ask questions requiring answers. Tell the learner if mastery is achieved and what to do next if it is not.

9. Enhancing retention and learning transfer

Present three to five additional concept instances, varied in form.
Example: Use sentences such as:
"Neoclassical expressions often supplant
mere platitudes."
Introduce review questions at spaced intervals.

Note. From "Planning and authoring computer-assisted instruction lessons" by R. M. Gagné, W. Wager, and A. Rojas, p.23. *Educational Technology,* September, 1981, Copyright 1981 by Educational Technology Publications. Reprinted by permission of the authors and copyright owners.

newly learned rule to examples, or originating solutions to novel problems. For these events of self-instruction, as well as for the judgment of correctness that gives an immediate sort of feedback, learners frequently must depend upon cognitive strategies available to them.

SUMMARY

This chapter is concerned with the events that make up instruction for any single performance objective as they may occur within a lesson. These are the events that are usually external to the learner, supplied by the teacher, text, or other media with which the learner interacts. When self-instruction is undertaken, as is to be more frequently expected as the learner's experience increases, instructional events may be brought about by the learner himself. However they originate, the purpose of these events is to activate and support the internal processes of learning.

The general nature of supporting external events may be derived from the information-processing (or cognitive) model of learning and memory, which is employed in one form or another by many contemporary learning investigators. This model proposes that a single act of learning involves a number of stages of internal processing. Beginning with the receipt of stimulation by receptors, these stages include (1) a brief *registration* of sensory events; (2) *temporary storage* of stimulus features in the short-term memory; (3) a *rehearsal* process that may be employed to lengthen the period of short-term storage to prepare information for entry into long-term memory; (4) *semantic encoding* for long-term storage; (5) search and *retrieval* to recall previously learned material; and (6) *response organization* producing a performance appropriate to what has been learned. Most theories include, either implicitly or explicitly, the process of (7) *reinforcement* as brought about by external feedback of the correctness of performance. In addition, this learning model postulates a number of (8) *executive control processes* that enable the learner to select and use cognitive strategies that influence other learning processes.

As derived from this learning model, instructional events are:

1. gaining attention
2. informing the learner of the objective
3. stimulating recall of prerequisite learnings
4. presenting the stimulus material
5. providing learning guidance
6. eliciting the performance
7. providing feedback about performance correctness
8. assessing the performance
9. enhancing retention and transfer

These events apply to the learning of all of the types of learning outcomes we have previously described. Examples are given to illustrate how each is planned for and put into effect.

The order of these events for a lesson or lesson segment is approximate and may vary somewhat depending on the objective. Not all of the events are invariably used. Some are made to occur by the teacher, some by the learner, and some by the instructional materials. An older, more experienced learner may supply most of these events by his own study effort. For young children, the teacher would arrange for most of them.

As these events apply to the various kinds of capabilities being acquired, they take on different specific characteristics (Gagné, 1985). These differences are particularly apparent in the following events from our list: Event 3, *stimulating recall of prior learning;* Event 4, *presenting the stimulus material;* and Event 5, *providing learning guidance.* For example, *presenting the stimulus* (Event 4) for the learning of discriminations requires conditions in which the differences in stimuli become increasingly fine. Concept learning, however, requires the presentation of a variety of instances and noninstances of the general class. Conditions of *learning guidance* (Event 5) required for the learning of rules include examples of application; whereas these conditions for verbal information learning are prominently concerned with linking to a larger meaningful context. For attitude learning, this event takes on an even more distinctive character when it includes a human model and the model's communication.

An example is given of using the events of instruction for the design of a computer-based lesson on a defined concept in English grammar.

REFERENCES

Anderson, J. R. (1985). *Cognitive psychology and its implications* (2nd ed.). New York: Freeman.

Estes, W. K. (Ed.), (1985). *Handbook of learning and cognitive processes: Introduction to concepts and issues* (Vol. 1). Hillsdale, NJ: Erlbaum.

Gagné, R. M. (1968). Learning and communication. In R. V. Wiman & W. C. Meierhenry (Eds.), *Educational media: Theory into practice.* Columbus, OH: Merrill.

Gagné, R. M. (1977). Instructional programs. In M. H. Marx & M. E. Bunch (Eds.), *Fundamentals and applications of learning.* New York: Macmillian.

Gagné, R. M. (1985). *The conditions of learning* (4th ed.). New York: Holt, Rinehart and Winston.

Gagné, R. M., Wager, W., & Rojas, A. (1981, September). Planning and authoring computer-assisted instruction lessons. *Educational Technology,* pp. 17–26.

Johnson, D. M. (1972). *A systematic introduction to the psychology of thinking.* New York: Harper & Row.

Klatzky, R. L. (1980). *Human memory: Structures and processes* (2nd ed.). San Francisco: Freeman.

Reynolds, J. H., & Glaser, R. (1964). Effects of repetition and spaced review upon retention of a complex learning task. *Journal of Educational Psychology, 55,* 297–308.

Tennyson, R. D. (1984). Artificial intelligence methods in computer-based instructional design: The Minnesota adaptive instructional system. *Journal of Instructional Development, 7*(3), 17–22.

Wittrock, M. C. (1966). The learning by discovery hypothesis. In L. S. Shulman & E. R. Keislar (Eds.), *Learning by discovery: A critical appraisal.* Chicago: Rand McNally.

11 SELECTING AND USING MEDIA

One of the essential decisions that must be made in instructional design is what medium to employ as a vehicle for the communications and stimulation that make up instruction.

In this chapter we describe the problem of media selection confronting designers of instructional programs and teachers. Some of the common features of media selection "models" will be described as well as the factors to be taken into account in selecting media. A method of media selection will then be outlined that indicates ways of incorporating desirable features. Some limitations of this method will be noted in comparison with other models.

SELECTION VERSUS DEVELOPMENT OF MEDIA PRESENTATIONS

When instructional design is developed from the very beginning, one expects that media presentations will be part of the design. The term media, however, is employed here in a very broad sense. Instructional delivery may be accomplished by means of the verbal speech of a teacher or by a printed text, as well as by way of vehicles of more complex technical material, such as sound and video recordings. Often, however, existing media presentations are *selected* as part of a larger instructional plan, rather than being separately *designed* and *developed*. Teachers, as well as teams of instructional designers, may carry out a comprehensive design of instruction that depends upon the selection of media.

Teachers as Instructional Designers

The teachers or instructors who deliver instruction often select and use a great variety of both print and nonprint media. Only infrequently do instructors develop such materials or media presentations themselves. In conventional procedures, teachers select materials that will enable learners to master the desired objectives. This selection function can be thought of as a part of lesson planning. Briggs (1977) has suggested that a completed lesson plan contain the following ingredients:

1. a statement of the objectives of the lesson;
2. a list of instructional events to be included in the lesson;
3. a list of media, materials, and activities by which each event is to be accomplished; and
4. notes on necessary teacher roles and activities, and communications to be made to the learners.

It will be recalled from earlier chapters that we view the purpose of communications to the learner as contributing desired instructional events for each lesson. Most of these events are needed in all of the domains of learning outcomes, depending somewhat on the experience level of the learners. One must choose *some* medium for delivery of instructional communications. The medium chosen may be the instructor's voice, a film, a book, or the learner's own study effort.

In selecting media and materials, the teacher is likely to discover immediately that available materials and media presentations are not *indexed* or *catalogued* according to the intended instructional objectives.

It is therefore suggested that the teacher make decisions about media in the process of lesson planning based on the following two questions:

1. What medium would I like to use for each intended learning outcome?
2. Where will I find the specific materials?

Actually, of course, there is a third question affecting the teacher's decisions, and that is how the selected materials will be used. But this chapter deals primarily with media selection, and a later chapter focuses on media utilization.

Although available materials will often contain much of the desired content (thus providing the desired instructional stimulus), they may not provide for other instructional events. For example, a film may deliver desired content, but it may not present the objective, guide semantic encoding, or contain pauses for student response and feedback. In that case the teacher must either create such events separately from the film, or stop the film at points so that questions and feedback can occur.

Most often, then, the teacher is faced with having to select media and materials that were not specifically designed for the exact objectives that have been adopted. Furthermore, the materials available often do not provide all the desired instructional events. Many materials are carefully organized to present information in a logical order but do not cover each instructional event appropriate for particular objectives. As a result the teacher cannot make all the media selections from catalogs; the actual materials must be previewed before final selection is made. Subsequently, the teacher's lesson plans are designed to indicate how the selected materials will be used and how the events not presented by the materials are to be accomplished.

On the brighter side, publishers and other suppliers of media are providing modules of instruction with increasing frequency. Such modules often give directions to both teachers and learners on how the materials and exercises can be used to reach the objectives, as well as how to know that they have been reached. Although such modules may not always be designed around the specific set of instructional events described in this book, they usually have been designed to implement a systematic strategy of instruction. Some modules contain the actual materials to be used, whereas others just list them. As the use of such modules becomes more widespread, teachers should experience fewer difficulties in lesson planning and in selecting media and materials.

MEDIA AND THE LEARNING SITUATION

Instruction is provided in a *learning situation*. The features of this situation impose constraints upon what media may be most effective. The kinds of media to be employed in a first-grade class, for example, are quite limited in variety in comparison with those appropriate for a class of high school students, or a class of adults refreshing their technical skills.

The following are features of the intended learning situations that need to be taken into account in selecting media:

1. communications to the learner delivered by the teacher (instructor) versus communications delivered via media for self-instruction
2. learners possessing sufficient verbal comprehension ability (see Chapter 6) to comprehend printed communications versus learners who have insufficient verbal comprehension ability
3. communications delivered directly to the learner or learners versus communications broadcast from a central station
4. the performance to be learned is such that errors are serious (that is, dangerous, as in the case of airplane emergency procedures) versus performances whose potential errors are not serious

Each of these features imposes some limitations on the kinds of media appropriate. Considering them in reverse order, the fourth feature would eliminate from consideration all media except those on which the desired performances can be directly practiced, such as the actual equipment or a simulator device. Media such as a lecture or a TV program might set the stage, but would surely be inadequate for the learning of a skill such as hose directing in firefighting, for example. The third feature also limits media selection severely; to be considered for this kind of learning situation, media must be those that involve the transmission of messages composed of sound, pictures, or both, and which do not provide for interactive responding by the learner.

These features of the learning situation lead to a classification of six general types of decisions about learning situations (Reiser and Gagné, 1983) as follows: (1) job competence decision (consequences of error); (2) central broadcast decision; (3) self-instruction with learner-readers; (4) self-instruction with nonreaders; (5) instructor with readers; and (6) instructor with nonreaders. Each of these choices of learning situation implies that certain media can be set aside as inappropriate. It may be noted, however, that for each type of learning situation, a number of media options continue to exist. As a general rule, most instructional functions can be performed by most media.

In practical instructional situations, the media employed are often chosen on grounds of availability, feasibility, and cost. Obviously these are important factors and must be considered in the final selection. What we suggest here, however, is that the question of appropriateness for learning be addressed *initially*. As we shall see, for most situations the types of media available for effective learning are more than a few. Giving initial consideration to learning support means that critical errors in media choice can be avoided. For example, one refrains from choosing: (1) a medium displaying printed discourse for learners who are nonreaders; (2) radio broadcast as a sole medium for teaching intellectual skills; or (3) an instructor lecture for the training of vehicle driving. These basic decisions based upon the type of learning situation lead to additional choices among a reduced list of learning-appropriate media. As factors in the final choice, cost, availability, and feasibility of use naturally come into play.

Media and Their Selection

In beginning to describe the details of the media selection process, we need first to define the term *medium*. A number of different definitions have been suggested, some reflecting sensory channels (audio-visual), others the nature of the content of the instructional message (print, picture). Here, we adopt Reiser's and Gagné's (1983) definition that instructional media are "the physical means by which an instructional message is communicated" (p. 5). By this definition, a printed text, an audio tape, a training device, a TV program, an instructor's talking, along with many other physical means, are all considered media.

Learning-effective media and the limitations on their choices occasioned by the type of learning situation are shown in Table 11-1.

TABLE 11-1.   Media and Media Decisions in Different Learning Situations

	Learning Situation	Media Decision
MEDIA		
Large equipment Simulator TV broadcast Radio broadcast	High performance competence (error consequence serious)	Exclude all media except (1) large equipment, (2) porta- ble equipment, (3) simu- lator
Portable equipment Training device	Central broadcast	Exclude all media except TV and radio broadcast
Computer Programmed text Interactive TV	Self-instruction with readers	All media are potentially ef- fective
Motion picture Slide/tape TV cassette Filmstrip Training aid Audio	Self-instruction with non- readers	Exclude media employing dis- cursive printed passages or complex audio messages
Overhead projector Slides	Instructor with readers	All media are potentially ef- fective
Instructor	Instructor with nonreaders	Exclude printed texts and complex instructor lectures

In examining the table, it should be borne in mind that its contents apply to the widest possible variety of learning situations. These include situations of the training of firemen, peace officers, and military personnel, as well as those of the typical school classroom. As a consequence, the listed media cover a range from equipment simulators (as a helicopter simulator) to the most primitive medium (as drawing on a chalk board). Of course, for the typical school classroom, the range of available media would be considerably shortened, and choices would be made among such devices as filmstrips, TV cassettes, and computer instruction modules. A more detailed description of the characteristics of the listed media may be found in Reiser and Gagné (1983).

Most of the media listed in Table 11-1 will be familiar to the reader. For a few, some words of explanation need to be made.

Real equipment and simulators   Real equipment may be employed as means of providing practice in skills. Equipment used for such a purpose may be *large,* as in the case of an earth mover, or *portable,* as in the case of a blood-pressure cuff and stethoscope. Sometimes, practice of skills is given by a *simulator,* a device that reproduces the operating characteristics of real equipment. Simulation may also be used to represent procedures, substituting symbols for real objects, as is often done with the computer.

Broadcast   Radio and TV broadcasts are used as instructional media when students are widely dispersed. Verbal knowledge of all kinds is readily transmitted in this manner. Intellectual skills have been successfully taught by broadcast radio and TV when provisions are made for student response.

Training devices and training aids   A device that permits practice of skills or of part skills, but which does not necessarily have the appearence or operating charateurisitics of the real equipment, is called a *training device.* A number of training devices are available for automobile driving, aimed at teaching the skills of braking, steering, and rapid response to unexpected events. Equipment that does not represent the operating characteristics of automobiles does not qualify as a simulator, strictly speaking, but is classified as a training device. A *training aid,* in contrast, is a device that exhibits the operation of some process (such as the circulation of blood) that the learner is expected to observe but not otherwise interact with.

Computers   The computer is a particularly useful and versatile instructional medium. It can display both print and diagrammatic pictures on a screen, and accept learner responses via a typewriter keyboard, joystick, pointer or other control device. By suitable design, computer-based instruction can be truly *interactive,* providing adaptive, informative feedback to learners in accordance with their responses, whether correct or incorrect. Because of this feature, computer-based instruction is particularly well suited for the learning of intellectual skills.

Interactive TV   Normally, the pictures and audio presentations of television are not learner-interactive. They may be made so, however, by using a TV tape cassette or video disc combined with a computer.

Motion picture and TV cassette   The medium listed as *motion picture* is to be understood as incorporating audio. Nowadays, motion and sound are typically combined in the *TV cassette,* although film is still widely used.

Instructor   The instructor characteristics employed as media are the voice in delivering speech, and the action in demonstrating procedures. Another important media function is the delivery of informative feedback adapted to learner responding. In addition to these various activities, of course, the instructor has the executive functions of planning and managing instruction.

MEDIA SELECTION PROCEDURES

If we consider the relationship of media characteristics to the type of learning situation, we discover a broad range of choices. This process of media choice, however, is only the first step. One must then proceed to narrow the field to the few media that are capable of supporting instruction most effectively. Most models for instructional media selection have the purpose of reducing the number of potentially effective media to a small number, with the final selection of one or two based upon factors of practical feasibility.

Media selection models have many features in common. For example, virtually all models call for consideration of the kind of objective being dealt with, the characteristics of the intended learners, and a host of practical factors such as size and composition of the group to be served, the range of viewing and hearing distances, and the like. Since separate summaries of the features of each model would be redundant, we will begin by listing some of the more common features.

Most writers on media selection models would initially agree that there is no one medium that is universally superior to all others for all types of desired outcomes and for all learners. This conclusion is also supported by research on media utilization (Aronson, 1977; Briggs, 1968; Briggs and Wager, 1981; Clark and Salomon, 1986). Most writers also agree that one cannot identify media that are particularly effective for a single school subject for a single grade level. Rather, careful design work and the results of media research both suggest that media are best selected for specific purposes within a single lesson. Thus, a motion picture may be an effective way to portray historical events in history lessons, but the teacher may be left to inform the learner of the objective, offer learning guidance, and provide feedback. It is not that a specially designed film could not provide all these events but rather that most available films do not do so. Films can also be used to portray theoretical events such as atomic particles in motion, or for enlarging, condensing, speeding up, or slowing down the portrayal of observable activities in nature or in manufacturing processes. Other examples of media usage, along with a listing of some advantages and limitations of various media, have been described by Briggs and Wager (1981).

Factors in Media Selection

Beyond the consideration of requirements of the learning situation, models of media selection typically include three categories of factors contributing to the narrowing of choices. These are (1) physical attributes of the media; (2) task characteristics; and (3) learner characteristics.

Physical Factors

Media differ from each other in terms of the physical characteristics of the communications they are able to display. Some media permit *visual* displays, and others do not. The property of visual display is obviously of use in teaching the identification of concrete concepts (shapes, objects) and spatial relationships (locations and distances). Generally, media are capable of presenting verbal displays, either as *printed text* or as *audio* messages. When printed text is otherwise appropriate, it is worth noting that print on paper is one of the least expensive media. The capability of presenting sound is a commonly occurring attribute of media. Not only may verbal messages play a role in media choice, but also the presentation of other varieties of sound, such as environmental noises and music. In connection with pictorial displays, *color* may be employed. Research findings indicate that color does not increase the effectiveness of instruction, except when it is an essential feature of what must be learned (Schramm, 1977).

A special category of media consisting of *real objects* is mentioned in Table 11-1 as *large equipment* and *portable equipment*. Most writers on media selection have recognized that real objects may sometimes be the most effective media, for example, in the learning of motor skills and complex procedures (piloting an airplane, putting air in a tire). Another characteristic of media of notable significance is pictorial *motion*. When the dynamic features of objects are a part of the essential content to be displayed, motion in pictures is virtually essential. Examples are the movements of storm clouds and the internal operations of a four-cylinder gasoline engine. Special uses of pictorial motion may be employed in the display of fast motion and slow motion. Of particular significance relating to motion is the potential of displaying the actions of living beings realistically, particularly humans. It would be difficult to over-emphasize the importance of this feature in such media as motion pictures and television.

The Learning Task

In choosing media for instruction, the type of performance expected of learners as a result of instruction (the *learning outcome*) needs to be considered with some care. Perhaps the most obvious media differences arise over the interactive quality. When intellectual skills are being learned, precise feedback to the learner about correctness and incorrectness of performance is a matter of great significance to learning effectiveness. When concrete concepts or rules involving spatial arrangements or spatio-temporal sequences are being learned, the presentation of pictures (as opposed to verbal descriptions) is an essential aspect of instruction. For example, the parts of a flower or the movement of a pendulum can be most effectively displayed in pictorial form, but only with difficulty by means of verbal presentations.

When verbal information is to be learned (for example, knowledge about a historical event), the medium to be chosen must have the capability of presenting verbal material, either in print or via audio channel as speech. Attitudes are best presented by media that make possible the display of a human model and the model's message about personal choices. Pictorial motion of human activities is therefore a major consideration in the choice of media for attitude learning or modification. Further details concerning the operation of these choices in a media selection model are described in the following section.

Learner Variables

Characteristics of learners must be considered when selecting media. Research on aptitude treatment interaction (ATI) has revealed the effects of instruction on learners differing in such traits as *anxiety* and *locus of control* (see Chapter 6). Some educators are convinced that learners differing in "learning styles" may benefit most from media presentations that match their styles. What these learning style differences are and whether they may be effective with different media has not been definitely established. If the styles were known, it might not be feasible or economical to provide enough parallel media packages for each lesson to accommodate all of the differing styles.

One kind of learner characteristic about which there should be little question in media selection is *reading comprehension*. Those learners who score low in this ability will undoubtedly have difficulty in learning from the printed page. Because this ability is closely correlated with performance in oral language understanding, low-ability learners are also likely to learn poorly from speech that is semantically complex. Two things can be done to alleviate these difficulties. One is to design the printed or oral prose passage to be low in reading difficulty, using familiar words and short sentences. The second is to use pictures and diagrams, to the extent possible, for the presentation of novel concepts, rules, and procedures. It will be evident that for those low in reading comprehension ability, pictures can convey such learning content more rapidly and efficiently than verbal expressions.

Given normal progression in reading skills for groups of learners, the learner's age is a useful matter to consider in media selection. In this connection, Dale's (1969) "cone of experience" is a useful tool. Dale listed 12 categories of media and exercises, in a somewhat age-related fashion. Thus, at level 1, "Direct purposeful experience," it is proposed that a child come into physical contact with objects, animals, and people, using all the senses to "learn by doing." As one goes up the age scale, pictorial and other simulated substitutes can be employed for some of the experiences. At the top of the cone is the use of "verbal symbols," which suggests learning by reading, an efficient method for sophisticated learners. When dealing with cognitive objectives—information, intellectual skills, and cognitive strategies—a rule of thumb previously suggested by Briggs and Wager (1981) is: "Go as low on the scale as you need to in order to insure learning for your group, but go as high as you can for the most efficient learning" (p.131). By considering the opposing factors of "slow but sure" (time-consuming direct experience) and "fast but risky" (typically occurring when learners are not skillful readers), one may decide just where on the scale is the best decision point for media selection. Dale's categories are as follows:

12. Verbal symbols
11. Visual symbols—signs; stick figures
10. Radio and recordings
9. Still pictures
8. Motion pictures
7. Educational television
6. Exhibits
5. Study trips
4. Demonstrations

3. Dramatized experiences—plays; puppets; role playing
2. Contrived experiences—models; mock-ups; simulation
1. Direct purposeful experience

For attitude objectives, Wager (1975) has suggested that Dale's age/media relationship becomes inverted, as compared to the relationship for cognitive objectives. Thus, a young child benefits from direct experience with real objects for cognitive objectives; he may acquire an attitude by hearing verbal statements from people he respects, as discussed in Chapter 5. Attitudes may be formed by listening to statements from respected models. Wager goes on to say that *changing* an already established attitude, on the other hand, may require real-life experiences for both children and adults. Briggs (1977) describes attitude formation and change as it takes place at various age levels by various methods: (1) classical conditioning, (2) hearing persuasive communications, and (3) human modeling. Clearly, different media would be involved in these different methods.

Some educators have sought to identify kinds of family backgrounds that are related to effectiveness of various media. Such factors may indeed be related to entering competencies and attitudes of pupils toward school learning. But it is not clear at present whether such data would be more useful in checking on the entering competencies of pupils in order to decide what they need to be taught or for selecting the media that relate to how they should be taught. However, a few indirect media selection implications are fairly likely. For example, children from homes where parents do not or cannot read are less likely than other children to have acquired a love for reading and skill in reading, other factors being equal.

The Assumed Learning Environment

Another set of factors in media selction is based upon administrative considerations rather than technical ones. The practicality of using media varies with such features of the learning environment as: (1) size of school budget; (2) size of class; (3) capability for developing new materials; (4) availability of radio, television, and other media equipment; (5) teacher capabilities and availability for an instructional design effort; (6) availability of modular materials for individualized, performance-based instruction; (7) attitudes of principal and teachers towards innovations; and (8) school architecture.

Many of the decisions related to these factors may have been made early in the instructional design process, as discussed in Chapter 2. However, these early decisions are intended to match the use of various delivery systems with media for specified school environments. Also, some of the factors clearly suit the situation where an established curriculum is being redesigned piecemeal as a gradual change strategy. These decisions are not at the same level of detail as media selection for specific lessons and instructional events with lessons.

The Assumed Development Environment

Obviously, it would be useless to plan to design a delivery system (and the attendant media) if the design and development resources were not adequate for the task. That is, the time, budget, and personnel available will influence the probability of success in designing specified delivery systems. The kinds of personnel available, for example, will determine the kinds of media that can be developed successfully. Beyond this, the personnel available will determine the kind of design model that is feasible for the situation. Carey and Briggs (1977) have discussed further how budgets, time, and personnel influence not only what instruction can be developed, but also what *design models* and what *team management* systems are appropriate to the task.

The Economy and the Culture

In designing an instructional system, one will wish to choose media that are acceptable to the users and within the budget and technology resources available. Attitudes toward various media may differ between urban and rural people or among ethnic or socioeconomic subgroups. Some countries or regions would not have the technological skills or the electric power to utilize radio and television, whereas these media would be practical and acceptable elsewhere. There may even be religious or cultural attitudes that help determine local reactions to various media. Print media may carry high prestige in one area, whereas radio and television have greater favor in another. All these factors should be considered if the media selected for a delivery system are to find acceptance.

Within boundaries of acceptance for various media, further consideration can then be given to cost effectiveness. Under one circumstance, cost may be the overriding factor, whereas under another circumstance a required level of effectiveness may be determined first, after which costs are considered. For example, designers of military training often separate those objectives that must be completely mastered at any cost because they are critically important from those objectives that are less critical. For the latter, lower performance

Designers will need to ascertain the intended user's status and intentions, in order to avoid selecting media that may be unacceptable or impractical. There are many ways of gathering such information, including visits to the users and the use of questionnaires. Perhaps the best way is to arrange to have some of the users become members of the design team. This practice may not only help assure acceptance of the media chosen but also enhance the effectiveness of the total instructional design.

Designers who serve as education consultants to other countries become aware of the need to avoid recommending a "United States solution" to problems in countries where such solutions are ill-suited. This refers not only to media selection but also to the total instructional approach. Even the translations of instructional materials and rather straightforward directions (as in teachers' guides) must be carefully reviewed to ensure clarity for the user. The importance of this point may be appreciated by recalling that even when designing materials for our own students we are not sure that the communications are understood until they are tested in use by those students. In short, what is perfectly clear to the writer may be very confusing to the reader.

Practical Factors

Assuming that the media under consideration are acceptable to the users and are within their capabilities, a number of detailed practical factors remain to be considered in order to select media that are effective and also convenient. A general discussion of such factors for each of several media may be found in books on media selection and utilization (e.g., Anderson, 1976; Bretz, 1971; Kemp and Dayton, 1985). The suitability of media for use in practical situations may depend on such factors as size of group, type of learner, response desired, type of stimulus presentation, simplicity of physical classroom arrangements, requirements for lighting or darkness in the room, and other environmental conditions.

Some of the practical factors to be considered in media selection are:

1. What size of group must be accommodated in one room on a single occasion?
2. What is the range of viewing and hearing distance for the use of the media?
3. How easily can the media be "interrupted" for pupil responding or other activity and for providing feedback to the learners?
4. Is the presentation "adaptive" to the learners' responses?
5. Does the desired instructional stimulus require motion, color, still pictures, spoken words, or written words?
6. Is sequence fixed or flexible in the medium? Is the instruction repeatable in every detail?
7. Which media provide best for incorporating most of the conditions of learning appropriate for the objective?
8. Which media provide more of the desired instructional events?
9. Do the media under consideration vary in probable "affective impact" for the learners?
10. Are the necessary hardware and software items obtainable, accessible, and storable?
11. How much disruption is caused by using the media?
12. Is a backup easily available in case of equipment failure, power failure, film breakage, and so on?
13. Will teachers need additional training?
14. Is a budget provided for spare parts, repairs, and replacement of items that become damaged?
15. How do costs compare with probable effectiveness?

You may be able to think of many items that could be added to this list of practical factors in media selection. Most teachers can report instances in which a burned-out bulb or a broken film has made necessary a sudden change in plans. Equally disruptive is the selection of a medium unsuitable for the size of the group, with the possibility that many persons cannot see or hear. Having a backup plan or an alternative lesson is one way to minimize the negative reaction when something "goes wrong" in the classroom.

A METHOD OF MEDIA SELECTION

Previous chapters have shown that there are many design steps to be taken prior to media selection, at least when following the general design model presented in this book. Some of these prior steps are: analysis of objectives, defining the objectives by domain of the learning outcome, and designing instructional sequences. Chapter 10, which dealt with the events of instruction, prepared you to apply those events to planning instruction for the particular objectives that represent the purposes of single lessons.

In acknowledging the importance of the *learning situation,* we propose that the identification of that situation be undertaken initially, as previously described. This is because the most prominent differences among instructional media are related to the learning situation. For example, in the learning situation involving "self-instruction with nonreaders," media that display *print* in the form of connected prose may be immediately excluded. Beyond this determination, the choice of media for learning effectiveness comes to depend particularly on the type of *learning outcome* intended. As an extreme example, the learning of a multistep procedure (and intellectual skill) cannot be adequately done via the medium of an instructor's lecture.

The purpose of this section is to outline a systematic method of media selection for each of six learning situations, beginning with the type of learning outcome implied by the lesson objective, and ending with the identification of a set of media that can be effective for the learning contemplated. From this smaller set, a final choice of one or more particular media may be made on the basis of practical considerations (cost, feasibility, etc.).

A Model of Media Selection

A method of media selection that proceeds from the identification of the type of learning situation to a consideration of the implications of the learning outcome is described by Reiser and Gagné (1983). Its main point will be briefly summarized here. It should be noted that this model is intended to be applicable to a broad range of situations, encompassing school education, adult education, technical and professional education, and training for specialized jobs in industry and the military services. Although acknowledging its broad applicability, we shall confine our attention mainly to the more "academic" kinds of instruction.

Reiser's and Gagné's model takes into consideration logistic factors (to what size group the instruction will be delivered), learner variables (readers vs. nonreaders), and task variables (task importance, type of learning outcome). At completion of the analysis, the media suggested as appropriate for teaching a particular type of task depend on their ability to deliver the *events of instruction* that will provide the necessary conditions of learning for a particular type of outcome.

An Example—Self-Instruction with Nonreaders

The model uses flowcharts to indicate the successive choices needed in media selection. An example applicable to the learning situation "self-instruction with nonreaders" is shown in Figure 11-1. The chart begins with a list of "candidate media" from which choices are to be made. As will be apparent, this list includes virtually all types of media except those that have been excluded because they meet only the requirements of other learning situations (such as broadcast media).

Entering the flow chart itself, the first question asks if the outcome is: "either an attitude or verbal information?" If the answer to this initial question is negative, then the model indicates that the learning outcome must be a skill (intellectual skill or motor skill). The next question asks whether "motor practice is necessary," indicating the identification of a *motor skill*. If indeed that is the learning outcome intended, the next critical question is whether suitable practice, with feedback, can be provided by one or both of the two media that suggest themselves: (1) portable equipment (e. g., a hand pump for inflating tires), or (2) a training device (e. g., a device for baseball batting practice). It may be noted that for this kind of learning situation, the identification of a motor skill as the learning outcome leads to the *exclusion of all media except these two*. These are the only two that make possible the direct practice required for motor skill learning. Should either of these media be judged inadequate to provide suitable feedback during practice, the chart says "go to box F.1." The meaning here is "go to the chart for the learning situation 'Instructor with Nonreaders'," with the implication that an instructor will provide the necessary feedback.

Now, returning to the point at which a skill was identified, suppose the question about motor practice was answered no. This would mean the skill to be dealt with is an *intellectual skill*. Decision 16 then seeks to answer the question about adequate feedback with respect to two media that have the quality of interacting with learner responses: (1) computer instruction, and (2) interactive TV. Why are such media as motion picture, slide/tape, and TV cassette not included here? Because these media do not permit the learner the response interaction that is essential for effective learning of intellectual skills.

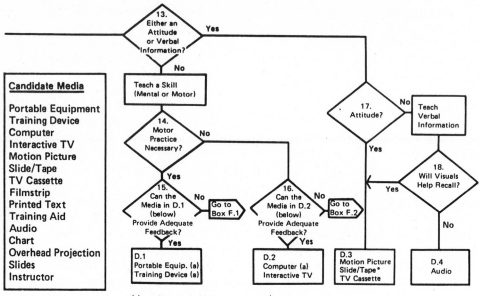

(a) supplemented with, or including, audio speech
* somewhat less effective

Explanation of Questions - Chart D

13. Either an Attitude or Verbal Information? Is the aim either to influence the
   student's values (attitudes) or to have the student learn to 'state' (rather than 'do') something?
14. Motor Practice Necessary? Does the skill to be learned require smooth timing of
   muscular movements (a "motor skill")?
15. Can the Media in D.1 Provide Adequate Feedback? Can the media in D.1 accept
   and evaluate the desired student responses and provide the type of feedback required?
16. Can the Media in D.2 Provide Adequate Feedback? Can the media in D.2 accept
   and evaluate the desired student responses and provide the type of feedback required?
17. Attitude? Does instruction aim to influence the student's values or opinions?
18. Will Visuals Help Recall? Is it likely that the use of visuals will help the student
   establish images that will aid recall of verbal information?

FIGURE 11-1   Portion of a flowchart for media selection, applicable to the learning
situation "Self-instruction with Non-Readers."
(From R. A. Reiser & R. M. Gagné, *Selecting media for instruction,* copyright 1983 by Educational Technology
Publications, Englewood Cliffs, NJ. Reprinted by permission).

Here is another instance, then, in which the appropriate media have been reduced from a rather long list to two. (The direction "go to box F.2" has the
same implication as that previously described for F.1)

To complete the explanation of this chart, we return to the question, "Is
this either an attitude or verbal information," and suppose that the answer is
yes. In either case, one seeks to identify media that do not employ printed
text and that do not necessarily provide for interactive feedback. Should the
learning outcome be an *attitude,* evidently motion pictures, slide/tape presentations, or TV cassettes will be effective media. This is because they can de-

scribe and picture the situations to which the attitude is applicable, and can realistically display a human model who will present the message conveying the desired attitude.

Should the answer to the previous question be no, the model identifies the learning outcome as *verbal information*. Again this implies the exclusion of media displaying printed discourse. The next question is "will visuals help recall?" If not, then a reasonable choice of media would appear to be one that presents the verbal information auditorially, as via an audiotape recorder. However, visuals in media often *do* help recall, by providing elaboration of the verbal information and additional visual cues (Chapter 5). If this possibility exists, the chart points again to the choices of media as (1) motion picture, (2) slide/tape, or (3) TV cassette.

By working through the flow chart for this particular learning situation (self-instruction with nonreaders), one can clearly appreciate the success of this method in reducing the candidate media to a relatively small number. The method is one that considers learning effectiveness as the basis for decisions, which in turn are derived from identifications of the five types of learning outcomes described in Chapters 3, 4, and 5. The media selected are those best able to put into effect the conditions of learning appropriate for each type of learning outcome.

Media Selection by Learning Outcome

It is possible in this chapter to present only one of the flow charts applicable to the six learning situations of the Reiser–Gagné model. Beyond referring the reader to that source (Reiser and Gagné, 1983), we can only summarize the main points of the model by describing the "media implications" of each learning outcome. This is done in Table 11-2. Recall that initial decisions leading to the exclusion of some media are made on the basis of the type of learning situation, as indicated in Table 11-1. Assuming that step has been taken, Table 11-2 lists the implications for media exclusion and media selection indicated by the type of learning outcome intended by the objective of the instruction.

The table may be read as follows. When the intent of the lesson under design has the learning outcome of one or more intellectual skills, *exclude* those media that make no provision for feedback and subsequent interactive responding by the learner. The reason is that the learning of intellectual skills cannot be adequately supported without this feature. Another exclusion, applicable to learners of low reading comprehension (here called nonreaders as an abbreviated expression), is that of semantically organized printed prose. Media *selection* for intellectual skills reflects complementary procedures. Select media that provide for feedback and learner interaction. For nonreaders, select media that communicate via pictorial and audio modes, either to supplement or supplant printed discourse.

TABLE 11-2. Implications for Media Exclusion and Media Selection of Type of Learning Outcome

Learning Outcome	Exclusions	Selections
Intellectual Skills	Exclude media having no interactive feature. Exclude printed discourse for nonreaders.	Select media providing feedback to learner responses. Select audio and visual features for nonreaders.
Cognitive Strategies	Exclusions same as for intellectual skills.	Select media with same features as those for intellectual skills.
Verbal Information	Exclude only real equipment or simulator with no verbal accompaniments. Exclude complex prose for nonreaders.	Select media able to present verbal messages and elaborations. Select audio and pictorial features for nonreaders.
Attitudes	Exclusions same as for verbal information.	Select media able to present realistic pictures of human model and the model's message.
Motor Skills	Exclude media having no provision for learner response and feedback.	Select media making possible direct practice of skill, with informative feedback.

Practical Considerations in Media Selection

Employment of the model described in the planning and designing of instruction is expected to assure suitable attention to learning effectiveness. In general, exclusion rules of the model will prevent consideration of media that are inappropriate for learning. Rules of inclusion will lead to the selection of media best suited for the presentation of communications that support learning. Following these rules will bring the instructional designer to consider the practical advantages of a relatively small list of media. Some of the principal practical questions in a media decision are described in the previous section on Practical Factors.

Adapting the Model to Design Circumstances

This chapter has presented a method of media selection that may be described as fine-grained and analytical. Emphasis is placed on making media decisions aimed at assuring learning effectiveness, based upon identification of the type of learning outcomes represented by the lesson objectives. Having excluded inappropriate media on this basis, a final choice can be made from a relatively small list of possibilities, on grounds such as feasibility, availability, and costs. Thus, explicit use of theory and research is provided for in the system, and subsequent empirical, formative, and summative evaluations are assumed.

The model of media selection and utilization presented here is therefore systematic, internally consistent, and related directly to the major theoretical orientation of this book. Actually, the model has been used with some ease and leads to expeditious media decisions (Reiser and Gagné, 1983). Designers trained in use of this model will have the detailed skills to fit the resource and constraints of specific design projects. Carey and Briggs (1977) have shown that the skilled design team leader must not only adopt, adapt, or design a specific model to fit the project budget and personnel available, but must also develop time periods for specific tasks, make personnel assignments, and monitor the entire management plan. A design model must be adapted to fit the circumstances of any particular design project relative to: budget, time, personnel, facilities, equipment, supplies, and institutional charcteristics of the developing agency and the using agency.

Surely the most significant aspect of this design model is its emphasis on the primacy of selecting media based on their effectiveness in supporting the learning process. Every neglect of the consideration of how learning takes place may be expected to result in a weaker procedure and poorer learning results. Experienced designers, of course, may do much of the analysis "in their heads," because the steps in problem solution have become familiar.

SUMMARY

This chapter begins with a brief account of how teachers may participate in the design of instruction by *selecting* media and the materials they display. Teams of designers may *develop* new materials for media. Much of the theorizing involved in the two functions, however, can be the same, according to the design model presented in this book.

Selection of media is determined by a number of factors, including the nature of the learning situation; the type of learning outcome expected; the environment for learning; the conditions for instructional development; the culture in which instruction will be given; and on various practical factors including accessibility, feasibility of use, and costs. The aim of learning effectiveness is to give primacy first to the learning situation (including the nature of intended learners), and following that to the kind of learning outcome expected.

A model of instructional media selection is described, based on the work of Reiser and Gagné (1983). This model requires, first, the identification of one of six kinds of learning situations, characterized by such features as the use or nonuse of broadcasts, instructors, and self-instruction. Following this determination, a flow chart exhibits successive choice points of media selection as they are influenced by the requirements of effective learning for the various learning outcomes. The model is based upon the conditions of learning pertaining to each of the learning outcomes described in Chapters 3, 4, and 5. A summary of implications of the model for media selection is presented in Table 11-2.

The method of media selection given here, along with the model, makes media selection a highly rational matter based upon theory and research pertaining to learning effectiveness. The method can be employed by teachers and designers of instruction as individuals or in teams. The model with its flow charts is easy to understand and use. The procedure is one that devotes initial attention to learning effectiveness, excludes inappropriate media, and applies practical considerations to a shortened list of candidate media.

REFERENCES

Anderson, R. H. (1976). *Selecting and developing media for instruction.* New York: Van Nostrand Reinhold.

Aronson, D. (1977). *Formulation and trial use of guidelines for designing and developing instructional motion pictures.* Unpublished doctoral dissertation, Florida State University, Tallahassee, FL.

Bretz, R. (1971). *The selection of appropriate communication media for instruction: A guide for designers of Air Force technical training programs.* Santa Monica, CA: Rand.

Briggs, L. J. (1968). Learner variables and educational media. *Review of Educational Research, 38,* 160–176.

Briggs, L. J. (Ed.) (1977). *Instructional design: Principles and applications.* Englewood Cliffs, NJ: Educational Technology Publications.

Briggs, L. J., & Wager, W. W. (1981). *Handbook of procedures for the design of instruction* (2nd ed.). Englewood Cliffs, NJ: Educational Technology Publications.

Carey, J., & Briggs, L. J. (1977). Teams as designers. In L. J. Briggs (Ed.), *Instructional design: Principles and applications.* Englewood Cliffs, NJ: Educational Technology Publications.

Clark, R. E., & Salomon, G. (1986). Media in teaching. In M. C. Wittrock (Ed.), *Handbook of research on teaching* (3rd ed.). New York: Macmillan.

Dale, E. A. (1969). *Audiovisual methods in teaching* (3rd ed.). New York: Holt, Rinehart and Winston.

Kemp, J. E., and Dayton, D. K. (1985). *Planning and producing audiovisual materials* (5th ed.). New York: Harper & Row.

Reiser, R. A., & Gagné, R. M. (1983). *Selecting media for instruction.* Englewood Cliffs, NJ: Educational Technology Publications.

Schramm, W. (1977). *Big media, little media.* Beverly Hills, CA: Sage.

Wager, W. (1975). Media selection in the affective domain: A further interpretation of Dale's cone of experience for cognitive and affective learning. *Educational Technology, 15* (7), 9–13.

12 DESIGNING THE INDIVIDUAL LESSON

The ultimate goal of instructional design is to produce effective instruction. When this goal is accomplished, it generally results in a lesson or set of lessons that may be delivered either by a teacher or by mediated materials. A mediated lesson is often called an instructional *module*. A lesson or module is generally planned to be a particular length in minutes, which usually means that any significant instructional curriculum is going to require more than a single lesson. In Chapter 9, we described the process of defining and sequencing lessons. In this chapter we will discuss the relationships among several different objectives within a lesson and the employment of the events of instruction in constructing such lessons.

Most of the characteristics of human capabilities that we discussed in Chapter 6 are used as a basis for planning the lesson, as are the events of instruction described in Chapter 10. These events apply to the design of all kinds of lessons, irrespective of the domain of the learning outcome intended. In this chapter, we emphasize the variations among lessons as they correspond to different domains of learning outcomes. These lesson variables are first considered in terms of their implications for designing sequences of instruction, and later with respect to the establishment of effective learning conditions for the different domains of learning outcome.

In designing a lesson one needs first to assure that the events of instruction are provided for. Additionally, it is necessary to classify the lesson objectives and arrange that specific events are placed in an appropriate sequence for the attainment of these objectives. The content of the events, or *instructional prescriptions,* are then written as the lesson content.

LESSON PLANNING AND MODULE DESIGN

Frequently, teachers select rather than develop instructional materials. In practice, teachers often "design as they teach"—that is, they may design sequences of lessons in advance but, perhaps, do not design all lessons for the course before the teaching begins. Because of these practical circumstances, teachers tend to plan each lesson in only enough detail so as to be "ready" for each lesson since they are able to improvise some of the details as the lesson progresses. This is not an entirely undesirable state of affairs because it gives the teacher flexibility to redesign "on the spot"—that is, to adjust procedures to the instructional situation and to responses of the learners.

As will be discussed in Chapter 14, the nature of instruction in large groups provides less precision of control in instruction than is possible in small groups. The unpredictability of student responses in a large group, coupled with restricted planning time, mean that instruction is often planned and carried out with only a moderate degree of control over the conditions of learning. Utilization of small group or individualized instruction modes makes possible greater precision of instruction. Adaptation to individual entering competencies and rates of learning is provided by instruction that permits self-pacing and self-correction for each learner. These functions are made possible in tutoring or small group modes and with materials that allow *branching* by the student to the most needed and helpful exercises contained in the instructional material. Such branching occurs in some learning modules in the form of programmed instruction, computer-assisted or computer-managed instruction, or by the frequent use of self-tests that allow the learner to use the instruction in an adaptive manner.

Individualized, Self-paced, and Adaptive Instructional Materials

These terms are often used synonymously, although there are shades of differences in their meanings. We define individualized instruction as that which takes into consideration the needs of students. Such instruction begins with an analysis of the entry skills of the learner, and the subsequent instruction is prescribed based on that individual's needs. Self-paced instruction is a phrase implying instructional management as well as the mediation of instruction. For example, videotaped or printed materials may be used in either group or self-paced instruction. However, in a self-paced system the learner can spend as much time as necessary to achieve the objectives. Self-paced instruction is generally associated with the procedures of *mastery learning,* in which achievement, rather than time, dictates the rate of the student's progress through the instruction. The term adaptive instruction usually refers to materials and management systems that constantly monitor the progress of the student and change the instructional content based on that student's progress. Generally speaking, adaptive instruction involves complex record keeping and decision making, and is facilitated by the use of computers. However, its procedures may be carried out manually for individuals or small groups. These types of instruction depend in some measure on mediated instructional materials, since all students in a class may be at different stages of learning at any particular point in time.

In summary, the object of instructional design is to produce a lesson or series of lessons that include consideration of the delivery system being used as well as the needs of the learners. The nature of the lesson will depend a great deal on how it is to be used. In a teacher-based system the lesson plan may be somewhat incomplete, because the teacher can fill in the gaps. In contrast, individualized or self-paced instruction must be more carefully planned and developed, since there is often no immediate teacher help available.

The remainder of this chapter will focus on how the principles of instructional design described in previous chapters may be applied to the development of either a teacher-led or a mediated lesson. Both these forms of instructional delivery retain the emphasis we have placed on these central themes:

1. Classifying objectives by the use of a taxonomy of learning outcomes
2. Sequencing objectives to take account of prerequisites
3. Including the appropriate events of instruction that apply to all domains of outcomes
4. Incorporating into the events of instruction the particular conditions of learning relevant to the domains of the objectives in the lesson

We now turn to further discussion of the sequencing of instruction and then to instructional events and conditions of learning. The chapter concludes with a discussion of steps in lesson planning and an example of a lesson plan that incorporates the form of the model usually adopted by an individual teacher who designs and conducts the instruction.

ESTABLISHING A SEQUENCE OF OBJECTIVES

In Chapter 9 we described top-down analysis of intellectual skills and the consideration of functional relationships among different domains of learning to which it leads. We indicated a way of diagramming these relationships through the use of instructional curriculum maps (ICMs) and showed how such maps can be used to identify different levels of curricula.

Within each ICM is an implied sequence of instruction. That sequence is based on principles underlying hierarchical prerequisite relationships and those that describe facilitative learning sequences. For example, Figure 12-1 shows a learning hierarchy that could also be an instructional curriculum map composed of objectives from only the intellectual skills domain.

Planning a Sequence for Intellectual Skill Objectives

We begin our account of lesson sequence planning with objectives representing *intellectual skills*. The subordinate skills required to attain an objective of this sort can be derived as a learning hierarchy. Suppose that one does indeed want to establish the skill of substracting whole numbers of any size, as diagrammed in Figure 12-1. The learning hierarchy for this objective lists 10 essential prerequisites, shown as boxes in the hierarchy. Let us assume that box I, simple subtraction facts, represents learning already accomplished earlier by the students. The teacher now needs to design a lesson or, perhaps more likely, a sequence of lessons to enable the learners to subtract any whole numbers they may encounter. Although there are several sequences of teaching the skills shown in boxes II through X that might be successful, the implication of the hierarchy is that the bottom row of boxes should be taught first, then the next higher row, and so on. It may also be inferred that a sequence going in numerical order from box II to box X might be the most effective sequence.

In summary, hierarchies are so arranged that there may be options in a sequence of boxes within a horizontal row, but there are not options in proceeding from the bottom row upwards, at least when we wish to adopt a single sequence for all learners in the group.

This is not to say that no student could learn the task by a sequence that violates the above rules of thumb. If a student did learn by such a nonhierarchical sequence, it might be because she already could perform the skills in some of the boxes or because she possessed sufficient cognitive strategies to discover some of the rules without having received direct instruction in their application.

Determining a Starting Point

Continuing with the example of learning to subtract whole numbers, it is possible that some students may have already learned some of the prerequisite skills in some of the boxes. One student may already be able to perform the skills in boxes II and III; another may be able to perform II and V. Obviously one needs to begin the instruction "where each student is." This is conveniently done in individualized instruction programs, as described in Chapter 15, but it can be done also with a group by arranging other activities for those students who do not need some of the instruction planned for the group. An alternative, of course, is for students to sit through some of the

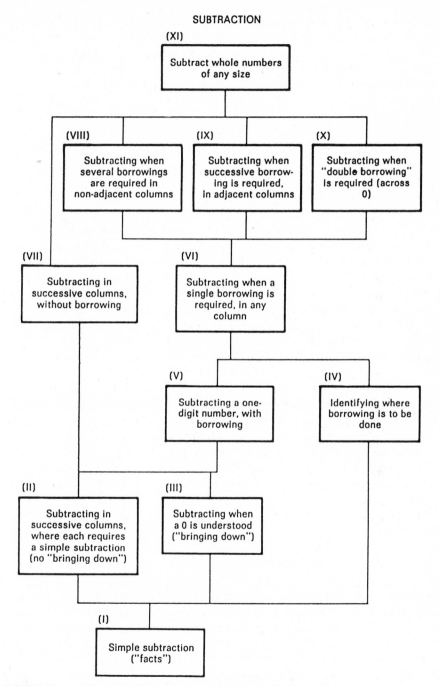

FIGURE 12-1   A learning hierarchy for subtracting whole numbers.

instruction as a review, although this may not always be the best solution. A review of earlier skills may be needed at the beginning of each lesson to be sure there is ready recall when the new learning is undertaken. Generally speaking, however, intellectual skills that are learned are recalled well, as compared with the recall of facts or labels, for example.

Specifying a Sequence of Lessons

A teacher must decide "how much" should be included in each lesson, as well as the sequencing of the lessons, following the implications for sequencing discussed earlier. It might often be convenient to teach some "boxes" as single lessons; other boxes may be combined within one lesson, as we have already shown in Chapter 9.

The hierarchy, then, implies several possible effective lesson sequences. The skill relationships that indicate essential prerequisites need to be maintained in planning such sequences—otherwise no particular sequence is implied. However, the teacher may choose to insert instruction relating to other domains of outcome into the sequence. Often a sequence of lessons is built around an intellectual skill objective in such a way as to include instruction concerning verbal information objectives, attitudes, and cognitive strategies (Wager, 1977; Briggs and Wager, 1981).

Achievement of Skills in Sequence

The planning of lessons designed to attain the final skill, XI, contains the assumption that each student will display *mastery* of prerequisite skills before being asked to learn the next higher skill. For example, before tackling skill X, requiring double borrowing across a column containing a zero, it must be ascertained that the learner can do skills VI and VII, requiring subtracting in successive columns without borrowing, and borrowing in single columns and multiple columns.

The notion of mastery must be taken with complete seriousness when one is dealing with intellectual skills. The lessons must be so designed that each prerequisite skill can be performed with perfect confidence by the learner before attempting to learn more complex skills in the hierarchy. Any lesser degree of learning of prerequisites will result in puzzlement, delay, inefficient trial and error at best, and in failure, frustration, or termination of effort toward further learning at the worst. For this reason we suggest that allowing the student to choose the sequencing is not likely to be the most efficient procedure.

Provisions for Diagnosis and Relearning

Lesson planning which utilizes the hierarchy of intellectual skills may also provide for diagnosis of learning difficulties. If a student has difficulty learning any given skill, the most probable diagnostic indication is that the student cannot recall how to perform one or more prerequisite skills. Any given lesson may provide for diagnosis by requiring that prerequisite skills be recalled. If one or more cannot be recalled, then relearning of these prerequisites should be undertaken. Thus the assessment of mastery for any given skill, occurring as a part of a lesson on that skill, may be followed by further assessment of prerequisite skills, in case mastery is *not* achieved. Following this, provision should be made for a "relearning loop" in the sequence of lessons, which gives the student an opportunity to relearn and to display mastery of the necessary prerequisites before porceeding.

Sequence in Relation to Cognitive Strategies

Since cognitive strategies are executive routines for information processing, it is often difficult to ascertain whether these skills have been learned. Usually, one cannot specify a particular sequence of prior learning leading up to the attainment of a cognitive strategy. What must be remembered, however, is that students already possess some kinds of cognitive strategies at the time instruction begins. These strategies are in the form of automatized rules for processing new information. When one speaks of teaching a new cognitive strategy, what is meant is introducing students to a new way of processing information. This means that they will have to learn to modify their existing strategy, or simply to forget it and adopt the new strategy.

The prerequisite skills essential for the learning of cognitive strategies are often simple skills established by prior learning. Examples are (1) associating unrelated names by using them in a sentence, and (2) breaking a complex problem into parts. Strategies such as these can usually be communicated by means of verbal statements to the learner. In addition, a sequence of instruction designed to improve cognitive strategies usually takes the form of *repeated opportunities* for the application of the strategies. Such occasions may be interspersed with instruction having other intended outcomes and typically are made to recur over relatively long periods of time. In this way it is expected that gradual improvement in the application of the new strategy will be possible. In the case of metacognitive strategies (see Chapter 4), it does not seem likely that observable amounts of improvement can occur with the time of a single lesson or two.

Planning a Sequence for Verbal Information Learning

As indicated in Chapter 5, the most important prerequisite for the learning of information is the provision of a meaningful context within which the newly learned information can be subsumed or with which it can be, in some meaningful sense, associated. The principles applicable to sequencing differ somewhat depending on whether the objective concerns learning a set of names (labels), learning an isolated fact, or learning the sense of a logically organized passage.

Names or Labels

The learning of a set of names (such as the names of a number of trees) is facilitated by the use of previously learned organized structures, which the learner has in memory. A variety of structures may be used by the learner to encode the newly acquired information. The encoding may take the form of a simple association, as when a new French word *la dame* is associated with the English word *dame,* which thus becomes an association for *lady.* Sometimes, encoding may involve the use of a sentence, such as that which associates *starboard* with *right* in "the star boarder is always right." Frequently, too, the method of encoding may involve the use of visual images, as would be the case if the learner associated an image of a crow with a person's name *Crowe.* Mnemonic techniques involving the use of images and the keyword method are reviewed by Pressley, Levin, and Delaney (1982). The imagery employed for encoding may be quite arbitrary, as when a learner uses the shops on a well-known street as associations for newly acquired names of things having nothing to do with the shops themselves (cf. Crovitz, 1970).

It is clear, then, that the learning of new names or labels calls upon previously learned entities stored in the learner's memory. In this kind of information learning, it does not seem reasonable that a *specific* content of previous learning, implying a sequence of instruction, can be recommended. Although it is possible to facilitate the learning of new labels by prescribing some particular "codes" for the learner to use, such a procedure is generally found to be less successful than permitting the learner to use an encoding system of his own. What the learner mainly needs to have learned previously, aside from the various meaningful structures that may already be in memory, is "how to encode." This is a particular kind of *metacognitive strategy.* The possibilities of long-term instruction designed to improve such a strategy have not been investigated.

Individual Facts

The learning of individual facts, as they might occur in a chapter of history text, also involves an encoding process. In this case, the encoding is usually a matter of relating the facts to larger meaningful structures—larger organized "bodies of knowledge" that have been previously learned.

Two kinds of procedures are available for instructional sequencing when one is dealing with factual information. Both of them should probably be employed, with an emphasis determined by other factors in the situation. The first is the prior learning (in a sequence) of what Ausubel (1968) calls *organizers.* If the learner is to acquire facts about automobiles, for example, an organizing passage may first be presented that informs the learner about the major distinctive categories of automobile description—body style, engine, frame, transmission, and so on. The specific facts to be learned about particular automobiles then follow.

The second procedure, not entirely unrelated to the first, involves the use of questions or statements to identify the major categories of facts for which learning is desired (cf. Frase, 1970; Rothkopf, 1970). Thus, if the learning of the names of persons described in a historical passage is the most important information to be learned, prior experience with questions about such names in a sample passage will facilitate the learning and retention of them. Should the lesson have the objective of stating dates, then dates could be asked about in a prior passage.

Organized Information

Most frequently, an objective in the category of verbal information is the expectation that the learner will be able to state a set of facts and principles in a meaningful, organized manner. For example, an objective in social studies might be to describe the process involved in the passage of a bill by the U.S. Congress. In this case a schema would likely include at least the essential steps, such as drafting the bill, introducing the bill, and so on. The learning of organized information of this sort is also subject to an encoding procedure that calls upon previously learned structure in the student's memory. Anderson (1984) describes such a memory structure as a *schema*. He defines schema as an "abstract structure of information . . . [which may be] conceived as a set of expectations" (p. 5). These expectations are considered to be slots in the learner's knowledge structure into which the new information can be integrated. Sequencing of organized knowledge should take into account existing schemas into which the new knowledge can be subsumed. The teacher should structure the new information so that it builds on what the student already knows. An example is cited in the work of Ausubel (1968); he speaks of the process of "correlative subsumption," occurring when information about Buddhism is acquired following what has previously been learned about a different religion, Zen Buddhism.

Planning a Sequence for Motor Skill Learning

The capabilities that constitute prerequisites for the learning of a motor skill are the part skills that may compose the skill to be learned, and the executive subroutine (the complex rule) that serves to control their execution in the proper order. Of course, the relative importance of these two kinds of prerequisite depends largely upon the complexity of the skill itself. To attempt to identify part skills for dart throwing, for example, would not be likely to lead to a useful sequencing plan; but in a complex skill such as swimming, practice of part skills is often considered a valuable approach.

Typically, the learning of the executive subroutine is placed early in the sequence of instruction for a motor skill, before the various part skills have been fully mastered. Thus, in learning to put the shot, the learner may at an early stage acquire the executive subroutine of approaching the line, shifting his weight, bending his are and body, and propelling the shot, even though at this early stage his performance of the critical movements is still rather poor.

The learning of particular part skills may themselves have important prerequisites. For example, in the skill of firing a rifle at a target, the concrete concept of a correct sighting picture is considered to be a valuable subordinate skill to the execution of the total act of target shooting. Accordingly, a plan of instruction for a motor skill must provide not only for the prior practice of part skills, when this is appropriate, but also on some occasions for a sequence relevant to the individual part skills themselves.

Planning a Sequence for Attitude Learning

As is true for other kinds of learned capabilities, the learning or modification of an attitude calls upon previously acquired entities in the learner's memory. A positive attitude toward reading poetry, for example, could scarcely be established without some knowledge of particular poems on the learner's part or without some of the language skills involved in interpreting the meaning of poetic writing. Thus, for many attitudes with which school learning is concerned, the planning of an instructional sequence must take into account these kinds of prerequisite learning.

The basis for an instructional sequence aimed at establishing an attitude is to be found in the particular verbal information and intellectual skills that become a part of the personal action that the teacher expects the learner to choose as a result of instruction. If the learner is to have a positive attitude toward associating with people of races different from his own, such an attitude must be based upon information concerning what these various "associations" (playing games with, working with, dining with, and so on) are about. Or, if the learner is to acquire a positive attitude towards the methods of science, this must be based upon some capabilities (skills) of using some of these methods.

An instructional sequence of learning an attitude, then, often begins with the learning of intellectual skills and verbal information relevant to that attitude. It proceeds then to the introduction of a procedure involved in establishing the positive or negative tendancy that constitutes the attitude itself, as described in Chapter 5. Since attitude learning may require prior learning of intellectual skills and verbal information, it is often necessary to consider learning domain interactions as described by Martin and Briggs (1986). These interactions can be analyzed by means of "audit trails," in which the attitude to be acquired is related to other skills that facilitate its acquisition. The audit trail may include other attitudes, verbal information, or intellectual skills, and it provides a guide for the sequencing of experiences leading to attitude change.

When the method of human modeling is employed for attitude modification, another prerequisite step in the sequence may be necessary. Since the "message" that represents the attitude needs to be presented by a respected source (usually a person), it may in some instances be necessary to establish or to build up respect for this person. A currently famous scientist is not likely to command the respect that a well-known scientist, such as Einstein, does; and Einstein as a pictured model is more likely to be respected if the learner knows of his accomplishments.

LESSON PLANNING FOR LEARNING OUTCOMES

The sequence of capabilities exemplified by the learning hierarchy (for intellectual skills) or by a set of identified prerequisites (for other types of outcomes) is used as a basis for planning a *series* of lessons. The implication it has for the design of a *single* lesson is that one or more prerequisite or supporting capabilities need to be available to the learner. Obviously, though, there is more than this to the planning of each lesson. How does the student proceed from the point of having learned some subordinate knowledge or skills to the point of having acquired a new capability? This interval, during which the actual learning occurs, is filled with the kinds of instructional events described in Chapter 10. These events include the actions taken by the students and teacher to bring about the desired learning.

Instructional Events and Effective Learning Conditions

The most general purpose for what we have called the events of instruction is that of arranging the external conditions of learning in such a way as to ensure that learning will occur. Instructional events are typically incorporated into the individual lesson. In a general sense these events apply to all types of lessons, regardless of their intended outcomes. Just as we have found it necessary to describe particular sequencing conditions that pertain to different lesson outcomes, we also recognize a need to give an account of the particular events that affect the learning effectiveness of lessons having different kinds of outcomes. This makes it possible to recall the *conditions of learning* for various classes of learning outcomes and to apply these principles to the arrangement of effective learning within a lesson. These conditions are described in greater detail by Gagné (1985).

Tables 12-1 and 12-2 are intended to consolidate several ideas that influence lesson design. First, they assume the general framework of instructional events, described in Chapter 10, without developing these ideas further. Second, they describe procedures for implementing optimal learning conditions that are specifically relevant to each class of learning objective. These have been referred to as the *external conditions* of learning. And third, they take account of the problem of lesson sequencing by representing the recall of pre-

TABLE 12-1.   Effective Learning Conditions for Incorporation into Lessons Having Objectives of Intellectual Skills and Cognitive Strategies

Type of Lesson Objective	Learning Conditions
Discrimination	Recall of responses Presentation of *same* and *different* stimuli, with emphasis on distinctive features Repetition of *same* and *different* stimuli, with feedback
Concrete Concept	Recall of discrimination of relevant object features Presentation of several instances and noninstances, varying in irrelevant object qualities Identification of instances by student, with feedback
Defined Concept	Recall of component concepts Demonstration of the concept, using its definition Demonstration of concept example by student
Rule	Recall of subordinate concepts and rules Demonstration of rule, using verbal statement Demonstration of rule application by student
Higher Order Rule	Recall of relevant subordinate rules Presentation of a novel learning task or problem Demonstration of new rule in problem solution by student
Cognitive Strategy	Recall of relevant rules and concepts Verbal statement or demonstration of strategy Practice of strategy in novel situations

requisite capabilities appropriate for each kind of learning outcome as *internal conditions*.

The result of this integrating exercise is a kind of checklist of distinctive conditions for effective learning that need to be incorporated into the general framework of instructional events in order to accomplish learning objectives. It may be noted that the checklist pertains only to the following events of instruction: Event 3, stimulating recall of prior learning; Event 4, presenting the stimulus; Event 5, providing learning guidance; and Event 6, eliciting performance. The remaining instructional events are described in Chapter 10.

Lessons for Intellectual Skill Objectives

Effective learning conditions for the varieties of intellectual skills that may be reflected in planning the events of a lesson are given in Table 12-1. Each list of conditions given in the second column begins with a statement designating the recall of a previously learned capability, often from a previous lesson in

TABLE 12-2. Effective Learning Conditions for Incorporation into Lessons Having Objectives of Verbal Information, Attitudes, and Motor Skills

Type of Lesson Objective	Learning Conditions
Verbal Information	
Names or Labels	Recall of verbal associations
	Encoding by student—relating name to image or meaningful sentence
	Using name in context of other knowledge
Facts	Recall of context of related meaningful information
	Performance of reinstating fact in larger context of verbal information
	Using fact in context of other knowledge
Knowledge	Recall of context of related information
	Performance of reinstating new knowledge in the context of related information
	Using knowledge by relating it to other facts and bodies of knowledge
Attitude	Recall of verbal information and intellectual skills relevant to chosen personal actions
	Establishment or recall of respect for "source" (usually a human model)
	Reward for personal action, either by direct experience or vicariously by observation of the human model
Motor Skill	Recall of responses and part skills
	Establishment or recall of executive subroutine (procedural rule)
	Practice of total skill

a sequence. The list then proceeds with conditions that are to be reflected in other instructional events (such as those of presenting the stimulus, providing learning guidance, eliciting the student performance, and so on). In interpreting the information in this column, the reader may find it useful to review the statements of internal and external conditions of learning for these types of objectives, as described in Chapters 4 and 5.

Lessons for Cognitive Strategy Objectives

Conditions designed to promote effective learning for cognitive strategies are listed in the bottom portion Table 12-1. The list pertains to the learning of strategies of learning, remembering, and problem solving. External and internal conditions for learning cognitive strategies were discussed previously in Chapter 3.

Lessons for Objectives of Information, Attitudes, and Motor Skills

The design of instructional events for lessons having one of the following objectives—verbal information, attitudes, or motor skills—needs to take into account the particular conditions for effective learning shown in the
corresponding portions of Table 12-2. These lists are derived from the fuller
discussion of internal and external conditions of learning contained in Chapter 4.

STEPS IN LESSON PLANNING

Assuming that a teacher has organized a course into major units or topics and
has further planned sequences of lessons for each, how does that teacher proceed with the design of a single lesson?

Following our emphasis upon providing for the events of instruction,
including the incorporation of effective learning conditions for the domain
represented in the objective of the lesson, we suggest that teachers employ a
planning sheet that will contain the following elements:

1. a statement of the objective of the lesson and its classification as to domain
   of learning outcome
2. a list of the instructional events to be employed
3. a list of the media, materials, and activities by which each event is to be accomplished
4. notes on teacher roles and activities (prescriptions for instruction)

Such a planning sheet might list the objective at the top, with a column
for each of the other three items in the previous list. After the planning sheet
is completed, a script for the lesson could be written. An example of a lesson
design is given in Table 12-3.

We will now describe some varieties of circumstances related to the four
elements of the planning sheet.

The Objective of the Lesson

As noted earlier, some lessons may have a single objective, whereas others
may include several related objectives. For example, a particular lesson may
be concerned with teaching a single enabling objective that appears in a learning hierarchy for a more complex intellectual skill objective. In the same lesson, however, the teacher may also incorporate one or more related
information objectives, and the total lesson may be intended to move the student a little closer to a longer term attitude or cognitive strategy objective. In
this sense, the instruction is "finished" in one lesson only for the enabling
objective and for the information objectives; it is not finished for the others.

TABLE 12-3. Lesson Planning Sheet for a Lesson on Sex-Linked Traits, Including Prescriptions for Media and Instructional Activities

Teaching Activity	Type of Stimuli	Possible Media	Prescription
(a)	spoken word pictures	teacher slides video tape	Gaining attention: Present a question to the student; e.g., "Why are more men color-blind than women? Why aren't more women bald? What determines if you will be bald when you get older?"
(b)	spoken word pictures	teacher slides video tape	Recall prerequisites: Review the concepts chromosome, trait, genes, recessive, and dominant, showing pictures with each definition.
(c)	spoken word written word	teacher handout video tape	Inform students of objective A and B: "In this lesson you will first learn what a sex-linked trait is, and how it is inherited."
(d)	spoken word written word	teacher handout video tape	State the definition of a sex-linked trait.
(e)	spoken word pictures	teacher video tape slides chalk boards	Present the objective, concept C, with new content showing relevant attributes of a chromosome pair with a sex-linked trait. Show three examples: X_cY X_cX X_cX_c. Indicate how each gene is carried on the X chromosome.
(f)	written word	worksheet	Elicit performance for concept C. "See if you can answer these questions: Does X_aY represent a sex-linked trait? Why? Does X_aX represent a sex-linked trait? Why? Does XY_a represent a sex-linked trait? Why?"
	spoken word writeen word	teacher handout	Review the correct answers with the students.
(g)	spoken word	teacher video tape	State objectives D and E: "Now you will learn why these traits are more common in men than in women; and how you can determine whether or not the trait will be *expressed*."

TABLE 12-3. *(Continued)*

Teaching Activity	Type of Stimuli	Possible Media	Prescription
(h)	spoken word illustrations	teacher handout chalk board	Show students how to develop a Pummet square for looking at possible outcomes of crosses:
(i)	spoken word	teacher handout	Present the Pummet square, filled in to show possible combinations of chromosomes and genes in offspring.
(j)	spoken word illustrations	teacher handout	Provide learning guidance, i.e., procedure for filling out the square, giving a number of different examples.
(k)	spoken word written word	teacher chalk board	Present the rule: "A sex-linked trait is always visible in a male because the Y chromosome does not mask the recessive gene on the X chromosome."
(1)	spoken word	teacher chalk board	Learning guidance: Show the Pummet square again. Demonstrate that the expression of a trait can be determined by looking at the cells. Give at least two examples from the following:
(m)	written word	worksheet	Have the students solve several problems applying the rule for determining expression of a trait. For any cross, ask: (1) Would a male exhibit the trait? (2) Would a female exhibit the trait? (3) What is the probability that a male or a female will not exhibit or carry the trait? For feedback, provide model answers to students.
(n)	written word	worksheet	Provide for retention and transfer by giving students word problems pertaining to sex-linked crosses. "If your mother's father was bald, what are your chances of being bald?" Discuss the relation of this lesson to work to be undertaken in subsequent lessons.

For activity (h), the Pummet square:

$$
\begin{array}{c|cc}
 & \multicolumn{2}{c}{\text{Female}} \\
 & X_c & X \\
\hline
\text{Male} \quad X & XX_c & XX \\
Y & X_cY & XY \\
\end{array}
$$

For activity (1), examples:

$$
\begin{array}{ll}
X_c\ X; & X_c\ \ Y \\
X\ \ X; & X_c\ \ Y \\
X_c\ X; & X\ \ Y \\
\end{array}
$$

Identify the concept *carrier*. Show that the recessive trait becomes expressed in the female only when both X chromosomes carry the recessive trait.

If some learners have not mastered the first two kinds of objectives by the conclusion of the lesson, it is not completed for them until further learning takes place.

As discussed in Chapter 9, it is useful to draw instructional maps (ICMs) in the process of planning sequences of lessons. These maps may be drawn at several levels, corresponding to the three levels at which the question of sequencing arises in the design of a course. Such maps may serve as a checklist of objectives in various domains so that none are omitted inadvertently in a series of lessons. Figure 12-2 illustrates a map for a lesson on the inheritance of sex-linked traits. In this lesson it is easy to see that the teacher would not

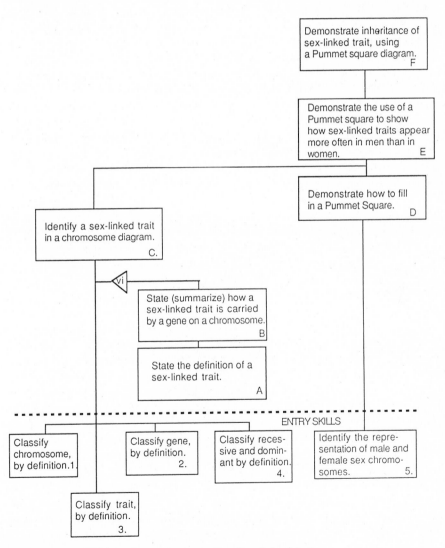

FIGURE 12-2   ICM for a lesson on genetics: Sex-linked traits.

blindly present all nine events of instruction for the information objectives and then all nine events for each intellectual skills objective. Rather, the teacher will probably group objectives and events of instruction into an instructional activity, and the lesson will be made up of many such instructional activities.

Listing the Instructional Events

The events of instruction are not all invariably found in a lesson, nor do they always appear in a fixed linear order. A teacher considers both the sophistication of the learners as self-directing learners and the nature of the objectives of the lesson.

Under some circumstances, an entire period may be taken for a single instructional event. Often a teacher will wisely spend an entire hour in attempting to establish motivation for a series of lessons on a task to be undertaken by the learners. Or, an hour may be required to present a complex objective to the students, including a discussion or a demonstration of its subordinate parts, to each of which the learners are to respond in some prescribed manner. For example, in a course in which objectives are written at the unit level, not at the lesson level, it may be reasonable to spend an hour or more clarifying the exact nature of the expected performance for each unit before the actual instruction for the unit is undertaken. This type of organization may be appropriate for a course whose major outcomes are problem-solving skills.

Although the events of instruction represent a crucial element in the design model presented in this book, it is also of great importance that the manner in which such events are planned reflect the best estimate of the capabilities and entering competencies of the students (See Chapter 6).

Choosing Media Materials and Activities

It is at this step in instructional design that the greatest differences may be noted between designing teacher-led and mediated instruction. Mediated instruction requires the designer to attend to all the events of instruction, as well as the manner in which they will be operationalized in the materials. The model for teacher-led instruction is much less exact, since the teacher fills in the gaps. However, the basic principle of planning the external events of instruction is the same in both. The question that must be answered is, "How will I accomplish this event with these students?" For example, when considering the event of gaining attention for an objective in a genetics unit in science for young children, the teacher might think, "I wish I could find a 16mm. film showing the various species of animals whose distinctive features are so exaggerated as to make them comical; I could use this to lead up to the objective about how genes determine these differences." Of course such a film may not be found; in that case the teacher would devise some other way to accomplish the event. Designers developing a module on the same topic would have to consider how they would accomplish the same event. Were they to decide upon the treatment using a film, they might consider producing that film.

Instructional Activities

The model for lesson design we propose has the teacher or designer determine a strategy for a lesson by grouping objectives and events of instruction into instructional activities. This is accomplished by listing the lesson's objectives vertically at the side of a sheet of paper. Across the bottom is a time line that depicts the desired length of the lesson. Within the matrix formed by the two axes of objectives and time are the events of instruction that make up the lesson, as shown in Figure 12-3. This figure pertains to a lesson on sex-linked traits, composed of six objectives labeled A through F.

Entry skills 1–5 are from previous lessons. They are to be reviewed to enhance retention and transfer (Event 9) for those skills and to recall prerequisites (Event 3) for the target objectives of the lesson. The event groupings are denoted by ellipses that enclose those events of instruction that are to be considered an *instructional activity*. Figure 12-3 shows these groupings, where the instructional activities are denoted by the lowercase letters (a)–(n) at the bottom of the figure, along a line denoting time.

As is evident from the figure, the first learning activity, (a), consists of instructional Event 1, gaining attention. The next instructional activity on the time line, (b), recalls the prerequisite skills learned in an earlier lesson. For these skills, the activity involved reflects Event 9, enhancing retention and transfer. The third activity, (c), informs the learners of the nature of objectives A and B. Notice that these are verbal information objectives. In this lesson, they are planned to be taught first, not because they are required as prerequisites, but because they provide a supportive context by aiding transfer to the learning of the main intellectual skill objectives (C, D, and E). The next activity, (d), presents the stimulus for the information objectives A and B.

Sometimes it is helpful to think of a group of events for a single objective as a single instructional activity, as illustrated for Event (f), in which furnishing learning guidance, eliciting performance, and providing feedback are planned to occur over a short time.

Sequencing Lesson Activities

Notice that the lesson described by the time and objectives matrix in Figure 12-3 can be described as a linear series of activities denoted by the lowercase letters (a)–(n). Within the framework of a lesson a teacher would have a fair amount of latitude with respect to which events to include, which to omit, and which to combine across objectives. However, sequencing considerations

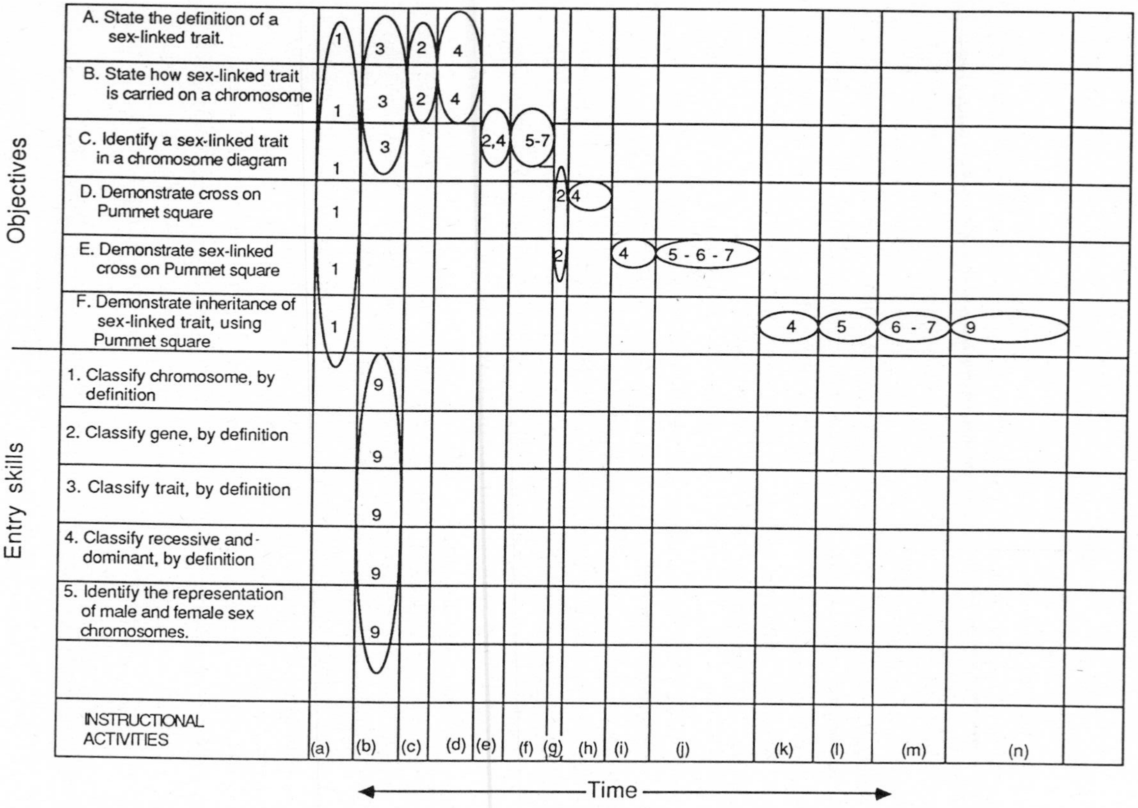

Objectives

A. State the definition of a sex-linked trait.

B. State how sex-linked trait is carried on a chromosome

C. Identify a sex-linked trait in a chromosome diagram

D. Demonstrate cross on Pummet square

E. Demonstrate sex-linked cross on Pummet square

F. Demonstrate inheritance of sex-linked trait, using Pummet square

Entry skills

1. Classify chromosome, by definition

2. Classify gene, by definition

3. Classify trait, by definition

4. Classify recessive and-dominant, by definition

5. Identify the representation of male and female sex chromosomes.

INSTRUCTIONAL ACTIVITIES

(a) (b) (c) (d) (e) (f) (g) (h) (i) (j) (k) (l) (m) (n)

←—— Time ——→

FIGURE 12-3   Time sequence of instructional activities (a-n) for a lesson on the inheritance of sex-linked traits, showing relationships of activities to instructional events 1 through 9.

related to prerequisite skills suggest that a most efficient strategy is that in which content will first be presented for the supportive objectives, followed by content related to target objectives.

The next step in our model for designing instruction is writing the prescriptions that will describe what will take place in each of the activities in the lesson. It is at this point that consideration must be given to the capabilities of media and delivery systems to provide the events constituting the lesson. Consistent with our model is the principle that the effectiveness of instruction depends upon the ability of the media employed to provide the events of instruction, in the manner required by the type of learning outcomes and by the sophistication of the particular learners. The fact that most media research shows no significant differences with regard to learning may speak more to the lack of consideration in the research of *when* media differences are important (that is, when the events being presented make a significant difference) than it does to whether media differences exist (Reiser and Gagné, 1983).

Prescriptions can be related to lesson activities and media on the lesson planning sheet. Table 12-3 illustrates a sample planning sheet that has the activities listed vertically on the left-hand side of the paper and the media and prescriptions listed in adjacent columns.

If the prescriptions are written before media of instruction are chosen, designers are left with a wide latitude within which they can make media decisions. This has been referred to as an open-media model of instructional design (Briggs and Wager, 1981). In contrast, if the medium is preselected, the prescription will have to take into consideration the limitations of the medium to provide the events that constitute the learning activity. Table 12-3 gives a brief example of how each of the prescriptions might be written in a classroom delivery system with a teacher. If this lesson were to be mediated, the teacher and film might possibly be replaced by videotape, and in this case the prescriptions would look very much the same. If the lesson were to be mediated in the form of printed text, however, the prescriptions would need to change substantially. As discussed in Chapter 11, media decisions become most important when attending to whether the learners are readers or nonreaders and the consequent requirements for visual display, practice, and feedback.

ROLES AND ACTIVITIES IN INSTRUCTIONAL DEVELOPMENT

There are a great many mediated instructional materials on the market. To the teacher, these items have value only in terms of the learning activities to which they may apply. To make the best use of these materials, the teacher needs to study them carefully. She must note particularly the instructional events they *do not* appear to address, so that plans can be made for such events in the lesson plan. The aim of this teacher activity is to produce a lesson plan in which all the needed instructional events occur.

When new instructional materials are developed, the instructional designer and the subject matter expert (SME) work together in analyzing the learning task, deciding on an appropriate delivery system and preparing prescriptions for the lessons in a course of study. In this process both the designer and the SME review existing materials and assess their appropriateness for use within an intended course. Then, like the teacher, they undertake to determine to which events, learning activities, and lessons these existing materials pertain. At this point the designer must determine how the remaining events or activities can be provided. Since the product of most instructional design projects is mediated instruction, the designer must be concerned with how the media chosen can be most appropriately used to support the instructional events. There is unlikely to be a "cookbook recipe" for developing instructional lessons. The best one may hope for is an understanding of the model summarized in Tables 12-1 and 12-2. This model indicates how particular kinds of lesson activities provide the events that support learning. Examples of applications of this model are given in Briggs and Wager (1981).

SUMMARY

This chapter has dealt with lesson planning as the accomplishment of two major activities: (1) planning for sequences of lessons within a course, unit, or topic, and (2) design of individual lessons in such a way that effective conditions of learning can be incorporated into the instructional events of each lesson.

The determination of sequences of lessons was discussed separately for each domain of learning outcome. The use of learning hierarchies was shown to be of central importance in the design of sequences of lessons for intellectual skill objectives, whereas other considerations enter into sequencing decisions for other kinds of outcomes.

To make each instructional event in the lesson successful, the conditions of learning relevant to the outcome (represented by the lesson objective) must be incorporated into the lesson. Although intuition, ingenuity, creativity, and experience are all valuable when planning lessons, reference to relevant conditions of learning can sharpen instruction and avoid neglect of some of the desirable functions of instruction.

Four steps in lesson planning were discussed. These include: (1) listing the objective(s) of the lesson, (2) listing desired instructional events, (3) choosing materials and activities, and (4) noting roles for teachers and designers. An example of a multiobjective lesson plan for a study of the inheritance of sex-linked traits provides an indication of the time scheduling of instructional events and prescriptions for teacher activities.

Up to the point of lesson planning, all stages of instructional design can be similarly done whether a team is designing an entire curriculum or a teacher is designing a course. At the point of lesson design, however, teachers must consider what they personally will bring to the lesson (and what role they will play), whereas designers of mediated materials must decide how to provide needed activities in a preplanned lesson. The purpose of both designs is the same—to incorporate effective conditions of learning into the instructional events for all lessons and modules.

REFERENCES

Anderson, R. C. (1984). Some reflections on the acquisition of knowledge. *Educational Researcher, 13,* 5–10.

Ausubel, D. P. (1968). *Educational psychology: A cognitive view.* New York: Holt, Rinehart and Winston.

Ausubel, D. P., Novak, J. D., & Hanesian, H. (1978). *Educational psychology: A cognitive view* (2nd ed.). New York: Holt, Rinehart and Winston.

Briggs, L. J. (Ed.). (1977). *Instructional design: Principles and applications.* Englewood Cliffs, NJ: Educational Technology Publications.

Briggs, L. J. & Wager, W. W. (1981). *Handbook of procedures for the design of instruction.* Englewood Cliffs, NJ: Educational Technology Publications.

Crovitz, H. E. (1970). *Galton's walk.* New York: Harper & Row.

Frase, L. T. (1970). Boundary conditions for mathemagenic behaviors. *Review of Educational Research, 40,* 337–347.

Gagné, R. M. (1985). *The conditions of learning* (4th ed.). New York: Holt, Rinehart and Winston.

Martin, B. L., & Briggs, L. J. (1986). *The affective and cognitive domains: Integration for instruction and research.* Englewood Cliffs, NJ: Educational Technology Publications.

Pressley, M., Levin, J. R., & Delany, H. D. (1982). The mnemonic keyword method. *Review of Educational Research, 52,* 61–91.

Reiser, R. & Gagné, R. M. (1983). *Selecting media for instruction.* Englewood Cliffs, NJ: Educational Technology Publications.

Rothkopf, E. Z. (1970). The concept of mathemagenic behavior. *Review of Educational Research, 40,* 325–336.

Wager, W. (1975). Media selection in the affective domain: A further interpretation of Dale's cone of experience for cognitive and affective learning. *Educational Technology 15* (7), 9–13.

Wager, W. (1977). Instructional design and higher education. In L. J. Briggs (Ed.) *Instructional design.* Englewood Cliffs, NJ: Educational Technology Publications.

13

ASSESSING STUDENT PERFORMANCE

Instruction is designed to bring about the learning of several kinds of capabilities. These are evidenced by improved performance on the part of the student. Although much learning goes on outside the school and much results from the student's own effort, the responsibility of the school is to organize and provide instruction directed toward specific goals—goals which might not be achieved in a less organized manner.

The outcomes of this planned instruction consist of student performances that show that various kinds of capabilities have been acquired. Five domains of such capabilites have been identified and discussed in previous chapters: intellectual skills, cognitive strategies, information, motor skills, and attitudes. Performance objectives in these categories, applicable to a course of instruction, may be further analyzed to discover the prerequisites on which their learning depends. These in turn can form a basis for deciding upon a sequence of individual lessons and for the design of the lessons themselves.

Both the designer of instruction and the teacher need a way to determine how successful the instruction has been, in terms of the performance of each individual student and entire groups of students. There is a need to assess student performance to determine whether the newly designed instruction has met its design objectives. Assessment may also be done to learn whether each student has achieved the set of capabilities defined by the instructional objectives. In this chapter we will discuss how both these purposes can be served by the development of procedures for assessing student performance.

PURPOSES OF PERFORMANCE MEASURES

In chapter 2, we pointed out that measures of student performance may have as many as five different purposes, when considered in relation to instruction in schools. These are briefly discussed in the following sections.

Student Placement

When students return to school after each summer vacation, they will have forgotten some skills learned the previous year, and they will have acquired new information, skills, and attitudes. Even in the unlikely event that the members of a group left school the previous term with highly similar capabilities, they will not all be at the same starting point in respect to the sequence of skills to be learned in a new school year.

Placement tests are used to determine just which skills in a sequence each student has learned and can recall at the time the tests are administered (usually soon after the beginning of a new term). The results of such tests show the pattern of each student's areas of mastery and nonmastery, for the purpose of identifying the starting points for instruction. Programs of individualized instruction (Chapter 15) are well designed for this purpose. Under group instruction the teacher will need to arrange some activities for students who need to catch up or who need to work ahead of the majority. The more suitable the provisions made for each learner, the more precise the instruction will be and the more likely the students are to experience success.

Diagnosis of Difficulties

Diagnostic tests may be constructed to measure the prerequisite skills revealed by learning hierarchies designed to represent the essential parts of total skills. These tests are especially helpful when some learners are seen to be "falling behind." A probable reason for falling behind, especially in group instruction, is that earlier skills in a sequence have not been mastered, thus making it difficult to learn superordinate skills. Based on the results of a diagnostic test, individual learners can be given remedial instruction on prerequisite skills. In some instances, of course, the remedial instruction may need to employ methods and materials different from those originally used in order to avoid a second failure at the same trouble point in a sequence of lessons.

Checking Student Progress

Performance tests are often administered after each lesson in a series to assure that each student is mastering each objective. Teachers learn to use such tests less often when an entire group consistently progresses well, and to use them more often when a number of learners are experiencing difficulties. Of course some progress checking is often done informally by the teacher, in making spot checks with a few learners on each occasion. But in individualized instruction programs, such as those presented by computer-based instruction, these tests are typically a part of each "module." Testing of this frequent sort soon shows that a learner is keeping up or falling behind. By using brief progress checks consistent with assessment adequacy, learners can receive assurance that they are progressing well. The results of such tests also represent dependable information for the teacher to use in planning the next steps in instruction.

For advanced learners, as in universities and colleges, progress checks are usually made less often. Some college instructors give weekly tests, but others may use only final examinations in their courses. In these settings also, the use of computer-based instruction introduces a desirably frequent progress-checking routine.

Reports to Parents

The use of performance measures not only assures both learners and teachers that all is well, it also constitutes a dependable basis for reporting progress to parents and administrators. The results of accumulated progress checks may provide a basis for promotion, for certification, or for admission to higher institutions of learning.

Evaluation of the Instruction

Another important purpose of performance testing is to evaluate and improve the instruction itself. In recent years it is not uncommon for the materials of instruction to have undergone *formative evaluation*—tryouts and subsequent revisions of the materials with individuals, with small groups, and with large groups in field test situations. For this purpose the total scores earned by each student on performance measures are of interest in indicating the overall degree of success attained. Even more important are item analyses showing which items are passed or failed by a majority of students. Item scores are particularly useful in deciding where the instruction needs to be improved. Techniques for formative evaluation are further discussed in Chapter 16 (see also Dick and Carey, 1985).

Performance tests are also used in conducting *summative evaluations* of the instruction. These evaluations are conducted after course revisions have been accomplished and after the resulting form of the course has been used for additional groups of learners. Procedures for summative evaluation are described in detail by Popham (1975), Dick and Carey (1985), and in Chapter 16.

The principles for preparing performance measures are similar whether applied to the construction of tests for each prerequisite skill in a learning hierarchy, for an entire topic, or for a larger unit of study. In the remainder of this chapter we will discuss tests in terms of their validity for measuring performance on a single objective. However, the single objective could be that of a course, a unit, a single lesson, or an enabling objective.

PROCEDURES FOR OBJECTIVE-REFERENCED ASSESSMENT

The phrase *objective-referenced assessment* is used with a literal meaning in the context of this book. It is intended to imply that the way to assess student learning is to build tests or other assessment procedures that directly measure the human performances described in the objectives for the course. Such measures of performance make it possible to infer that the intended performance capability has indeed been developed as a result of the instruction provided. Similar tests can be administered before the instruction is given (pretests), and provisions made to allow students to bypass instruction they do not need. Normally, a teacher tests only for the "assumed entering capabilities" before introducing the instruction and assesses performance on the objective itself only following instruction (that is, by a posttest). A convenient compromise practice may be for the teacher to permit any student who thinks he has mastered the objective before instuction to take the test reflecting that objective as a pretest, and to excuse the student from that portion of the instruction if he passes it.

The performance objective is the keystone in planning assessment of performance. We have indicated the critical importance of the *verbs* in the statement for correctly describing the objective (Chapter 7). The verbs are equally crucial as a basis for planning the performance assessment. Such verbs tell what the student should be asked to *do* when taking the performance assessment test. Note that the *capability* verb refers to the capability that is *inferred* to be present in the student's repertoire, when the student has successfully performed as stated in the *action* verb in the objective. The capability verb is the *intent* of the objective; the action verb is the *indicator* that the intent has been achieved by the learner.

Congruence of Objective and Test: Validity

The objective-referenced orientation to assessment greatly simplifies the concept of *validity* in performance measurement. This approach to assessment results in a direct rather than an indirect measure of the objective. Thus it eliminates the need to relate the measures obtained to a criterion by means of a correlation coefficient, as must usually be done when indirect measures are used or when tests have been constructed without reference to any explicit performance objectives. Accordingly, one can address the matter of test validity by ansering this question: "Is the performance required during assessment the same performance as that described in the objective?" If the answer is a clear yes, then the test is valid. In practice, it is desirable for more than one person to make this judgment and for consistency to be attained among these judges.

Validity is assured when the assessment procedure results in measurement of the performance described in the objective. This occurs when the test and the objective are *congruent* with each other. A caution may be interjected here, however. This method of determining validity assumes that the statement of the objective is itself valid, in the sense that it truly reflects the purposes of the topic or lesson. The procedures described for defining objectives in Chapter 7 are intended to ensure that this is the case. Nevertheless, there may be an additional need to recheck the consistency of specific objectives and more broadly stated purposes. Sometimes, inconsistencies become apparent when statements of objectives are transformed into tests of student performance.

It should be recognized that the word *test* is used here in a generic sense to mean any procedure for assessing the performance described in an objective. Thus the use of this word can cover all forms of written and oral testing as well as procedures for evaluating student products such as essays, musical productions, constructed models, or works of art. We choose the term *assessment* rather than the alternative *achievement testing,* to refer to the measurement of student performance. The latter term is often associated with norm-referenced measurement, which will be the subject of a later discussion in this chapter. At this point, however, *test* and *assessment* are used to refer to objective-referenced performance measurement.

Some of the performance objectives given in Chapter 7 can be used to illustrate how judgments about test validity may be made. Initially, we shall be concerned primarily with two of the five parts of an objective statement, the two verbs that describe the *capability* to be learned and the *actions* the student takes in demonstrating this capability. Later on, other parts of the objective will be related to performance assessment.

First, consider the example of *generating* a letter by typing. The word *generate* is the clue that in the test situation the student must compose his own letter, rather than typing a different form of a letter composed by someone else. It is clear that the learner must use his capability of generating a particular kind of letter within the constraints of the situation described in the objective. In the alternative objective relating to typing, the learner receives a written longhand letter composed by someone else. These two objectives relating to business letters are very different. One requires only the skill of typing a letter already composed, whereas the other requires also the problem-solving capability of composing the letter. Thus two domains (motor skills and intellectual skills) are sampled.

In a second example drawn from Chapter 7, the student must *demonstrate* the use of a rule by supplying the missing factor in an equation. Simply copying the missing value from a book or remembering the value from having seen the problem worked before would not constitute a valid test for this capability. In designing a test, care must be taken to use different examples for testing than those used for teaching, so as to minimize the chance that the correct response can be supplied by any means other than the intended intellectual process.

In any example of demonstrating mastery of a *concept,* the learner may *identify* the concept by printing the first letter of the concept (name) in a blank. This is not the same as either copying the first letter, or spelling the name of the concept. It is also different from the performance of explaining how the concept may be used. Any or all of these latter instances may be useful performances, but they do not reflect the intent of the objective, either as to the capability required or the action signifying that the capability is present.

Exercises on judging the validity of test items by comparing them with the corresponding performance objectives are given by Briggs and Wager (1981) and by Dick and Carey (1985).

Designing the Test Situation

The form of performance objectives described in Chapter 7 serves as the basis for the test situation. It will be recalled that the five components of an objective statement are given as: (1) situation, (2) learned capability, (3) object, (4) action, and (5) tools and constraints. An objective statement also provides a description of the situation to be used in testing.

For certain types of objectives and for learners who are not too young, the change of only a few words may convert the objective statement to a test. For example, one could give the objective on generating and typing a letter to the learner as "directions for taking the test." All that would be needed in addition would be to supply the "received letter," and to provide an electric typewriter, typing paper, and carbon paper. The person administering the test would further be instructed to ensure a favorable (monitored) environment and to record and call "time." For the objective of demonstrating the procedure of short division, about all the test administrator (teacher) would need to do would be to supply a division expression in the form *abc/d* and make clear where the students were expected to write their answers.

It is clear, then, that the more closely objectives follow the outline given in Chapter 7, the fewer decisions remain to be made in planning the test and the fewer directions must be given to the student. Statements of objectives prepared for the use of the instructional designer or teacher are also used to define most of the test situation for the student. Of course, both objectives and test items derived from them have to be presented in simpler terms for young children, either for communicating to them the purpose of the lesson or for testing their performance after the lesson is completed.

Some Cautions

In using objectives to plan tests, a few cautions should be noted. The more incomplete the statements of objectives, the more these cautions may be needed, because more must be "filled in" in moving from the objective to the test situation.

1. Avoid substituting verbs that change the meaning of either the *capability* or the *action* described in the objective. When synonyms or more simple explanations are needed to translate the objective into a test, these restatements must be reviewed for agreement with the intent of the objective. Particular care should be taken not to change from an answer the student must somehow construct or develop for himself, to an answer he must merely choose, select, or recall. If an objective says "generate a position and a defense for the position," he can only do this orally or in writing—not by selecting answers from a multiple-choice test. Avoidance of amibiguity in "guessing at" what vague verbs mean in poorly stated objectives can be achieved by using the standard verbs from the Table 7-1. But careful attention needs to be given to deciding upon unambiguous meanings for verbs such as summarize, describe, list, analyze, and complete, except as *ing* verbs denoting the particular action expected. Review of an objective in these terms sometimes reveals that the objective itself needs to be changed. In that case, it should be changed before planning the instruction and before using the statement either as a lesson objective or as a part of the directions for a test.

2. Changes in other elements of the objective should be avoided, except when needed to simplify directions for the student on how to take the test. That is, unless a deliberate change is intended, the situation, the object, and the tools and other constraints, as well as the two verbs denoting the capability and the action, should be congruent between the objective and the test. It is possible that changes might be so great as to make the test call for capabilities the students have not yet been taught. In a "worst possible" mismatch between objective and test, capabilities in different domains of learning outcomes might be specified in the objective and in the test. In such a situation, if the teaching were to be directed to an objective in still a third domain, there would be maximum incongruence among the three anchor points. It might be revealing to ask teachers or designers, on three separate occasions, to produce their *objectives*, their *examinations*, and their *lesson plans*. It is conceivable that the objectives might call for "appreciation" while the teaching contains "facts," and the examination calls for the "use of concepts and rules".

3. Tests should not be made either easier or more difficult than the objectives. These terms need not enter into testing of the objective-referenced variety. The aim is one of accurately representing the objective, rather than one of estimating how to make tests sufficiently difficult.

4. The test should not try to achieve a large range in scores or a normal distribution of scores. The aim of such testing is not that of discriminating among the students. That is to say, testing does not have the purpose of finding that

one student scores higher or lower than another. Rather, its purpose is to discover which objectives both students have learned.

THE CONCEPT OF MASTERY

The introduction of the idea of *mastery* of learning outcomes (Bloom, 1968) requires a change in viewpoint towards the conduct of instruction, as well as towards its assessment. In conventional instruction, both the teacher and the students expect that only a few students will learn so well as to receive an *A* in the topic or course. The rest will either do fairly well, as represented by a *C*, for example, or they will fail. When test scores are plotted as frequency distributions, a normal curve is formed, and certain percentages of students are assigned to various letter grades.

In commenting on the impact of this system of assessment, Bloom, Hastings, and Madaus (1971, p. 43) observe that the expectations so established tend to fix the academic goals of teachers and students at inappropriately low levels, thus reducing both teacher and student motivation. The particular educational practice that produces these effects is "group-paced" instruction, in which all students must try to learn at the same rate and by the same mode of instruction. When both pace and mode are fixed, the achievement of each student becomes primarily a function of his aptitude. But if both mode and rate of instruction can vary among learners, the chances are that more students can become successful in their learning (Block and Anderson, 1975).

It is easier to set up means by which the *rate* of learning is allowed to vary among learners than it is to predict the *mode* of learning which will benefit each student the most. And of course there are economic and other limits— one cannot provide a different mode for every single student. Modularized, individualized instruction can largely take care of the rate problem, and to some extent (when alternative materials or modes are available) the problem of learning style as well. The diagnostic features of individualized assessment also make it possible to help a student redirect his efforts properly.

Mastery learning means essentially that if the proper conditions can be provided, perhaps 90 to 95 percent of the students can actually master most objectives to the degree now only reached by "good students." Thus the mastery learning concept abandons the idea that students merely learn more or less well. Rather, an effort is made to find out why students fail to reach mastery, and to remedy the situation for such students. The resolution of a learning problem by a student usually requires one of the following measures: (1) more time for learning, (2) different media or materials, or (3) diagnosis to determine what missing prerequisite knowledge or skills he must acquire to master the objective. Within this context, the personal knowledge of the teacher can be added to form decisions concerning students whose performance is exceptional even when these methods have been fully utilized. The general aim implied by the notion of mastery includes the resolution to provide materials and conditions by means of which most learners can be successful at most tasks, in a program that is reasonable for each individual.

Determining Criteria for Mastery

How can it be known when a student has performed satisfactorily or attained mastery on a test applicable to any particular objective? The student needs to be told he was successful, so that he can then move on to work toward achieving the next objective he chooses or has assigned to him. In case he has not been successful in attaining the objective, the teacher needs to determine what remedial instruction is needed.

A remedial decision for an objective in the intellectual skill domain can best be made by administering a diagnostic test over the capabilities subordinate to the objective. In other instances, the teacher may use oral testing methods to find out where in the teaching sequence the failure to learn first began. When instruction is individualized, the individual lessons often include such diagnostic tests on subordinate capabilities. For a known slow learner, such diagnostic tests of subordinate competencies may be used as assessments of performance so that the learner is known to have mastered each capability before he goes on to the next. This procedure detects small failures before they accumulate into large failures of entire lessons, topics, or courses. Certainly, consistent use of frequent testing could often prevent the year-after-year failures, or at least alert the school earlier to a need to reappraise the program for a particular student.

When *mastery* is defined for a test assessing performance on an objective, this also defines the criterion of success for that objective. The first step is to define *how well* the learner must perform on the test to indicate success of that objective. Then a record is made of *how many* students have reached the criterion (mastery). This makes it possible to decide whether the instruction for that objective has reached its design objective. Later, at the end of an entire course, the percentage of students who reached the criterion of mastery of all the objectives (or any specified percentage of the objectives) can be computed. From such data, one can determine whether the course design criterion has been met. A frequently used course design criterion is that 90 percent of the students achieve mastery of 90 percent of the objectives, but other percentages than these may of course be used. Sometimes three design criteria are set, with one indicating minimal acceptable success, and the others representing higher degrees of success. In general, this means of representing course design criteria can be used to give *accountability* for the performance of students following instruction.

The administration of tests applicable to course objectives and the definition of mastery level for each objective provide the means for evaluating both the course itself and the performance of individual students. Thus students can be promoted on the basis of such tests, and the test results can be used in the *formative evaluation* of the course, showing where revisions are needed, if any (see Chapter 16). This built-in capability for course improvement is compatible not only with fair promotion standards for students, but also with the individualization of instruction and with the development and evaluation of entire instructional systems.

Although the action of defining mastery on each objective, when objective-referenced tests are employed, is intended primarily for the purpose of monitoring student progress and for discovering how successful the course is, data from the same tests can be used for assigning grades when that is required by the school.

CRITERIA FOR OBJECTIVE-REFERENCED ASSESSMENT

The question to be addressed next concerns the matter of deciding upon criteria of mastery for each kind of learning objective. Typical procedures for each domain of learning outcomes are described in the following section. More extensively described procedures for criterion-referenced testing may be found in Berk (1984).

Intellectual Skill Objectives

Problem Solving

As an illustration of criteria for assessing performance for this type of learning outcome, we begin with an objective for the acquisition of a capability in *problem solving,* briefly described in Table 7-1. The statement of this objective is "generates, by synthesizing applicable rules, a paragraph describing a person's actions in a situation of fear."

To score such a paragraph as acceptable, a list of features the paragraph should include would be prepared. For this kind of objective, no verbatim key is possible, and mechanical scoring does not appear to be feasible. Since no grammatical requirements are included in this abbreviated statement of the objective, it may be assumed that an adequate description need not be error-free in grammar and punctuation. If several teachers are using the same objective, they might work together to define assessment criteria and to agree upon how many actions must be described and what aspects of fear are to be included. The minimum number of rules to be synthesized in the solution could be agreed upon, with the application of some rules being mandatory and others optional.

The test for a problem-solving objective is not to be based on a judgment such as "8 out of 10 questions correct." The criteria to be employed may be both qualitative and quantitative in nature. Whatever the checklist for scoring may contain, its application will require judgment, not a clerical checking of answers with an answer key. Consequently, degree of agreement among teachers in applying the checklist to determine acceptable or unacceptable paragraphs is a relevant factor in determining the reliability of the measure of performance obtained. The criteria employed for judging such a performance might be (1) facial expression, (2) bodily reaction, and (3) two statements of rules governing emotional expression in behavior.

Rule Learning

For the learning of a rule, the example given in Table 7-1 is "demonstrates, by solving verbally stated examples, the addition of positive and negative numbers." To examine the matter of performance criteria more exactly, one needs to begin with an expanded version of this objective: "Given verbally stated examples involving physical variables that vary over a range of positive and negative values, demonstrates the addition of these values by writing appropriate mathematical expressions yielding their sum." Obviously, this more complete statement adds to the specification of the situation, and therefore to the adequate formulation of a test item. Such an item, for example, might say, "The temperature in Greenland on one day was 17°C during the day and decreased by 57° during the night. What was the nighttime temperature?"

Thus the *situation* part of the objective statement defines the class of situations from which particular test items are to be drawn. Suppose the objective is: "Given a verbal statement defining values of length and width of a rectangularly shaped face of an object, finds the area of the face." From such a statement, an item such as the following could readily be derived: "A box top is 120 cm in length and 47 cm in width; what is its area?" It may be noted that the statement of the objective in this case implies that the performance will be assured in a situation including a verbal statement of the problem. A different statement beginning "Given a diagram of a rectangle with values of length and width indicated . . ." would of course imply a different form of test item.

A remaining decision pertaining to the criterion of performance measurement has to do with the question of how many items to employ. Obviously, the aim is to achieve a genuine measurement of "mastered" versus "not mastered." It may need to be determined empirically how many items must be used in order to make such a decision correctly. By convention, 10 or 20 items might be considered necessary as a number of examples for a test of the learning of an arithmetic rule. However, Lathrop (1983) has shown that, by using some reasonable assumptions of sequential analysis (Wald, 1947), decisions about mastery and nonmastery can be made on the basis of a sequence of as few as three correct items. The aim in using multiple examples is primarily one of avoiding errors of measurement, which may arise because of one or more undesirable idiosyncratic features of a single item. Additional procedures for determining desirable test length are described in the book edited by Berk (1984).

Defined Concepts

To derive an illustration of performance criteria for the measurement of a *defined concept,* the following example of an objective may be used: "Given a picture of an observer on the earth and the sky above, classifies the *zenith* as the point in the sky vertically above the observer." Again it is apparent that the situation described in this statement may be directly represented in the form of a test item. For example, such an item might first depict (in labeled diagram) the earth, the sky, and an observer standing on the earth. Going on, it could say: "Show by an angular diagram the location of the *zenith.*" For answer, the student would draw a vertical line pointing from the observer to the sky, indicate that it made a 90 degree angle with the earth's surface at the point at which the observer was located, and label the point in the sky to which the line was directed as the zenith.

An item of this type would not be highly dependent on the verbal abilities of the student and might be a desirable form of measurement for that reason. Alternatively, providing you could assume the student's verbal facility, an item might be based upon a differently stated objective, as follows: "Asked to define, classifies *zenith* as the point in the sky vertically (or 90° to the surface) above an observer on the earth by stating a definition orally." It is evident that measurement in this case is subject to distortion. Unless one is entirely convinced that the student has mastered the subordinate concepts (earth, sky, observer, 90°), the resulting response of the student may have to be interpreted as a memorized verbalization. Nevertheless, it is noteworthy that verbal statements (preferably in the learner's own words) are often employed as criteria for assessment of defined concepts.

Concrete Concepts

The assessment of *concrete concept* learning involves the construction of items from an objective statement like the following: "Given five common plants and asked to name the major parts, identifies for each the root, leaf, and stem by pointing to each while naming it." For such an assessment to be made, the student would be provided with five plants laid out on the table and, in response to the teacher's question, would point to and name the root, leaf, and stem for each plant. Of course, an objective with a somewhat different statement of the situation would lead to a corresponding difference in the test item. For example, the objective statement: "Given pictures of five common plants, identifies the root, leaf, and stem of each by placing labels bearing these names opposite the appropriate part," implies a specifically different kind of test item. Whereas the example assumes only that the oral responses *root*, *leaf*, and *stem* can be made without error, the latter example requires the assumption that the labels containing these words can be read.

A simple example of assessment for concrete concept is provided by the task of identifying a common geometrical shape, as it may occur in an early grade. The objective statement might read: "Given a set of common geometrical shapes and the oral directions 'show me the circles,' identifies the circles by pointing." From this statement an assessment item may be derived that involves giving the student a piece of paper on which figures such as the following appear:

○ □ △ ○ ○ □ ◯ ▭ ○ △ ○

Upon being given the oral directions, "Point to each one that is a circle," the student would make the appropriate response to each circular figure and not to other figures, in order to be counted as having attained the concept.

Discrimination

Assessing discrimination requires the presentation of stimuli to which the learner responds in a way that indicates *same* or *different*. The example in Table 7-1 is "discriminates, by matching the French sounds of *u* and *ou*." To represent this objective as a test item, it would be necessary to present the sounds of a number of French syllables or words containing these vowels (as in *rue* and *roux*) and ask for an indication of *same* or *different* by the student.

An example of visual discrimination would be provided by items that presented figures such as the following to be matched to the model:

Model: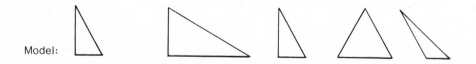

The directions for an item of this sort would be "Circle the figure or figures that match the model." It may be noted that discrimination tasks are purely perceptual; they do not require that the learner name the stimulus or identify its attributes. What is assessed is simply perceiving a difference or no difference.

Cognitive Strategies

In contrast to the techniques of assessment for intellectual skills, the indicators of cognitive strategies are somewhat indirect and often require a longer chain of inference. For example, if the strategy of Table 7-1, "adopts the imagining of a U.S. map to recall the names of the states," is employed, the observed performance will be a list of the states. However, such a list could be given by a learner who used quite a different cognitive strategy, possibly one that was less efficient (such as a strategy of reporting the states systematically by the beginning letters of their names). Thus, the performance by itself fails to indicate the adoption and use of a particular strategy. Assessment of the imaging strategy would require the additional observations that the states were named in a sequence indicating regional locations and also that this strategy yielded an efficient performance.

Several different strategies of solving geometric problems involving relations between angles of a complex figure were studied by Greeno (1978; see also, Gagné, 1985, pp. 143–145). Here, too, the strategies could not be revealed simply by the successful solution of the geometric problem. Instead, they were indicated by verbal reports of the students themselves, who were asked to "think aloud" while working on the problem.

Although it would seem desirable to extend the notion of mastery learning to all domains of instructional objectives, its application to the measurement of cognitive strategies cannot readily be achieved. Whether the strategies we deal with are those that primarily control the processes of attending, encoding, retrieving, or problem solving, it is apparent that the *quality* of the mental process is being assessed, and not simply its presence or absence. Sometimes, novel problems have many solutions, rather than a single solution. In such instances, cognitive strategies will have been used by the student in achieving the solution, whatever it may be. Accordingly, assessment becomes a matter of judging how good the solution is, and it is unlikely that a "pass-fail" decision will be made.

It is noteworthy that standards of originality and inventiveness are applied to the assessment of such student products as theses and dissertations in university undergraduate and graduate education. Besides being thorough and technically sound, a doctoral dissertation is expected to make "an original discovery or contribution" to a field of systematic knowledge. Exact criteria or dimensions for judging this quality are typically not specified. Varying numbers of professionally qualified people usually arrive at a consensus concerning the degree of originality exhibited by a dissertation study, and its acceptability as a novel contribution to an area of knowledge or art.

Productive Thinking

The measurement of productive thinking, and by inference the cognitive strategies that underlie such thinking, has been investigated by Johnson and Kidder (1972) in undergraduate psychology classes. Students were asked to invent novel hypotheses, questions, and answers in response to problem statements which go beyond the information obtained from lectures and textbooks. The problems employed included (1) predicting the consequences of an unusual psychological event, (2) writing an imaginative sentence incorporating several newly learned (specified) concepts, (3) stating a novel hypothesis related to a described situation, (4) writing a title for a table containing behavioral data, and (5) drawing conclusions from a table or graph. When items such as these were combined into tests containing 10 to 15 items, reasonably adequate reliabilities of *originality* scores were obtained. Quality was judged by two raters whose judgments were found to agree highly after a short period of training.

Assessments of originality can presumably be made of students' answers, compositions, and projects at precollege levels. In fact, such judgments are often made by teachers incidentally, or at any rate informally, concerning a variety of projects and problems undertaken by students in schools. It seems evident that systematic methods of assessment can be applied to cognitive strategies at these lower levels of the educational ladder, although this has not as yet been done.

It should be pointed out that the assessment of cognitive strategies or the originality of thought as an *outcome* of learning does not necessarily have the same aim nor use the same methods as those employed in the measurement of creativity as a *trait*. Creativity has been extensively studied in this latter sense (Torrance, 1963; Guilford, 1967; Johnson, 1972), and the findings go far beyond the scope of the present discussion. When assessment of the quality of thought is to be undertaken as a learning outcome, two main characteristics must be sought. First, the problem (or project) that is set for the student must require the utilization of knowledge, concepts, and rules that have been recently learned by the student, rather than calling upon instances of skills and information that may have been acquired an indeterminate number of years previously. Second, it must be either assumed or, preferably, shown that students have in fact learned relevant prerequisite information and skills before the assessment of originality is undertaken. This condition is necessary to ensure that all students have the same opportunity to be original and their solutions are not handicapped by the absence of necessary knowledge and intellectual skills.

Verbal Information

In this domain, the concept of mastery must be related to a predetermined set of facts, generalizations, or ideas, an acceptable number of which the student can state in acceptable form or degree of completeness and accuracy. The conventional *norm-referenced* measurement is often closely related to assessment of information. The fundamental distinction to be held in mind, however, is that of *objective*-versus *content-referenced* measurement. The aim of assessment is to determine whether certain objectives have been attained, rather than to discover whether some content has been covered.

Objective-referenced assessment may be achieved for the information domain of learning outcomes by specifying what information is to be learned as a minimum standard of performance. Objectives pertaining to information should state clearly *which* names, facts, and generalizations should be learned. They thus differentiate the core content of information to be recalled from the incidental information that may be in the book and that some students may be able to recall, but that represents learning beyond the required level.

It would be a mistake to make the objectives in the information domain so exhaustive as to leave no time for objectives in other domains. Instead, one should deliberately seek out and identify those informational outcomes that are likely to contribute most to the *attainment of objectives in other domains*. Although masses of information should be acquired over years by the well-educated person, this goal should not be allowed to interfere with the attainment of objectives in the areas of intellectual skills and problem-solving strategies.

Typically, assessing the learning of verbal information means measuring quantity (Gagné and Beard, 1978). The intention is to assess *how much* the student knows about some particular historical event or era, or about a natural phenomenon like earthquakes. How much do students know about the varieties of oak trees or about the cutting of logs into lumber? Answers to quantity questions come from items selected from a domain that is more or less well defined. It may be as precisely defined as a specific longer prose passage. Or it may be more loosely defined as the declarative knowledge a student is expected to learn from lectures, text, and other references available on a particular subject.

A variety of methods have been proposed for the analysis of prose texts displaying verbal knowledge (Britton and Black, 1985). Some of these proposals suggest the possibility that the *quality* of knowledge can be assessed as a learning outcome. It is possible that some kinds of memory organization resulting from learning represent "deeper understanding" of verbal information. It is of course possible to distinguish main ideas from subordinate ideas by the use of these methods. But the fuller meaning of quality or depth of knowledge remains to be shown by research and theoretical development, before measures of this aspect of verbal information can be developed.

Examples of Verbal Information Items

Some typical items for assessment of verbal information are:

1. Describe at least three of the causes of the American Revolution, as discussed in the textbook.

2. State the chemical name for the following substances: baking soda, blue vitriol, chalk, . . .
3. Write a paragraph summarizing how a president is elected when the electoral college fails to elect.
4. Name any 15 of these 20 animals from their pictures.
5. What is guaranteed by the Fourth Amendment to the U.S. Constitution?
6. Read this report and write a summary of the four main themes developed in the report.

As these examples indicate, objective-referenced testing of information requires the exact indentification of what information is to be learned and retained. If the listing of names or dates is to be acquired, this should be made clear. Alternatively, if the substance of a passage is to be recounted, this objective should be made equally apparent to the student. These procedures make learning for mastery feasible, as well as fair and reasonable.

Attitudes

As Chapter 5 has indicated, attitudes vary in the intensity with which they influence the choice of personal actions. Since the strength of attitudes is what one wishes to assess, it is evident that mastery cannot be identified. The assessment of stength of an attitude toward or against a class of choices of action may be obtained in terms of the proportion of times the person behaves in a given way in a sample of defined situations. For example, attitude toward using public transportation might be assessed by observing the likelihood of a student's choosing various forms of public (rather than private) transportation in the various situations in which such choices are made. The observed incidents would be the basis for inferring the degree to which the person *tends* to use or not to use public conveyances.

In assessing an attitude such as "concern for others," it is evident that no pass-fail criterion of mastery can be set. However, a teacher might adopt the objective that all her second-grade pupils will improve in this attitude during a year's period. In addition, it would be possible to adopt the standard that each child will exhibit concern for others, either in verbal expression or overt actions, more times per month in May than during the previous October. Anecdotal records may be kept recording such actions, and reports of "improvement" or "nonimprovement" made at the end of the school year. Such reports can be quantified in terms of number of positive actions and in terms of proportion of positive to total (positive plus negative) actions. Behaviors representing neither kind of action would simply not be recorded, in recognition of the fact that some of the child's time is spent in study periods offering little opportunity for behaving either way toward other people.

Attitudes are often measured by obtaining self-reports of the likelihood of actions, as opposed to direct observations of the actions themselves. As is well known, the most serious limitation in the use of questionnaires for this purpose is the possibility of bias resulting from the students' attempts to answer questions so as to win approval rather than reflecting their choices accurately. There appears to be no simple solution to the problem of obtaining truly accurate information from self-reports, although many investigations have been carried out for this purpose (cf. Fishbein, 1967). Best results appear to be achieved when students are first assured that the assessment being done is not intended as an adversary process; that is, that they need not report only what will (they think) be approved. When questionnaires are administered to groups, the additional precaution is frequently taken to ensure that responses are anonymously recorded.

As previously indicated, attitudes are best conceived and measured as a consistency in *choices of personal action* toward some class of objects, persons, or events (Chapter 5; see also Gagné, 1985). A domain of assessment items that defines these choices may be carefully specified along several dimensions (Triandis, 1964). For example, in assessing the choices made by whites in accepting "social contact with Negroes," items were chosen from a domain that included the dimension of sociopersonal characteristics of Negroes (occuption, age, and so on). Of course, the specific content of the Triandis instrument reflects the prevailing values of an earlier age. But the method, or a variant of it, can presumably be used to define a set of choices of personal action that make possible an acceptable quality of objective-referenced scores for attitudes.

Motor Skills

Motor skills have for many years been evaluated by comparison with standards, as in the case of handwriting. Many years ago, a familiar device in the elementary school room was the Palmer Scale for grading handwriting. A sample of the student's writing was compared with ideal samples on a chart containing various degrees of "correct" penmanship, each having a numerical value such as 90, 80, 70, and so on, indicating the standard for each level of skill in writing. This was a criterion-referenced form of grading, in that standards were stable and teachers could say that 60 was "passing" at the third grade, 70 at the fourth grade, and so on.

The standards for assessment of motor skills usually refer to the *precision* of the performance, but often also to its *speed*. Since motor skills are known to improve in either or both of these qualities with extended practice, it is unrealistic to expect that mastery can be defined in the sense of learned or not learned. Accordingly, a standard of performance must be decided upon in order to determine whether mastery has been achieved.

Typing skill provides a good example of assessment methods in this domain. A number of different standards of performance are set at progressively higher levels for practice that has extended over increasingly long periods of time. Thus a test standard of 30 words per minute with a specified minimum number of errors may be adopted as a reasonable standard in a beginning course, whereas 40 or 50 words per minute may be expected for an advanced course after more time has been given for additional practice.

Reliability of Objective-Referenced Measures

Choosing criteria for items and tests designed to accomplish objective-referenced measurement requires the selection of standards of performance appropriate to the stated objective, as the preceding discussion has shown. In addition, the items employed for assessment need to yield measurement that is *dependable*. This latter feature of the assessment procedure is referred to as *reliability*, and it has two primary meanings.

Consistency

First, reliability is *consistency* of measurement. It is necessary to determine that the student's performance in answering or completing one particular item designed to assess his performance on an objective is consistent with his performance on other items aimed at the same objective. A pupil in the second grade may be asked by one item to demonstrate his mastery of an arithmetic rule by means of the item: $3M + 2M = 25$; $M = ?$ Obviously, the purpose of assessment is to discover whether he is able to perform a *class* of arithmetic operations of this type, not simply whether he is able to do this single one. Accordingly, additional items belonging to the same class (for example: $4M + 3M = 21$; $5M + 1M = 36$) are typically employed in order to ensure the dependability of the measurement.

In informal testing situations, as when the teacher probes by asking questions of one student after another, single items may be employed to assess performance. However, it is evident that no measure of consistency is available in such situations. On any single item, a student may make a successful response because he happens to have seen and memorized an "answer." Alternatively, his response may be incorrect because he has inadvertently been misled by some particular characterisitic of the item. The single item does not make possible a confident conclusion that the student has mastered the performance implied by the objective.

In those instances in which the class of performances represented by the objective is well defined (as in the arithmetic example previously given) the procedure of selecting additional assessment items of the same class is fairly straightforward. It is essential to bear in mind that the conclusion aimed for is not "how many items are correct?" but rather "does the number correct dependably indicate mastery?" Although two items are obviously better than one, they may yield the puzzling outcome, half right–half wrong. Does this mean that the student has attained mastery, or does it mean he got one item right only because he somehow managed to memorize an answer? Three items would seem to provide a better means of making a reliable decision about mastery. In this case, two out of three correctly answered leads to a certain confidence that reliability of measurement has been achieved. More items can readily be employed, but three seems a reasonable minimum on which to base a reliable assessment of mastery.

When cognitive strategies are the aim of assessment, the item selected for the purpose of assessment may actually be a rather lengthy problem-solving task. For example, such a task might be to "write a 300-word theme on a student-selected topic, within one hour." Assessing performance consistently may require several items, since it is necessary to disentangle the prior learning of information and intellectual skills from the quality of original thought. A number of occasions must be provided on which the student can display the quality of his performance within this domain of learning outcome. The aim is to make it unlikely that a student could meet the criteria set for such tasks without possessing a genuine, generalizable capability of writing original themes on other topics.

Temporal Dependability

The second meaning of reliability is dependability of the measurement on temporally separated occasions. One wishes to be assured that the student's demonstration of mastery of the objective as assessed on Monday is not different from what it would be on Tuesday, or on some other day. Is his performance an ephemeral thing, or does it have the degree of permanence one expects of a learned capability? Has his performance, good or bad, been determined largely by how he felt that day, by a temporary illness, or by some adventitious feature of the testing situation?

Reliability of measurement in this second meaning is usually determined by a second testing separated from the first by a time interval of days or weeks. This is the test–retest method, in which good reliability of the tests is indicated by a high degree of correspondence between scores obtained by a group of students on the two occasions. Often this procedure is used in the formative evaluation of the test, but it may also be employed in practical assessment to determine whether what has been learned has a reasonable degree of stability.

NORM-REFERENCED MEASURES

Tests designed to yield scores that compare each student's performance with that of a group or with a norm established by group scores are called *norm-referenced*. Characterisitically, such tests are used to obtain assessments of student achievement over relatively large segments of instructional content, such as topics or courses. They differ from objective-referenced tests in that they typically measure performance on a *mixture* of objectives, rather than being confined to assessment of single, clearly identifiable objectives. Thus, a norm-referenced test is more likely to have the purpose of assessing "reading comprehension" than it is to measure the attainment of the individual skills involved in reading, considered as specific objectives.

Because of this characteristic of comprehensive coverage, norm-referenced tests are most useful for *summative* kinds of assessment and evaluation (see Chapter 16). They provide answers to such questions as, "How much American history does a student know (compared to others at his grade level)?" "How well is the student able to reason using the operations of arithmetic?" "What proficiency does the student have in using grammatical rules?" Obviously, such assessment is most appropriate when applied to instruction extending over reasonably long periods of time, as in mid-course or end-of-course examinations.

At the same time, the characteristics of norm-referenced measures imply some obvious limitations, as compared with objective-referenced tests. Since their items usually represent a mixture of objectives, often impossible to identify singly, they cannot readily be used for the purpose of diagnostic testing of prerequisite skills and knowledge. For a similar reason, norm-referenced tests typically do not provide direct and unambiguous measures of what has been learned, when the latter is conceived as one or more defined objectives.

Often a norm-referenced test presents questions and tasks that require the student at one and the same time to utilize learned capabilites of intellectual skills, information, and cognitive strategies. In so doing, they make possible assessments of student capabilities that are "global" rather than specific to identifiable objectives. For this reason they are particularly appropriate for assessing outcomes of learning in a set of topics or in a total course. Since the scores obtained are also representative of a group (a single class, or a larger "referenced" group such as ten-year-old children), the score made by each student may conveniently be compared with those of others in the group. Percentile scores are often used for this purpose; the score of a student may be expressed, for example, as "falling in the 63rd percentile."

Teacher-Made Tests

Tests constructed by teachers are sometimes of the norm-referenced variety. The teacher may be interested in learning how well students have learned the content of a course, which may represent a number of different objectives and several categories of learning outcome. Mid-course and end-of-course examinations often have this characteristic of mixed purposes of assessment. These may also be conceived as being aimed at testing the student's *integration* of the various skills and knowledge he is expected to have learned.

At the same time, a norm-referenced test makes possible the comparison of students' performances within a group, or with a referenced group (such as last year's class). Often, such tests are refined over periods of years, using methods of item analysis to select the most "discriminating" items (cf. Hills, 1981; Payne, 1968). This means that items that do not discriminate—those that many students answer correctly and those that few answer correctly— are progressively discarded. Tests refined in this manner tend increasingly to measure problem-solving and other cognitive strategies. They may also, in part, measure intelligence, rather than what has been directly learned. Although this may be a legitimate intention when the aim is to assess the total effects of a course of study, it is evident that this quality of norm-referenced tests makes them very different from objective-referenced tests.

When assessment is aimed at the outcomes of individual lessons or parts of lessons, little justification can be seen for the use of norm-referenced tests. When such tests are used to assess student performance resulting from the learning of defined objectives, they are likely to miss the point of assessment entirely. When instruction has been designed so as to ensure the attainment of objectives, tests should be derived directly from the definition of the objectives themselves, as indicated in the earlier portion of this chapter. Unless objective-referenced tests are used for this purpose, two important purposes of assessment will likely be neglected: (1) the assessment of mastery of the specific capabilities learned; and (2) the possibility of diagnostic help for students in overcoming particular learning deficiencies by retrieving missing prerequisite skills and knowledge.

Standardized Tests

Norm-referenced tests intended for broad usage among many schools within a school system, a region, or in the nation as a whole may have norms that are *standardized*. This means that the tests have been given to large samples of students in specified age (or grade) groups, and that the resulting distributions of scores obtained become the standards to which the scores of any given student or class of students may be compared. Sometimes the standard norms are expressed as percentiles, indicating what percent of the large sample of students attained or fell below particular scores. Often, too, such standards are expressed as grade-equivalent scores, indicating the scores attained by all children in the group who were in the first grade, the second grade, and so on. Procedures used in the development and validation of standardized tests are described in many books on this subject (cf. Cronbach, 1984; Thorndike and Hagen, 1986; Tyler, 1971).

Standardized tests are generally norm-referenced tests; the development of objective-referenced tests has not yet proceeded to the point of availability for a variety of objectives and for a variety of levels of instruction. Accordingly, standardized tests typically exhibit the characteristics previously described. They are usually mixed in their measurement of particular objectives since their items have not been directly derived from such objectives. Their items are selected to produce the largest possible variation in scores among students, and thus their scores tend to be rather highly correlated with intelligence, rather than with particular learning outcomes. With a few exceptions, they fail to provide the identification of missing subordinate capabilities that is essential to diagnostic aims.

Obviously, then, standardized tests are quite inappropriate for use in the detailed assessment of learning outcomes from lessons having specifiable objectives. Their most frequent and most appropriate use is for the purpose of summative evaluation of total courses of several years of instruction. When employed for these purposes, standardized tests can provide valuable information about the long-term effects of courses and of larger instructional programs.

SUMMARY

Up to this point, we have been concerned primarily with goals and performance objectives, with the domains of learning they represent, and with the design of lessons that employ instructional events and conditions of learning suitable for the chosen objectives. In this chapter we turn our attention to the assessment of student performance on the objectives. Thus we proceed from the *what* and the *how* to the *how well* aspect of learning.

For the purpose of assessing student performance on the planned objectives of a course, *objective-referenced tests employing a criterion-referenced interpretation* constitute the most suitable procedure. Such tests serve several important purposes:

1. They show whether each student has mastered an objective and hence may go on to study for another objective.
2. They permit early detection and diagnosis of failure to learn, thus helping to identify the remedial study needed.
3. They provide data for making improvements in the instruction itself.
4. They are *fair* evaluations in that they measure performance on the objective that was given to the student as an indication of what he was supposed to learn. This kind of testing is consistent with the honesty of the relation of teacher to learner.

Objective-referenced tests are direct rather than indirect measures of performance on the objectives. They deal with each objective separately, rather than with very large units of instruction, such as an entire year of study. For this reason they have diagnostic value, as well as value for formative evaluation of the course.

The *validity* of objective-referenced tests is found by determining the congruence of test with objective. *Reliability* is obtained by measuring the consistency of the performance assessment and its dependability over time. The concept of *mastery* is relevant for objective-referenced tests in the domains of intellectual skills, motor skills, and information. For these types of learning outcomes, mastery levels can be defined as error-free performances. In the case of cognitive strategies and attitudes, since assessments deal with *how well* or *how much,* the use of criteria of mastery is less clearly applicable. Examples are given of how criteria of performances can be chosen for each learning domain.

Another type of test is called *norm-referenced.* Such tests do not measure separate, specific objectives of the course. Rather, they measure mixtures or composite sets of objectives, whether these are identified or not. When a norm-referenced test is a standardized test, it has been carefully designed and revised to yield high variability of scores. The interpretation of the scores is made by reference to norms, which represent performance on the test for large groups of learners. Such tests permit comparison of a score of one pupil with that of others; they also permit comparing the average score for a group with the scores of a larger norm group.

REFERENCES

Berk, R. A. (1984). *A guide to criterion-referenced test construction.* Baltimore, MD: Johns Hopkins University Press.

Block, J. H., & Anderson, L. W. (1975). *Mastery learning in classroom instruction.* New York: Macmillan.

Bloom, B. S. (1968). Learning for mastery. *Evaluation Comment, 1*(2), 1–5.

Bloom, B. S., Hastings, J. T., & Madaus, G. F. (1971). *Handbook on formative and summative evaluation of student learning.* New York: McGraw-Hill.

Briggs, L. J., & Wager, W. W. (1981). *Handbook of procedures for the design of instruction* (2nd ed.). Englewood Cliffs, NJ: Educational Technology Publications.

Britton, B. K., & Black, J. B. (1985). *Understanding expository text.* Hillsdale, NJ: Erlbaum.

Cronbach, L. J. (1984). *Essentials of psychological testing* (4th ed.). New York: Harper & Row.

Dick, W., & Carey, L. (1985). *The systematic design of instruction* (2nd ed.). Glenview, IL: Scott, Foresman.

Fishbein, M. A. (Ed.). (1967). *Attitude theory and measurement.* New York: Wiley.

Gagné, R. M. (1985). *The conditions of learning* (4th ed.). New York: Holt, Rinehart and Winston.

Gagné, R. M., & Beard, J. G. (1978). Assessment of learning outcomes. In R. Glaser (Ed.), *Advances in instructional psychology* (Vol. 1). Hillsdale, NJ: Erlbaum.

Greeno, J. G. (1978). A study of problem solving. In R. Glaser (Ed.), *Advances in instructional psychology* (Vol. 1). Hillsdale, NJ: Erlbaum.

Guilford, J. P. (1967). *The nature of human intelligence.* New York: McGraw-Hill.

Hills, J. R. (1981). *Measurement and evaluation in the classroom.* Columbus, OH: Merrill.

Johnson, D. M. (1972). *A systematic introduction to the psychology of thinking.* New York: Harper & Row.

Johnson, D. M., & Kidder, R. C. (1972). Productive thinking in psychology classes. *American Psychologist, 27,* 672–674.

Lathrop, R. L. (1983). The number of performance assessments necessary to determine competence. *Journal of Instructional Development,* 6(3), 26–31.

Payne, D. A. (1968). *The specification and measurement of learning outcomes.* Waltham, MA: Blaisdell.

Popham, W. J. (1975). *Educational evaluation.* Englewood Cliffs, NJ: Prentice-Hall.

Thorndike, R. L., & Hagen, E. (1986). *Measurement and evaluation in psychology and education.* New York: Wiley.

Torrance, E. P. (1963). *Education and the creative potential.* Minneapolis: University of Minnesota Press.

Triandis, H. C. (1964). Exploratory factor analyses of the behavioral component of social attitudes. *Journal of Abnormal and Social Psychology, 68,* 420–430.

Tyler, L. E. (1971). *Tests and measurements* (2nd ed.). Englewood Cliffs, NJ: Prentice-Hall.

Wald, A. (1947). *Sequential analysis.* New York: Wiley.

PART FOUR

DELIVERY SYSTEMS FOR INSTRUCTION

14 GROUP INSTRUCTION

A great deal of instruction is done with learners assembled in a group. When instruction is delivered in this way, one has to bear in mind constantly that learning still occurs within individuals. Older learners, to be sure, may attain a high degree of control over the managment of instructional events, to the extent that their learning depends on self-instruction. For learners of whatever sort, the attempt is usually made in group instruction to ensure that each instructional event is as effective as possible in supporting learning by all members of the group.

Groups assembled for instruction are of various sizes. The group sizes that seem of particular significance for instructional design include, first, the *two-person group,* which makes possible the *tutoring* mode of instruction. A second commonly distinguished kind of group is simply the *small group,* containing roughly three to eight members, a size that favors *discussion,* as well as what may be called *interactive recitation.* In this latter mode, the performances of individuals are affirmed or corrected by other members of the group. The third kind of group is a *large group,* with fifteen or more members. The most commonly used mode of instruction in such a group is the *lecture,* which may of course incorporate such other presentations as projected or televised pictures and demonstrations. Another mode of instruction with large groups is *individual recitation,* commonly used with such subjects as language, both native and foreign, and sometimes with other subjects as well.

Although these three group sizes clearly appear to have different implications for instructional delivery, the distinctions among them are not hard and fast. What can be said, for example, of the group containing between eight and fifteen members? Sometimes instruction proceeds as with a small group (by discussion, for example), but at other times a large-group mode (a lecture) might be employed. Division into small groups is also possible. Considering the other end of the scale of group size, one can distinguish a *very large group* (a hundred or more students). In such cases, however, instructional factors differing from those of the large group are likely to be logistic in character, pertaining to seating arrangements, acoustics, and others of this general nature. Although these factors have their own peculiar importance, we do not attempt to discuss them here. Otherwise, large-group modes of instruction are assumed to be relevant and applicable to groups that are very large.

CHARACTERISTICS OF GROUP INSTRUCTION

How shall the instructional features of different kinds of instructional groups be characterized? One way would be to describe in detail the various *modes* of instruction, such as the *lecture,* the *discussion class,* the *recitation class.* These various modes do indeed have different characteristics and vary in their feasibility of use with different sizes of groups. There are, however, variations in what is done by the lecturer, in what happens during a discussion, and in what occurs in a recitation class that appear to be of particular significance for an understanding of the effects of instruction on learning. Systematic knowledge of several of these instructional modes is summarized in a volume edited by Gage (1976).

Rather than describing features of the different modes of instruction, our approach in this chapter is to consider how such varieties of instruction can be planned for delivery to different sizes of instructional groups—the two-person group, the small group, and the large group. Our discussion is concerned with questions of what instructional arrangements (including instructional modes) are *possible* and are likely to be *most effective* with each of these types of groups.

Patterns of Interaction in Instructional Groups

The size of an instructional group is an important determinant of the environment in which learning occurs. Some patterns of interaction among teachers and students are more readily attained with small groups, some with large. Owing to differences in the way they are perceived by students, these patterns may influence the outcomes of learning. Some classroom interaction patterns, similar to those described by Walberg (1976), are depicted in Figure 14-1.

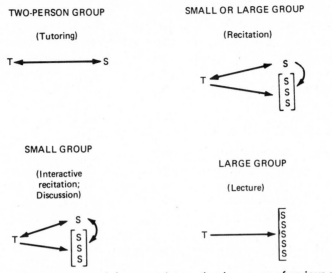

FIGURE 14-1 Some patterns of classroom interaction in groups of various sizes. Arrows indicate the direction of interactions.

As the figure shows, communication between teacher and student flows in both directions during instruction in a two-person group. When recitation is the adopted mode, with either a small or a large group, mutual interaction occurs between the teacher and one student at a time, while the other students are recipients of a teacher communication. Interactive recitation and discussion occur in a small group when there is interaction among students as well as between teacher and student. In the lecture mode of instruction, used typically with a large group, the communication flow is from teacher to students.

Variations in Instructional Events

Any or all of the events of instruction (Chapter 10) may be expected to vary with group size, both in their form and in their feasibility of use. For example, the event of gaining attention can obviously be rather precisely managed in a two-person group, whereas it can only be loosely controlled for the individual learners in a large group. Learning guidance, in a two-person group, is typically under the control of the instructor (tutor), whereas the semantic encoding suggested by a lecturer is likely to be modified in a number of individual ways by the strategies available to individual learners. When feedback consists of information indicating correct or incorrect student answers, it can often be controlled in a large group with about as much precision as that provided to a single student. However, when the feedback includes information about the causes of incorrect responding, it will vary with the individual student.

The primary factors that appear subject to variation in different types of instructional groups, then, are those pertaining to the events of instruction. The size of group not only determines some of the necessary characteristics of these events but also sets limits upon their effectiveness in supporting the processes of learning. It is these features of group instruction that we shall be considering in the following sections.

Diagnosing Entry Capabilities of Students

Another factor of some importance in its influence on instructional effectiveness is the assessment of *entry capabilities* of students (cf. Bloom, 1976). Procedures for accomplishing such diagnosis are not instructional events themselves, but they do make possible the design of certain of these events. The ways of conducting diagnostic procedures are likely to vary with the size of the instructional group.

Entry capabilities of students may be assessed at the beginning of a course of study or at the start of a semester or school year. Student capabilities may be assessed in a finer sense, and weaknesses or gaps diagnosed just prior to the beginning of each new topic of a course. Diagnosis of this latter sort is commonly done, for example, in the period devoted to teaching prereading skills to children or at various stages of beginning reading. Other examples of the application of fine-grained diagnostic procedures are sometimes seen in mathematics and in foreign language study. Such diagnoses are likely to be most successful if based upon learning hierarchies. Simple tests or probes can be administered to students to reveal specific gaps in enabling skills. Diagnosis in this form is not severely affected by group size, since testing for gaps in enabling skills can usually be done as effectively with large groups as with small.

Designing and executing the instruction implied by diagnosis of entry capabilities is, however, greatly affected by the size of the instructional group. This fact provides the basic rationale for instruction that is "individually prescribed." The effects of group size on instruction are particularly critical when the concern is with diagnosis relevant to each *lesson* or, in other words, with the *diagnosis of immediate prerequisites*. Depending on what has happened previously to prepare students for a particular lesson, a different individual pattern of prerequisite capabilities may conceivably appear for each student in a group. Thus, one might discover 20 different patterns of capability in enabling skills within a group of 20 students. Such patterns of instructional needs may readily be managed in a tutorial situation; in fact, this feature is often considered the most prominent advantage of the tutoring mode. The same finding about individual patterns of immediate prerequisites, however, presents a considerable challenge to the teacher of a 20-person group, for example. Some further implications of this circumstance will be discussed in a later section.

INSTRUCTION IN THE TWO-PERSON GROUP

Instructional groups of two persons consist of one student and one instructor or tutor. Groups of this sort may, however, be composed only of students, one of whom assumes the tutoring role. In schools, the tutoring of younger students by older ones is not an uncommon practice. However, peer-tutoring has also been successfully done, even in early grades (Gartner, Kohler, and Riessman, 1971). The alternation of student-tutor roles by pairs of older students or of adults is sometimes chosen as a mode of instruction. Regarding any of these possible arrangements, it is worthy to note that learning gains are about as frequently reported for tutors as they are for students (Ellson, 1976; Devin-Sheehan, Feldman, and Allen, 1976; Sharan, 1980).

As noted in the previous chapter, systems of individualized instruction are usually designed so that diagnostic tests of student weaknesses (or gaps) will be followed by prescriptions of specific instruction designed to fill these gaps. In such systems, teachers are essentially behaving as tutors when they follow up the prescription with oral instruction. Individualized instruction, then, although it frequently calls upon the learner for self-instruction, often also involves tutoring in a two-person group.

Events of Instruction in Tutoring

The group composed of a single student and a single tutor has long been considered a kind of ideal situation for teaching and learning. The primary reason for this preference would appear to be the opportunities the two-person group provides for the *flexible adjustment of instructional events*. Thus, the tutor can employ just enough stimulation to gain the attention of the student, or can increase the amount if a first attempt fails. The tutor can suggest a number of alternative schemes for the encoding of information to be learned; if one doesn't work well, another can be employed. The student's comprehension of a new idea and his storing of it can be assessed immediately, and again after a lapse of time, in order to affirm its learning and to reinforce it.

Some of the main features that exemplify flexible adjustment of instructional events for a two-person group may be described as follows:

1. *Gaining attention* Assuming that the student participates willingly in the tutorial situation, the gaining of attention (in the sense of alertness) may readily be accomplished. The tutor may demand attention by giving a verbal direction, and watch for the overt signs that are typical of an attentive state. Obviously, immediate adjustments of the stimulation necessary to bring about attention can be made if the student's attention wanders.
2. *Informing the learner of the objective* In the case of this event, flexibility may typically be achieved by repeating the objective in different terms, or by demonstrating an instance of the performance expected when the learning is

completed. Of course, if the objective is already known by the student, an event such as this may be entirely omitted.

3. *Stimulating recall of prerequisite learnings*   The possibility of flexible determination of this event gives the tutoring mode a definite advantage in the support of learning. Assuming that diagnosis of the previously learned prerequisites has been made, the tutor can proceed to fill in the gaps of missing student capabilities, should that be necessary. Being assured that prerequisites have been acquired, the tutor can then proceed to require recall by the student. These acts of the tutor in making prerequisite skills accessible in the working memory will do much to ensure that learning proceeds smoothly.

4. *Presenting the stimulus material*   Here, too, there is a great flexibility of choice available to the tutor. Selective perception may readily be aided: The tutor can give emphasis to lesson components by changes in oral delivery, by pointing, by drawing a picture, and in many other ways. If a foreign language is being learned, for example, the tutor can choose just the right oral expression to illustrate the grammatical rule to be taught. If varied instances are required, as in the teaching of a new concept, the number and varied features of these instances can be carefully chosen to meet the student's need, as indicated by an immediately preceding performance.

5. *Providing learning guidance*   This event is also one in which the flexibility of the two-person situation results in an important advantage. In fact, it is in this connection that the phrase "adapting instruction to the needs of the learner" has its clearest meaning. The tutor can employ a variety of means to encourage *semantic encoding* on the part of the learner. Furthermore, the tutor can try such means one after another, if necessary, until one is found that works best. Rule applications can be demonstrated; pictures can be used to suggest visual imagery; organized information can be provided as a meaningful context for the learning of new knowledge. The tutoring mode offers many opportunities for the selection of effective communications by the tutor, all aimed at supporting the learning processes of the student.

6. *Eliciting the performance*   In the two-person group, learner performance can be elicited with a degree of precision not possible in larger groups. On a moment-to-moment basis, the tutor is usually able to judge by the learner's behavior that the necessary internal processing has occurred and that the learner is ready to show what he has learned.

7. *Providing feedback*   The provision of feedback is also capable of greater precision in a two-person group than in other groups. Precision in this case pertains not primarily to the timing of feedback but to the nature of the information given to the learner. The learner can be told, with a high degree of accuracy, what is right or wrong with his performance and given directions that permit correction of errors or inadequacies.

8. *Assessing the performance*   Flexibility in assessment is available to the tutor, in the sense that the performance may be tested at various intervals following the learning. The testing of learner performance may also be repeated as many times as deemed necessary for a reliable decision to be made.

9. *Enhancing retention and transfer*   The management of this kind of event may be done with considerable flexibility in a two-person group and, therefore, with a good deal of precision. The tutor can select cues that, according to

past experience, work effectively to facilitate retrieval in a particular learner. Just enough varied examples can be chosen to aid the transfer of learning. Spaced reviews can be conducted to the extent needed to ensure long-term retention for the particular student, based upon previous experience with that student in the tutoring situation.

The Flow of Instruction in Tutoring

It is evident that the two-person instructional group permits maximal control of instructional events by the tutor. As the manager of instruction, the tutor can decide which events to employ, which to emphasize, and which to assign to the learner's own control. The determination of timing for these events can operate to make each act of learning optimally efficient. In addition, the flexibility of choice of exactly how to select and arrange each event makes it possible for a tutor to provide instruction meeting the needs of the individual student. In the tutoring mode, instruction can most readily be *adapted* to the instructional needs of each student.

In practice, tutoring has taken a variety of forms (Gartner, Kohler, and Riessman, 1971). The advantages it appears to offer have often been shown to yield favorable results in student achievement, although not always (Cloward, 1967; Ellson, 1976). The evidence appears to show that the advantages of tutoring do not result from the individual attention provided to the student in the two-person situation. Rather, tutoring works best when the instruction it makes possible is highly *systematic* (Ellson, 1976). In other words, tutoring can be a highly effective mode when advantage is taken of the flexibility it provides in achieving *precision* in the arrangement of instructional events. The freedom made possible by the two-person instructional group does not produce good results if it leads to sloppiness in instruction; on the contrary, favorable outcomes appear to depend upon careful control of the events of instruction.

Typically, an episode in tutoring runs somewhat as follows. We assume a task in beginning reading, having the objective, "Given an unfamiliar printed two-syllable word of regular spelling, demonstrates phonics rules by sounding the word." The printed word is *plunder*. The tutor is a volunteer, mother of one of the children in a first grade, tutoring a six-year-old girl pupil.

First, the tutor gains the child's attention, and tells her of the objective by saying, "Here is a word you probably haven't seen before in your reading *(plunder)*. I want you to show me how you can sound it out." Should the pupil sound out the word immediately, either by correctly using rules or by recognizing the printed word, the tutor says "Good!" and goes on to another printed word. Otherwise, the tutor encourages the child to sound out the first syllable *(plun)*, then the second one, and the both together.

Actually, the procedure is one of combining reminders of what the child already knows (recall of prerequisites), such as the sounds of *pl* and *un*, and learning guidance that suggest the strategy called "blending." Thus, the tutor may tell the child to place her finger over the last part of the word, leaving the letters *p* and *l* exposed, and then ask "What sound does *pl* make?" If the pupil answers correctly, positive feedback is given. If she gives an incorrect response, the child is told what the correct response is and asked to repeat it. Then the procedure is followed again for each sound and for successive combinations of sounds, until the entire word can be sounded correctly. At that point, the child is asked to repeat the word, and some acknowledgement is made of her accomplishment. (In fulfilling a secondary objective, the meaning of the word would probably also be explained to the child.)

The systematic steps in this tutoring situation can be seen to be those of repeating, as necessary, the events of instruction calling for the stimulation of recall of prerequisites, presenting the stimulus material, providing learning guidance, eliciting and assessing the performance, and providing feedback. Essentially the same steps would be followed in the tutoring of older students or adults in the learning of an intellectual skill, except that somewhat greater dependence might be placed on encouraging the student to institute these events himself. Tutoring at the university level, of course, usually consists almost entirely of self-instruction—the tutor's activities being largely confined to assessing performance and to suggestions of means the student may employ to enhance retention and transfer of learning.

INSTRUCTION IN THE SMALL GROUP

Instructional groups of up to eight students are sometimes found in formally planned education. The university teacher, or the teacher of adult classes, meets with a small group of students on some occasions. More frequently, such groups may be formed by deliberate division of larger ones. In the elementary and middle grades, the schoolteacher may find it desirable to form small groups from an entire class of students, in order to instruct students who have progressed to approximately the same point in their learning of a particular subject.

The employment of small groups for instruction is a common practice in such elementary school subjects as reading and mathematics. In the first grade, for example, a teacher may find that some pupils have not yet mastered the oral language skills of reading readiness; others may be just beginning to learn to sound the letters and syllables of printed words; still others may be reading entire printed sentences without hesitation. Obviously, these different sets of students need to be taught different sets of enabling skills. It would accomplish nothing to present pages of print to pupils who are still struggling with oral language. Nor is it likely to be of advantage to those pupils who already read pages of print to have to suffer through lessons that require them to make oral descriptions of objects shown in pictures. The practical solution is to divide the class into a number of small groups.

Classes of older students or adults are sometimes divided into several sets constituting small groups. The groups thus formed may meet on separate occasions, as in "quiz sections," or they may meet in separate small groups for a portion of the time of a scheduled class. In either case, the aim is to attain some of the advantages of small-group instruction and to provide some added variation in the forms of instruction possible with large groups.

Events of Instruction in Small Groups

The control of instructional events in the small group (three to eight students) can best be compared to what is possible in the tutorial situation. This kind of arrangement of teacher and students might be described as "multistudent tutoring." The characteristics of the instructional situation resemble those of the two-person group and are rather unlike those of the large group. In the small group, the teacher typically attempts to use tutorial methods, sometimes with single students, sometimes with more than one, and most often by "taking turns." The general result is the management of instructional events in a way that applies to each individual student in the group, but with some evident loss of flexibility and precision.

Procedures of diagnosis may have been used to select the members of a group for small-group instruction. As previously noted, this is typical practice for small groups in elementary reading, language, and mathematics. During an instructional session with a small group, it is also possible for the teacher to diagnose each student's attainment of *immediate prerequisites*. In fact, this may be seen as one of the important features of small-group, as contrasted with large-group, instruction. By suitable questioning of each student in turn, the teacher is able to judge with a fair degree of accuracy that the necessary enabling skills are present in all students. In this way the estimate of students' readiness for taking the next step in learning can be made to approximate the degree of precision available in the two-person instructional group.

The possibilities of control of the events of instruction in the small group are discussed in the following paragraphs:

1. *Gaining attention*   In a small group, arranged so that the teacher can maintain frequent eye contact with each member, gaining and maintaining student attention poses no major difficulties.
2. *Informing the learner of the objective*   This event can also readily be managed in a small group. The teacher can, as necessary, express the objective of the lesson and ensure that it is understood by each member of the group. Of course, it may take a bit more time to ensure understanding of objectives for eight students than it does for only one (as in the two-person group).
3. *Stimulating recall of prerequisite learning*   By questioning several students in turn, the teacher is able to be fairly sure that necessary enabling skills and relevant items of supportive information are accessible in the working memories of *all* students. Using best judgment, the teacher may direct questions that require selected students to recall relevant items. The same questions

have the added effect of reminding other students of material that is already available to them.

4. *Presenting the stimulus material*   The materials for learning may be presented in ways appropriate to the objective, but without necessarily being tailored to individual student characteristics. For example, features of the oral presentation may be given emphasis by voice changes. In pictures or diagrams, particular features of objects and events may be appropriately made prominent. For this particular event, the degree of lessened flexibility compared with that of the two-person group appears to be minimal.

5. *Providing learning guidance*   Here the choice is either to present a communication to the group or to members of the group in turn. With the first of these alternatives, the teacher is behaving as though in a large-group setting; with the second, the event is managed as in the tutoring mode, involving a teacher interaction with one student, then with another, and so on. Obviously, the more students there are in the group, the more time the latter procedure takes. It is not uncommon for the teacher of a small group to alternate between these two approaches, judging one to be more appropriate at one time, one at another. In any case, the function of the event remains the same—providing cues and suggesting strategies that will aid the semantic encoding of material to be learned.

   Quite a different kind of learning guidance is provided by the small group which is engaged in *discussion*. In such groups, the discussion may be managed and led by the teacher. In other instances, the small groups formed out of large classes may have designated students as discussion leaders. The learning guidance provided by discussion is of various sorts, and its function in support of learning depends on the nature of the objective. More generally, it would appear that discussions place a fairly high degree of dependence on the self-instructional strategies of the individual students. In adult discussion groups, the members are, to a large extent, using the strategies of deciding what they want to learn, and selecting these elements from what they hear or say as part of the discussion.

6. *Eliciting the performance*   It is clear in the case of this event that the only way of eliciting *each* student's performance in a small group is to do it one by one. Since this uses up a good deal of time, it is not always done. Instead, the teacher usually calls on one or two students to show what they have learned, and assumes that the learning has been equally effective for those not called upon. As the lesson proceeds, other students take their turns being called upon. Obviously this procedure aims at approximating that of the two-person group, but the precision of the event is considerably reduced. The teacher comes to depend upon a probabilistic estimate of learning outcomes, rather than a precise determination of them.

7. *Providing feedback*   Since this event is tied to the occurrence of student performance, it is subject to the same kinds of limitations in the small group. For the student who is called upon, feedback may be precisely provided. For the other students, it is only probable because it depends on which of them has made the same response (perhaps covertly).

8. *Assessing the performance*   Performance assessment may also lose some precision of control, since only one student's performance can be assessed at

one time by oral questioning. The other students must wait their turns; this means that a sample of performances will be assessed for each student, but not the entire repertoire he is supposed to have learned. Of course, a test covering an entire lesson or topic may later be given to the entire group of students (a technique equally applicable to large groups).

9. *Enhancing retention and transfer* For instructional groups in the elementary grades, the teacher is able to estimate the desirability of varied examples and additional spaced reviews in providing favorable conditions for retention and learning transfer. Such estimates are made by a kind of averaging of the group's performance, and thus do not have the precision afforded in the tutoring situation of the two-person group. In the case of older students and adults, the conduct of discussions has as one of its major purposes the enhancement of retention and transfer.

Instruction in the Small Group—Recitation

Suppose that a teacher has assembled a small group of pupils who are to learn the skill of adding fractions with dissimilar denominators. Since one of the steps in this procedure is "finding the least common multiple of the denominators," it may be assumed that diagnostic tests have indicated that students already possess such prerequisite concepts as *numerator, denominator, factor,* and *multiple,* as well as rules for multiplying and dividing small whole numbers, and the rule for adding fractions with identical denominators.

Once the attention of all members of the group is gained, the teacher tells the students the objective of the lesson, using an illustration such as $2/5 + 4/15$. Calling upon one or two members of the group in turn, the teacher stimulates recall of prerequisite concepts and rules. For example, students may be asked to add $2/13 + 5/13$, to obtain $7/13$. Having assurance that prerequisite skills are readily accessible, the teacher presents a single example such as $2/5 + 3/7$ as stimulus material. The next step is to provide suitable learning guidance for the learning of the rule having to do with the finding of a least common multiple. This may be done by demonstrating the multiplication of denominators $(5 \times 7 = 35)$; alternatively, a discovery method may be employed, initiated by such a question as, "How might we make it possible to change these into fractions that could be added?" In this case, one student of the group is called upon for an answer while others await a later turn. A different student may then be called upon for performance, that is, in arriving at the changed expression $14/35 + 15/35$, and in supplying the sum $29/35$. Feedback, in the form of affirming the correct response or correcting a wrong one, is then provided.

The subsequent instructional events are conducted in such a group by calling on different students, using different examples. Thus, performance is followed by appropriate feedback for the students in the group by taking turns. The immediate performance of each student is assessed in this manner. The varied examples used serve the function of enhancing retention and transfer, assuming that students other than the one called upon are responding in a covert manner while learning.

Instruction in Discussion Groups

Instruction that takes the form of discussion is said to be characterized as "interactive communication" (Gall and Gall, 1976). One student speaks at a time, and is listened to by the entire group. The order in which students initiate or respond to speech is not predetermined. Often, one student is responding to the remarks or questions introduced by another student. The teacher may interpose remarks or questions, and sometimes may call upon individual students to speak. Of course, small groups of this sort may be organized with students as discussion leaders.

Three kinds of objectives are often considered appropriate for instruction via group discussion: subject matter mastery, attitude formation, and problem solving (Gall and Gall, 1976). It is not unusual for a class discussion to have more than one of these types of objectives.

The formation and modification of attitudes is usually the major aim of *issue-oriented discussion,* examples of which are found in the "jurisprudential model" and the "social inquiry model" described by Joyce and Weil (1980). The discussion may be initiated by the account of an incident illustrating a social issue (such as freedom of speech or job discrimination). The teacher or group leader may then ask for one or more opinions about the issue. Comments are made about these opinions, either by the discussion leader or by other students. As the discussion proceeds, the leader attempts to achieve progressive sharpening and clarification of the issue by introducing different examples and by encouraging statements by various group members. Often, what is aimed for is a group consensus, as represented by a set of statements to which no major disagreements remain. This attitude-forming situation may be conceived as a particular kind of learning guidance, namely one involving communications from a number of human models. These models are members of the group and its leader. This kind of learning guidance, particularly effective in attitude formation, is followed by performance (choice of action) by the individual students, and by feedback in the form of group consensus.

Problem solving is also a commonly adopted goal for discussion groups (Maier, 1963). It appears that the kinds of problems that provide the most effective instruction in discussion groups are those with multiple solutions and those that include attitudinal components. Maier (1971) points out that small-group divisions of large college classes can increase the opportunities for student participation and can be used to form discussion groups for problem-solving and other related purposes. Maier suggests the presentation of problems or issues that capture student interest and emotional involvement, as a means of enlisting motivation. With this kind of objective, small groups have the chance to practice both communication skills and problem-solving strategies. Obviously, this type of instructional group is one that depends very largely upon the students to manage instructional events for themselves. The students must provide themselves with the stimulation to recall relevant knowledge and must employ their own cognitive strategies of encoding and problem solving. Attitude change objectives are a secondary, although not necessarily less important, outcome of this type of discussion session.

LARGE GROUP INSTRUCTION

In instructing large groups, the teacher employs communications that do not differ in function from those employed with two-person groups or with small groups. For a large group, the teacher initiates and manages the events of instruction that are specifically relevant to the primary objective of the lesson. Because the teacher's cues for timing and emphasis come from several (or many) sources rather than from a single student, there is a marked *reduction in precision* in the management of instructional events. Teachers of large groups cannot be sure they have gained the attention of *all* students; they cannot always be sure that *all* students have recalled prerequisites or that the semantic encoding they suggest will work well with *all* students. The strategy of instruction in a large group is therefore a *probabilistic strategy*. Instruction so designed will be effective "on the average" but cannot by itself be ensured as effective for each individual learner (cf. Gagné, 1974, pp. 124–131).

It may be argued that this pattern of large-group instruction is the way instruction should be designed in general. The instruction itself (that is, the communications of the teacher) is "good," and it is up to the student to profit from it. Students, in this view, must do a great deal of organizing of the events of instruction themselves—it is up to them to infer the objective of instruction, to remind themselves to recall prerequisite skills, to choose a method of encoding, and so on. Such a view appears to be widely held, and widely employed in college and university teaching. It may be noted, also, that this conception of instruction runs contrary to the notion of *mastery learning* proposed by Bloom (1974, 1976). Bloom's conception relates the quality of instruction to the occurrence of events described as providing *cues, participation, reinforcement,* and *feedback/correctives*. This set of instructional features closely resembles the instructional events we have described. It is evident that mastery learning requires the management of events that go beyond the "giving of information" by the teacher.

Instructional Events in the Large Group

As is the case with the small group, the influence of instructional events in the large group is only *probable*. The teacher's communications reach individual students with different probabilities, and their effects on individual students also cannot be monitored with certainty. Since the degree of instructional readiness, the intensity of motivation and alertness, the appropriateness of the semantic encoding suggested, and the accessibility of relevant cognitive strategies are all factors likely to vary with the individuals who make up the group, instruction, as delivered, is relatively imprecise. Any lack of effectiveness instruction may have for one individual may, of course, be overcome by the student's own efforts at self-instruction. For example, what some students fail to learn from a lecture they may learn later by employing their own encoding strategies on notes of the lecture. Other students may find this kind of encoding ineffective and may prefer to process the information in its oral form as originally given.

1. *Gaining attention*   This event, as all teachers know, is highly important for the effectiveness of instruction delivered to a group. It is surely no more than a probable occurrence in a class of young people, and often little more likely in a class of older students. The occasional use of demonstrations and audiovisual media can aid the gaining of attention at times when other critical instructional events are to follow.

2. *Informing the learner of the objective*   The objective can readily be stated and demonstrated to a large group. It will probably be comprehended by all students, when suitably presented.

3. *Stimulating recall of prerequisite learning*   As indicated previously, this event may be of critical importance for learning. It is also, perhaps, one of the most difficult events to accomplish with reasonable probability in a large group. Typically, the teacher calls upon one or two students to recall relevant concepts, rules, or information. Obviously, though, the necessary retrieval may not be achieved by other students, many of whom are hoping to avoid being called upon. As a result, the management of this event may often be inadequately accomplished. Those students who have not recalled prerequisite skills will probably not learn the relevant objective. The cumulative effects of this inadequacy are therefore quite serious. Various means (such as "spot quizzes" for the entire group) are employed to improve the operation of this event. It appears to deserve a great deal of attention by instructional designers.

4. *Presenting the stimulus material*   The content to be learned can be presented in a way that emphasizes distinctive features. This means that the presentation can be made optimally effective, on the average.

5. *Providing learning guidance*   In a large group, learning guidance can be provided in a way that works, in a probabilistic sense, for most members of the group. For example, the encoding of a historical event can be suggested by a picture or dramatic episode, which may be generally effective in the group as a whole. The particular encoding suggested, however, cannot be adapted to the individual members of the group, as it can in smaller groups.

6. *Eliciting the performance*   Control in obtaining the learner's performance is much weakened in the large group. Whereas a tutor can expect several occasions during which the student exhibits what he learns in a single lesson, the

teacher of a group cannot manage this for each student in the group. Instead, in a typical class, the teacher calls on one or two students at a time. Other students in the group may occasionally be responding covertly, but this is not a highly likely possibility. Accordingly, it may be seen that the student response has a low degree of precision as an instructional event in the large group.

Quizzes and tests are often given in an attempt to overcome the difficulty of eliciting student performance. To be most effective as instructional events, quizzes should be frequent. Even daily quizzes, however, cannot approximate the frequency with which the tutor is able to ask for student performances that reflect capabilities learned in an immediately previous moment.

7. *Providing feedback*   Since this event is inevitably tied to the occurrence of performances by the students, it is subject to the same limitations as those occurrences. Feedback to students in a large group occurs with low frequency and is likely to be confined to results of tests covering a number of different learning objectives.

8. *Assessing the performance*   Similar comments may be made concerning this event in the instruction of large groups. The more frequent and regular assessment (followed by corrective feedback) can be, the better will be the outcome of learning. For example, regularly scheduled quizzes following segments of study material are considered to be the most valuable feature of some computer-managed courses in college subjects (Anderson et al., 1974). When the computer is used for assessment, this event can be managed with a degree of precision that is impossible for the teacher of a large group.

9. *Enhancing retention and transfer*   Events of this nature can be accomplished by the teacher of a large group, again in a probabilistic sense. That is, the teacher can use the varied examples and spaced reviews that have been found to work best on the average, but she is unable to adapt these techniques to differences in individual learners.

The Lecture

Surely the most common mode of instruction for the large group is the lecture. The teacher communicates orally with students assembled in a group. The oral communication may be accompanied by occasional demonstrations, pictures, or diagrams; and these may be presented in various media, including the chalkboard. The students listen, and some take notes, which they may use later for recall or as a means of generating their own semantic encodings.

As pointed out by McLeish (1976), the lecture can accomplish some positive instructional purposes. In particular, the lecturer can: (1) inspire an audience with his own enthusiasm; (2) relate his field of study to human purposes (and thus to student interests); and (3) relate theory and research to practical problems. The lecture attains these goals with the utmost economy, which doubtless accounts for its preservation as an instructional mode over two thousand years of higher education.

McLeish's interpretation implies that the good lecture can attain certain instructional objectives very well, because it is able to implement certain instructional events effectively. For example, "inspiring students with his enthusiasm" implies that the lecturer often functions as a human model in establishing positive attitudes toward the subject of study. The motivational effects of lecturing are also incorporated in the idea of relating a specialized field of study to the more general concerns of human living. As for the concept of relating research findings to practical problems, this purpose of the lecture functions to provide a context of cues that will aid retention and learning transfer.

As pointed out in the previous section, the communications delivered to groups of learners via the lecture can be aimed at optimizing the effectiveness of many of the events of instruction in a probabilistic sense. For example, attention can be gained by dramatic episodes; instructional objectives can be simply and clearly stated; suggested encoding of material to be learned can be provided by summary statements, visually presented tabular arrays, or diagrams; and so on. It is clear that many of the events of instruction can be appropriately presented within a lecture, although they cannot be managed with precision. Their momentary effects cannot be ensured for all students; neither can their specific forms be adapted to individual differences in students.

Viewed from the standpoint of instructional events, perhaps the weakest features of the lecture reside in its lack of control over (1) the recalling of prerequisites, and (2) the eliciting of student performance, with its succeeding provision of corrective feedback. The lecturer can remind students of what they need to recall as prerequisite knowledge, but he cannot take the steps necessary to ensure such recall. Again, the lecturer can encourage students to practice the capabilities they are learning, but he cannot require them to exhibit the performances that reflect what they have learned. Thus, when the lecture is used as the sole mode of instruction, there is heavy dependence upon the student to institute these events for himself. This degree of self-instruction is a common expectation in college and university instruction, as well as in adult education. It is worth noting that quizzes and tests are able to overcome this limitation of the lecture only to a small degree, since they are typically both infrequent and "coarse-grained" in their assessment of specific learning objectives.

The Recitation Class

Another form of large-group instruction, more frequently used with some subjects than with others, is the recitation class. This mode of instruction partially overcomes some of the limitations of the lecture, at least in a probabilistic sense. In a recitation class, the teacher calls on one student after another to respond to questions. In a class in foreign language, for example, one student at a time may be answering questions posed in that language or otherwise continuing a conversation in the foreign tongue.

The teacher's *questions* in a recitation class may represent several different instructional events at different times during a single lesson. A question may be designed to stimulate recall of a prerequisite capability, bringing it to the forefront of the student's memory. Or, a question may be one that asks the student to perform—to show what he has learned. A different kind of question may be used to suggest a question of thought to the student (in the manner of "guided discovery"), and thus to guide his learning in the sense of semantic encoding. Still another kind of question may require the student to think of examples of application of a newly learned skill or body of knowledge—a process that will contribute cues useful for recall and learning transfer. For instance, having learned the concept of homeostatic control, for instance, questions might direct students to describe several examples of practical devices of this category.

In recitation classes, some instructional events are often left for the student to manage. This is typically the case when recitation follows a homework assignment. In such instances, it is usually expected that events such as control of attention, gaining information about the objective, semantic encoding, and the provision of corrective feedback will be managed by the student himself as he does his homework. These events are obviously relevant to the student's study activities in reading his textbook, practicing his newly learned skills in examples, or rehearsing the statement of organized information. Good study habits are, in these circumstances, the determiners of effective learning.

The control of instructional events in the large recitation class is decidedly imprecise, so far as their effects on individual students is concerned. When questions are asked, for whatever purpose, there is time for only a few students to respond. Should the teacher call upon students who are typically well-prepared, and thus engage in relative neglect of students who may be less able to guide their own learning? Or should the teacher call upon the less able students, and through the necessity of supplying corrective feedback, bore those who have already learned correctly? It is clear that what usually happens in the use of recitation with a large class is that the necessary events of instruction affect only a few students on any one occasion. Time does not permit the teacher to allow everyone to take a turn. All too frequently, students learn to resort to the game of avoiding being called upon to recite. This, of course, is the wrong game so far as the learning of lesson objectives is concerned.

FEATURES OF TUTORING IN LARGE GROUPS

Methods of large-group instruction, including lecture and recitation, may be combined in various ways with small-group, two-person group, or individualized instruction in restoring some of the advantages of the tutoring situation. One rather simple scheme is to divide the large group into small groups for part of its meeting time, or into classes of smaller groups for meetings subsequent to a large-group lecture or recitation. Either of these arrangements is intended to make possible a degree of precision in the control of instructional events which surpasses that of a large group.

Mastery Learning

An outstanding system of teaching that directly attempts to introduce precision in the management of instructional events is called *mastery learning* (Bloom, 1974; Block and Anderson, 1975). Generally speaking, this method supplements large-group modes of teaching with *diagnostic progress testing* and *feedback with correction* procedures.

In using this system, the teacher divides a course of study into units of approximately two week's length, each unit having clearly defined objectives. Following the teaching of the unit, students take a test to determine who has mastered the objectives. The test diagnoses which objectives have or have not been acquired. Those students who exhibit mastery are permitted to engage in self-instructional enrichment activities. For those who have not yet shown mastery, additional sessions of instruction are provided, such as small-group study, individual tutoring, or additional self-study materials. These students are again tested when they believe themselves prepared, with the intention that all will eventually show mastery of the objectives. The addition of progress and diagnostic and corrective feedback procedures makes a distinct contribution to instructional precision. Evidence of the system's effectiveness has been reviewed by Block and Burns (1976).

Tutoring versus Other Methods

Bloom (1984) has described a series of studies, conducted by students under his direction, that provide direct contrasts of the effectiveness of several methods of instructional delivery. As compared with conventional instruction in groups of 30 students per teacher, the use of mastery learning procedures resulted in improvements in achievement amounting to a rise from the 50th percentile to the 84th percentile. (This is an increase of one standard deviation, or 1 sigma, as Bloom describes it). When tutoring was employed as a method, the increase in achievement was from the 50th to the 98th percentile, or 2 sigma.

These striking effects of the tutoring method, extending even beyond those of mastery learning in 30-student groups, raised the general question: What aspects of tutoring can be incorporated into large-group instruction so as to increase its effectiveness? Can strategies of instruction be used with 30-person groups that can raise achievement from the 50th percentile to the 98th percentile, the tutoring level?

One technique of instruction that was investigated in 30-person groups was called "enhanced prerequisites." Actually this was the same as the instructional event "stimulating recall of prerequisites," since it involved helping students review and relearn the prerequisites they lacked. The subjects being learned were courses in second-year French and second-year algebra. The achievement resulting from his treatment, using otherwise conventional instruction, was from the 50th to the 76th percentile. In other comparable classes, the enhanced prerequisite technique combined with mastery learning procedures brought achievement to the 95th percentile.

Other aspects of tutoring-related instruction investigated by Bloom's students included combinations of (1) enhanced cues and student participation and (2) enhanced cues, participation, and reinforcement (corrective feedback). The meaning of enhanced cues in these studies was the provision of explanations of the concepts and rules being learned; participation was encouraged by having students note the frequency of their participation in learning and the problems they had in understanding the instruction. Both of these combinations of supplementary instructional events brought about substantial increases in student achievement. The greatest effects, beyond the 96th percentile, were found when these added techniques were combined with the procedures of mastery learning.

These studies confirm and reconfirm the effectiveness of mastery learning procedures, particularly those that inform students of correct and incorrect performances and permit restudy until success is achieved. Beyond this, they show that certain instructional events that normally characterize the tutoring situation can be employed satisfactorily in groups of about 30 students. These "enhanced" instructional events include (1) assuring review of prerequisites, (2) employing student participation as a part of learning guidance, and (3) enhancing cues to retrieval by employing elaboration of concepts and rules (as in using explanations). In the absence of one-on-one instruction, it appears that particular attention to these events of instruction can accomplish in large groups much of what is expected from tutoring.

SUMMARY

The nature of instruction delivered to groups is determined in many important respects by the size of the group. For purposes of distinguishing the characteristics of instruction, it is useful to consider three different group sizes: (1) two-person groups; (2) small groups containing approximately three to eight students', and (3) large groups of fifteen or more members.

The characteristics of instruction applicable to groups of these three different sizes can be understood in terms of the degree of precision with which instructional events can be managed by the teacher. Generally speaking, the two-person situation, consisting of a tutor and a student, affords the greatest degree of precision for instructional events. As the size of the group increases, control over the management of instructional events grows progressively weaker. That is to say, the effects of necessary instructional events on individual learners decreases from near certainty to lesser degrees of probability as the group size grows larger. Learning outcomes, accordingly, come to depend increasingly upon the self-instructional strategies available to the individual learner.

A particular feature of the instructional situation that is typically more difficult to manage as group size increases is the diagnosis of entering capabilities. Means of assessing what the individual students know or do not know at the beginning of each lesson are readily available to the tutor but become more difficult to accomplish with larger groups. This factor is of particular importance for the execution of the instructional event *stimulating recall of prerequisites* since students will obviously be unable to recall something they have not previously learned. The control of this event is thus likely to grow weaker with larger groups, and the result may be a cumulative deficit in student learning.

The two-person group, using tutoring as a mode, makes possible relatively precise management of instructional events, from early ones such as gaining attention to the late ones providing for retention and learning transfer. In the small group, precision of control of instructional events is attained largely by multiperson tutoring, that is, by initiating each instructional event for the different members of the group in turn. In such circumstances, some events become only probable (rather than certain) for some students on some occasions. With the aid of self-instructional strategies of individual students, small-group instruction can attain considerable effectiveness. Small groups are frequently formed by dividing larger groups. Examples of small groups formed in this manner are those for instruction in basic skills in the elementary grades, and student-led discussion groups in college classes.

Large-group instruction is characterized by weak control of the effects of instructional events by the teacher. The gaining of attention, the cuing of semantic encoding, the eliciting of student performance, and the provision of corrective feedback can all be instituted as events, but their effects upon the learning processes of students are only probable. Sometimes, indeed, the effects of these events on the individual learner have quite low probabilities. Learning from large-group instruction therefore depends to a considerable degree on the learner's own strategies of self-instruction. This circumstance is more or less expected in college students and adult groups.

Typical modes of instruction in large groups are the lecture and the recitation class. A number of techniques have been suggested for overcoming the weaknesses of these large-group instructional methods. Frequently, large groups are divided into smaller groups, and sometimes into two-person groups, in order to bring about some of the advantages of increased precision of control over instructional events. One system for the improvement of large-group instruction is mastery learning, in which units of instruction are managed so that diagnosis and corrective feedback follow the learning of each unit until mastery is achieved.

Studies have also shown that aspects of the tutoring situation, such as enhancing the learning of the prerequisites, encouraging student participation in learning guidance, and adding elaboration cues for retrieval, can bring about substantial improvements in achievement in large-group situations.

REFERENCES

Anderson, T. H., Anderson, R. C., Dalgaard, B. R. Wietecha, E. J., Biddle, W. B., Paden, D. W., Smock, H. R., Alessi, S. M., Surber, J. R., & Klemt, L. L. (1974). A computer-based study management system. *Educational Psychologist, 11,* 36–45.

Block, J. H., & Anderson, L. W. (1975). *Mastery learning in classroom instruction.* New York: Macmillan.

Block, J. H., & Burns, R. B. (1976). Mastery learning. In L. S. Shulman (Ed.), *Review of research in education, 4.* Itasca, IL: Peacock.

Bloom, B. S. (1974). An introduction to mastery learning theory. In J. H. Block (Ed.), *Schools, society and mastery learning.* New York: Holt, Rinehart and Winston.

Bloom, B. S. (1976). *Human characteristics and school learning.* New York: McGraw-Hill.

Bloom, B. S. (1984, June). The 2 sigma problem: The search for methods of group instruction as effective as one-to-one tutoring. *Educational Researcher,* pp. 4–16.

Cloward, R. D. (1967). Studies in tutoring. *Journal of Experimental Education, 36,* 14–25.

Devin-Sheehan, L., Feldman, R. S. & Allen, V. L. (1976). Research on children tutoring children: A critical review. *Review of Educational Research, 46,* 355–385.

Ellson, D. G. (1976). Tutoring. In N. L. Gage (Ed.), *The psychology of teaching methods* (Seventy-fifth Yearbook of the National Society for the Study of Education). Chicago: University of Chicago Press.

Gage, N. L. (1976). (Ed.). *The psychology of teaching methods* (Seventy-fifth Yearbook of the National Society for the Study of Education). Chicago: University of Chicago Press.

Gagné, R. M. (1974). *Essentials of learning for instruction.* New York: Dryden Press-Holt, Rinehart and Winston.

Gall, M. D., & Gall, J. P. (1976). The discussion method. In N. L. Gage (Ed.), *The psychology of teaching methods* (Seventy-fifth Yearbook of the National Society for the Study of Education). Chicago: University of Chicago Press.

Gartner, A., Kohler, M., & Riessman, F. (1971). *Children teach children.* New York: Harper & Row.

Joyce, B., & Weil, M. (1980). *Models of teaching* (2nd ed.). Englewood Cliffs, NJ: Prentice-Hall.

Maier, N. R. F. (1963). *Problem-solving discussions and conferences.* New York: McGraw-Hill.

Maier, N. R. F. (1971). Innovation in education. *American Psychologist, 26,* 722–725.

McLeish, J. (1976). The lecture method. In N. L. Gage (Ed.), *The psychology of teaching methods* (Seventy-fifth Yearbook of the National Society for the Study of Education). Chicago: University of Chicago Press.

Sharan, S. (1980). Cooperative learning in small groups: Recent methods and effects on achievement, attitudes, and ethnic relations. *Review of Educational Research, 50,* 241–271.

Walberg, H. J. (1976). Psychology of learning environments: Behavioral, structural, or perceptual? In L. S. Shulman (Ed.), *Review of research in education, 4.* Itasca, IL: Peacock.

15 INDIVIDUALIZED INSTRUCTION

Teachers have long sought ways to make teaching more precise while adjusting both the objectives and the methods of learning to the needs and characteristics of individual learners. These efforts have largely been frustrated because teachers have had no *delivery systems* designed to adjust instruction to the individuals in a group of 25 or more learners. Although teachers have traditionally divided their time between working with individuals and with groups of varying sizes, this arrangement often leaves some pupils unoccupied and unable to progress for some periods of time. Little is accomplished by teachers merely asserting a determination to adjust their teaching to the individual needs of pupils. They need a delivery system designed to achieve such a purpose.

In the previous chapter, we described some of the attempts made by Bloom and his students (1984) to overcome some of the difficulties inherent in group instruction. Bloom maintains that one-to-one tutorial instruction is the most effective form, and that an average student in a tutorial program achieves more than 98 percent of students in conventional classroom instruction. (Bloom calls the difference in effectiveness of instructional mode the 2-sigma problem, referring to the fact that the achievement difference is two standard deviations in size). We note here, however, that besides the tutoring mode of instruction, there is another set of solutions for the 2-sigma problem. These are centered on the use of instructional materials that address the student directly, without depending upon a teacher for delivery, and that are tailored to individual student needs.

Early efforts to design delivery systems to assist teachers in individualizing instruction were made before the advent of more recent technology such as the current models of the design of instructional systems represented, for example, by this book. Also, the early efforts were largely local ones because there was no marketing agency concentrating on development and diffusion of special materials and media to support the plan for individualization. Many early experiments in schools thus left too great a burden on the teacher. In short, a total support system for individualized instruction was lacking.

In recent years, universities and private research and development agencies have developed comprehensive delivery systems for individualized instruction. The design and development phases for the delivery system were often supported in part by federal or state funds, and private funding made possible the diffusion of the developed system. The purposes of these systems, broadly defined, were:

1. to provide a means for assessing the entry skills of students;
2. to assist in finding the starting point for each student in a carefully sequenced series of objectives;
3. to provide alternative materials and media for adjustment to varying learning styles of students, including choices between print and nonprint materials;
4. to enable students to learn at their own rates, not at a fixed pace for the entire group; and
5. to provide frequent and convenient progress checks so that students did not become "bogged down" with cumulative failures.

In total, these measures were to enable all pupils to work each day on objectives within their individuals needs, capacities, prerequisite skills, and rates of learning. This was accomplished, in part, by designing the learning materials and media that could carry more of the support for more of the instructional events. Team efforts were often employed to design, develop, evaluate, and diffuse the learning materials as a component of the entire delivery system. In short, a systems design model was employed to provide teachers with a total delivery system to support the classroom activities.

NATIONALLY DIFFUSED SYSTEMS

Three individualized-instruction delivery systems have been used widely in the elementary and middle schools throughout the nation. The three systems vary somewhat in subject areas covered and in age range of pupils. Generally speaking, all were eventually intended to include the areas of reading, mathematics, sciences, and social studies from kindergarten through the upper elementary and sometimes to the secondary levels. The three programs referred to have been described by their designers in books edited by Weisgerber

(1971) and by Talmage (1975). The names of the programs are: Program for Learning in Accordance with Needs (Project PLAN); Individually Prescribed Instruction (IPI); and Individually Guided Instruction (IGE). Other widely diffused systems for individualizing instruction are also described in the book edited by Weisgerber (1971).

The three nationally diffused systems are the subjects of an evaluation report (EPIE, 1974). Other reports describe the operations of the programs in the schools and provide guidance for school administrators and teachers concerning their potential selection, adoption, or adaptation to meet local needs (Edling, 1970; Briggs and Aronson, 1975). Reiser (1987) has recently reviewed the substantive objectives of each of these plans and the fate of each. In spite of their proved effectiveness, all three of these programs have declined in use or have been discontinued when federal funding ceased.

LOCALLY DEVELOPED SYSTEMS

Some schools or school districts have undertaken either to make adaptive changes in one of the widely marketed systems of individualized instruction or to develop entirely new systems locally. Thus some schools have adopted available systems with few modifications, some have made major modifications of available systems, and some start at the beginning to develop their own systems. When the objectives of a school coincide closely with those of an available system, it seems wasteful to repeat the entire instruction design when it has already been carefully done. Of course, schools sometimes purchase certain materials or components of a system and develop others locally.

In visits to two nonoverlapping samples of schools employing one or another systems of individualized instruction, great variations in specific classroom applications were found, although all were directed generally toward the five purposes previously described. Edling (1970), who visited 46 schools, highlighted both the common and the unique features of operation. Briggs and Aronson (1975), who visited 42 schools, described typical operations in selected schools and summarized the various factors to be considered by schools when plans are made to initiate a new program. These factors pertain to information needed by school boards, parents, administrators, teachers, and pupils.

The development of locally developed systems is sometime facilitated by publishers who produce materials that identify skills in a sequence. Schools may adopt a reading series based on objectives keyed to test items furnished to the publisher. However, few publishers have actually validated the effectiveness of their material. Usually, they have simply added objectives and test items onto existing materials. The State of Florida has recently recognized this deficiency and has implemented a set of textbook selection criteria that require publishers to validate the effectiveness of their materials. It is possible that other states will follow suit.

VARIETIES OF ACTIVITIES

Once a system of individualized instruction is in operation in a school, what typical activities might be observed? The following description (adapted from Briggs and Aronson, 1975) represents a typical hour in an individualized reading program in a classroom with one teacher, one teacher's aide (who might be a paid paraprofessional or volunteer parent), and 25 third-grade pupils.

Several children are still learning to visually discriminate the letters of the alphabet. They are working individually with programmed instruction booklets. Each page shows a letter at the top; the child is to underline one of the two letters placed lower on the page that matches the letter at the top; feedback is provided when the page is turned. These children learn slowly, but they are learning and so still experience success rather than failure during this hour. These children were in a regular classroom last year.

Two other children can discriminate the letters, and now they are learning to pronounce the names of the letters. They are taking turns running cards through a machine; each card has a letter printed on it, and a sound recording pronounces the name. The children imitate the pronounciation.

Five children are sitting in a corner of the room with the teacher. These children can read, pronounce, and give their own definitions of many words, and they can read some sentences. However, they need help on word attack skills for unfamiliar printed words. After some instruction in this small group, the teacher will assign various lessons to be completed at another corner of the room. There each child will work with a booklet accompanied by a sound tape describing the exercises and providing feedback. The tape is paced to give instruction, to pause for pupil responses, and to give correct-answer feedback.

Four children are listening to "read along tapes" while silently reading from a printed text.

One child is taking an oral test administered by the teacher's aide. Another child has in his hand a completed written test for an objective; he is waiting for it to be checked by the aide. One child is checking off an objective for himself on a record sheet on the wall; he has passed a self-graded test, using an answer key. Another child is at the materials file, looking up the materials indicated by a sheet giving him a new objective for study.

The noise level in the room is higher than in a conventional classroom, but it is mainly productive noise and the children are no longer distracted by it. The teacher pauses from her small-group work to reprimand one boy who is annoying a classmate who is trying to read.

Over in the far corner of the room, one boy is lying on the floor reading a sixth-grade level book.

How can a teacher arrange for all these activities and conduct some direct instruction with individuals and small groups? The diagnostic and placement tests are keyed to objectives in a sequence, which in turn are keyed to varieties of materials available for the objectives. The materials are arranged in files that the students have learned to use to gain access to materials. They also have learned to return nonexpendable materials to their proper places in the file.

Still the children do not spend a major portion of their time working alone. In the next hour "show and tell" is scheduled for everyone, since they can all communicate at a common level of oral speech although they vary widely in reading ability. Also, in this particular school, only reading and mathematics instructions are individualized. Other subjects are taught by conventional methods. The children are able to operate the variety of equipment available. The equipment is simple to operate, and instruction on its operation was given at the beginning of the year.

IMPLEMENTING INSTRUCTIONAL EVENTS

As may be inferred from such a sketch of a typical hour of activity, media are often employed for some of the events of instruction. Careful attention is also given to the sequencing of objectives and to each child's progress in the sequence.

Taking Account of Prior Learning

Diagnostic and placement tests are given at the beginning of the school year to determine just which skills, in a carefully ordered sequence of objectives, each student already has mastered and can recall at the time of testing. The results of such tests determine which objective represents the starting point for an individual student. Frequent subsequent testing helps to update individual records of mastery of prerequisites for later objectives in the series. In the case of reading and mathematics, a single sequence of objectives is often adopted for all learners, but the materials used and the pace of learning vary among learners. In science and social studies, there are often core objectives assigned to all learners, with enrichment or "excursion" objectives selected by the student according to personal interest.

Gaining Attention

This event usually presents fewer problems in an individualized program than in large-group instruction. Each student is usually eager to begin a new objective, having achieved success on the previous objective. As soon as students obtain the materials (and in some cases, the equipment) for a new objective, they usually turn at once to the learning task. It is rare that more than one student at a time is not actually working on the assigned objective.

Maintaining attention also is seldom a problem. The systematic cycles of presenting a problem, requiring a response, and providing feedback, which are built into the material, tend to maintain attention. In elementary grades, children are often encouraged to "turn off the machine" or to "read for fun" or to turn to something like clay modeling when they tire of a task. They then will usually return to the task without prompting.

Informing the Student About Objectives

Owing to the highly structured nature of much of the learning materials, the objective is often evident to the learners. However, each objective in the sequence usually carries both a number and a name—the number to facilitate the filing of material, and the name as a shortened form of the objective. The students thus become aware of the various objectives in the series. In small-group sessions, when the teacher undertakes to initiate a new skill or to verify completion of objectives, the objective is made evident if it is not already known to the members of the group. It may be noted that under individually paced programs, the composition of small groups shifts constantly. A group of five who are all at the same point of progress on one day may not be at the same common point on another day.

Although in general there is no reason why objectives should not be given to learners in terms that they can understand, this event does not appear to have as much importance for highly structured material as it does for loosely structured material. In the latter type of material, the student needs the objectives to determine which portions of the material are most relevant, so that selective reading and review may be undertaken.

Stimulating Recall of Prerequisites

Recall of prerequisite learning also may be easy to achieve in a carefully sequenced, highly structured program. The careful sequencing makes recall of immediately prior learning highly probable, and the structure of materials makes more probable the initial mastery of prerequisites. Also, the frequent progress checks (usually after study of each objective) prevent cumulative forgetting that could slow further learning.

Presenting the Stimulus Materials; Eliciting the Response; Providing Feedback

Since individualized programs for elementary schools make much use of self-instructional materials, it follows that there are built-in cycles of presenting the stimulus information, requiring a response, and providing feedback. This feature is commonly found in the various print and non-print media used for individualized instruction. It may be the appropriate and precise management of these events of instruction that represents one of the strongest features of individualized programs. This strength, it may be noted, is primarily a feature of the *materials*. The primary strengths of the *teacher* in such programs lie in managing and monitoring the entire system in the classroom and in assuring that personal guidance is available when provided materials and tests fail to function adequately for an individual learner.

Providing Learning Guidance

Much of this event is also designed into instructional materials in the form of prompts, cues, and suggestions to the learner. This function may be blended in with the event of *providing the stimulus material* in a somewhat more precise manner than can be provided by a teacher, except when using the tutorial mode of instruction.

Teachers, however, are often able to give a more general form of guidance than the provided encoding cues. Teachers discuss with pupils which alternative materials may be best for them, and which enrichment or elaboration objectives they may wish to choose. One of the advantages of individualized delivery systems is that they give the teacher time to spend with individual pupils. Once the pupils learn the basic procedures for pretesting, learning, and posttesting, and how to locate materials and equipment, the basic system runs itself. This does not mean that the teacher is not busy. Often there are times when students must wait to see the teacher for guidance or testing. But gradually pupils learn to signal the teacher of their needs and to turn in the meantime to enrichment activities.

Individualized systems can be a great boon to slow learners, protecting them from inappropriate tasks and inappropriate instruction, and hence helping them to avoid failure. These systems can also free fast learners from the boredom of unnecessary instruction, allowing time for more challenging activities. It is the teacher, of course, who must see that learning arrangements are suitable for each learner.

Assessing the Performance

For some objectives, learners may test themselves by use of an answer key. But at least periodically, and more often than in large-group instruction, the teacher makes the assessment with or without the use of written tests. Again, time freed from other instructional events is available for this assessment by the teacher.

Enhancing Retention and Transfer

Once the teacher has assessed performance and found it to be satisfactory in terms of mastery or adequate according to acceptable standards, attention may be turned to enhancing retention and transfer of learning. As shown in Chapter 14, group instruction is often suitable for this purpose. When several learners have been assessed and found to be at the same point of progress in mastery of given objectives, discussion led by the teacher can be lively, interesting, and effective. Students can hear each other's applications of skills or information, and at the same time progress may be made toward terminal objectives pertaining to attitudes or cognitive strategies. Other techniques for enhancing retention and transfer may be designed as part of the materials or may take the form of projects selected to match student interests.

Because the events of instruction can be arranged in the ways described, learning in an individualized program is often perceived to be less difficult (and perhaps more enjoyable) than in a large-group situation. Although it is frequently mentioned that an individualized program requires self-management and self-instruction, it may also be the case that given adequate motivation, such a program is more precisely helpful to the learner than instruction in a large group. Therefore, it need not be concluded that individualized instruction is effective only for mature learners. Owing to the assured appropriateness of each lesson in the series and to the precision of the instruction, it might be said that this form of instruction is needed more by less mature learners. Adults and college students, on the other hand, are expected to be able to discern appropriate from inappropriate materials and to provide many of the instructional events for themselves. For mature learners, the form and structure of individualized programs may be expected to differ from those designed for younger students.

SYSTEMS FOR OLDER STUDENTS

In college-level instruction, several individualized methods have been developed and fairly widely adopted. Two of the best known are the Keller Plan (Keller, 1966; Ryan, 1974) and the Audio-Tutorial Approach (Postlethwait, Novak, and Murray, 1969).

As might be expected considering the maturity of the learners, the individualized methods employed at the college level are not as fine-grained and precisely controlled as those for elementary schools. Generally, the procedures are designed to utilize an economical combination of large-group lecture sessions, smaller quiz sessions, and independent study. Tests are employed to enable students to participate in the most appropriate activities. Insofar as possible, the mastery learning concept is adopted, within the limits of administrative policies concerning the pursuit of courses beyond scheduled academic terms.

In college programs of this sort, students concentrate their study in the areas needing the most work, as indicated by progress checks. Less use is made of the small-step variety of learning materials. Instead, laboratory assistants are available to provide some individual tutoring when other methods fail. The typical combination of conventional and special procedures results in a degree of instructional precision that is perhaps intermediate between that obtained in large groups and that described previously as appropriate for elementary programs. This may represent a reasonable degree of control in trying to enhance learning for adult learners without undue costs.

An earlier strategy for individualizing learning at the college level was developed by Pressey in the 1920s. He later termed the procedure *adjunct autoinstruction* (Pressey, 1950). The strategy was quite straightforward; a regular college textbook was employed, along with sets of practice test questions for each chapter. Mechanical devices were used to provide "right-wrong" feedback after each response to a practice test question. This procedure was employed in regular classrooms, in independent study programs, and in special classes for superior students (Pressey, 1950; Briggs, 1947, 1948). Adjunct autoinstruction was not widely adopted as a regular classroom procedure. Interestingly enough, questions about various ways of placing the adjunct questions in an overall study procedure have become prominent in recent years (Frase, 1970; Hiller, 1974; Rothkopf and Bisbicos, 1967).

VARIETIES OF INDIVIDUALIZED INSTRUCTION

Apart from the types of delivery systems described earlier for elementary and secondary schools and for college instruction, the term individualized instruction has been used in reference to a diverse array of educational methods. Some of these may be described as follows:

1. *Independent study plans,* in which there is agreement between a student and a teacher on only the most general level of stated objectives to indicate the purpose of studying. Students work on their own to prepare for some form of final examination. No restrictions are placed upon students as to how they may prepare for the examination. A course outline may or may not be provided. The task may be described at the course level in such terms as "preparing for an examination in differential calculus," or at the degree level as in honors programs in English universities. A similar procedure is used in the United States in preparing for the doctoral comprehensive examination in many fields.
2. *Self-directed study,* which may involve agreement on specific objectives but with no restrictions upon how the student learns. Here the teacher may supply a list of objectives that define the test performances required to receive credit for the course; the teacher may also supply a list of readings or other resources available, but the student is not required to use them. If a student passes the test, he or she receives credit for the course.
3. *Learner-centered programs,* in which students decide a great deal for them-

selves within broadly defined areas—what the objective will be and when to terminate one task and go to another. This degree of openess is sometimes found in public schools and has been the customary style of operation for a few private, special schools. Usually in public schools, learner choice is permitted only for enrichment exercises, and then only after certain required or "core" skills have been mastered. Often such activities are offered as an incentive to the student to learn the core skills.

4. *Self-pacing,* in which learners work at their own rates, but upon objectives set by the teacher and required of all students. In this case all students may use the same materials to reach the same objectives—only the rate of progress is individualized.

5. *Student-determined instruction,* providing for student judgment in any or all of the following aspects of the learning: (a) selection of objectives; (b) selection of the particular materials, resources, or exercises to be used; (c) selection of a schedule within which work on different academic subjects will be allocated; (d) self-pacing in reaching each objective; (e) self-evaluation as to whether the objective has been met; and (f) freedom to abandon one objective in favor of another.

This description implies the possibility of more than 20 different ways in which instruction may be said to be individualized or "learner-determined" if various permutations and combinations of elements controlled by the learner are considered.

Clearly, most of the varieties of individualized instruction tend to place greater responsibility upon students for providing the events of instruction than is the case for the delivery systems discussed in previous sections. For the most part, these varieties also permit greater freedom of choice as to what the objectives will be and how the objectives may be attained. For these reasons, the latter methods have been employed mainly for selected groups of students.

THE MANAGEMENT OF INDIVIDUALIZED INSTRUCTION

Systems of individualized instruction may provide a large number of materials, each separate item of which is keyed to an objective. Alternatively, use may be made of existing materials to form a module of instruction by giving printed directions on how to use the materials to achieve one or more objectives. For younger children, often an objective will be so limited in scope that it can be mastered in an hour or less. For older students, a module may require a week or two of study, and evaluation may be made for a cluster of objectives rather than for a single objective.

The management of the day-to-day progress of pupils is related closely to the frequent progress checks that are keyed either to single objectives or to groups of objectives representing modules. Thus the frequency of formal checks upon pupil progress tends to decrease with the age of the learners. In a similar vein, the frequency of responding (followed by feedback) may often also be decreased for older learners; this feature is adjusted either by the way self-instructional materials are designed or by the way a module is assembled. The management of learning, then, is usually centered on single objectives for young children and on modules for older children.

Sometimes the module will contain all the instructional materials needed to pass a test on the objective. It usually also contains practice tests that the students can use to judge their readiness for taking the actual test. In the event that materials and resources physically independent of the module itself are to be used, directions for how to locate and use them are included. Thus the module and its directions for using related materials allow the learner to go about the learning task without directions from another person, except when difficulties are encountered.

We now give further consideration to the nature of materials for individualized procedures that are applicable to young learners, and to those applicable to adults.

Components for Young Learners

A typical individualized program of instruction for children may be expected to have somewhat different components than a program for adults. The procedures for using the materials will also be different. The following are brief descriptions of typical components of modules designed for children at about the sixth-grade level, who are assumed to have some reading ability.

A List of Enabling Objectives

Often the learner may benefit from seeing both the target objective for the module and the prerequisite capabilities to be acquired. These may be shown simply as a list, or they may be in the form of a learning hierarchy (described in Chapter 12).

A Suggested Sequence of Activities

In part, the sequence of activities may be derived from the sequence of enabling objectives, and in some instances alternative sequences may be chosen. The sequence as a whole needs to make suitable provision for the enhancement of retention and transfer. Sometimes alternative materials, resources, or exercises may be offered as options. Students may be encouraged to find for themselves which materials seem suitable. One student may prefer or profit most from a programmed text; another may find a slide-tape presentation more effective.

A Menu of Modules

Some programs contain only required modules. The total menu, however, should be designed to meet the needs of fast learners as well, and not solely those of slow learners. Alternatively programs may offer both core and enrichment modules, and still others may consist entirely of student-chosen modules. The student-selected modules may be designed to provide only self-evaluations of student performance, since the objectives represent what the student wants to learn.

Programs may be designed to make use of the principles of contingency management—using a preferred (high-reward) activity as an inducement to undertake a study activity. Often such programs include procedures that give the student opportunities to make "contracts," with some required minimum number of modules to be completed by each student. The student may receive a number of points at the outset, which he can "spend" to negotiate time to complete a module; and he may, in turn, earn points for successful completion within the contract period. The earned points, within limits, may then be spent to earn free time for preferred new learning or for other kinds of preferred activity.

A more extreme curriculum philosophy holds that there should be no modules and no objectives. According to this view, the learner would simply be put into a learning environment that includes learning resources, laboratory materials, supplies, and so on, perhaps attractively arranged to induce interest, but with no requirements, points, or other rewards, other than the intrinsic reward of enjoyment of learning.

Opinions differ greatly on whether the student should be required to learn or even to try to learn anything he does not voluntarily undertake. Opinions also differ on how specific objectives for individual learning should be. Those who dislike specific objectives usually shun the use of modules, preferring an open environment that permits the student to choose what is to be learned. It would seem, however, that society must take responsibility for teaching children how to live in our culture as productive, happy, responsible, adult citizens. Since it is difficult to determine the exact nature of human capabilities a child will need to achieve his goals and to solve problems not yet foreseen by today's adults, emphasis needs to be placed on intellectual skills and problem-solving strategies, rather than simply upon presently known "facts" (see Rohwer, 1971). Programs in science and social studies are often designed to emphasize the attainment of "process" objectives, such as, "the student will arrange numerical data in tabular form."

Alternative Materials for Single Objectives

It is evident to most teachers that some children master a given learning objective better by using one book, medium, or exercise, than by using another that may have equally good content. In some cases the reason may be obvious; a poor reader will understand a tape-slide series better than a book. In less obvious cases, individual *learning styles* are cited, although the specific meaning of this phrase is not entirely clear. Research studies, however, have identified few intellectual and personality characteristics that can be related to success with specific forms or media of instruction (Briggs, 1968; see also Chapter 6). This finding may result either from the existence of differing entering capabilities that match the specific content of various materials or from the fact that instructional events more appropriate to some individuals occur to a greater extent in certain materials. It may be that features of the instructional materials—small versus large steps; inductive versus deductive; concrete versus abstract; or other characteristics of this sort can be shown to be differently effective for different learners. At any rate, providing several versions of a module is often worthwhile. One version might have a simpler vocabulary; one might employ a shorter sentence length; and another might combine an outline or advance organizer with a presentation that is technically complex.

The alternative-module concept clearly raises an economic question. Research is needed to assess the extent of advantages of alternative materials so that these can be considered in light of costs. Similar data are needed relevant to the instance in which one form of materials is superior to other, less costly materials for most learners.

A Feedback Mechanism

For young learners and lengthy modules, it may not be wise to wait until a test is given to provide feedback. Feedback after small increments of study is usually desirable. Feedback at frequent intervals is a built-in feature of both text-based and computerized programmed instruction. The effectiveness of many media, such as television or film, can be improved by building in explicit provisions for learner response and feedback. In addition to enhancing learning, such response and feedback may suggest the need for diagnostic testing and remedial instruction or for restudy of the module when performance is poor.

Mechanical devices and chemically-treated answer sheets can be used to provide feedback after each response to practice questions. Television teachers can pose questions after a brief lecture segment, pausing for the viewer to write his answer or just to think of the answer; after the pause, feedback can be given. Questions used with live lectures have also been found to benefit learning and usually retention also. Classroom devices can be employed to provide immediate automatic recording of students' answers (to multiple-choice questions) for viewing by the teacher who can reprogram the lecture on the spot.

When a module is brief, or when the learner typically succeeds readily, a parallel form of the formal test on the objectives can be used to give feedback after learning is completed. Some form of self-testing or response with corrective feedback can usually be devised as part of each module. This enhances learning and saves time for the teacher, who can then do individual remedial teaching when all other means fail—a valuable activity for which teachers have too little time when employing conventional group instruction. Such a procedure provides an added bonus—the teacher has time to guide thinking and give added feedback on an individual basis when most needed.

Components for Adult Learners

The nature of both the materials and the procedures may properly be less highly structured for college students or other experienced adult learners.

Objectives

Course objectives for adults may sometimes be quite precise and specific. However, it may still be assumed that evaluation of learner performance can be made at less frequent intervals than would be the case for children. Whether one broad objective or many more specific ones are employed in modules, checks on the performance of the adult learner are typically not made until after a rather long period of study.

Directions

Directions for pursuing study may also be greatly abbreviated for adult learners. The learners may be provided with a list of resources or simply told to "use the library and laboratory." The objective itself may be the main source of directions.

Learning Materials

Materials for learning may be highly structured, as in a programmed text; semistructured, as in an outline or laboratory guide; or unstructured, as would be the case when the student does library research on a topic.

Evaluation of Performance

Adult students may have a few weeks or an entire semester to complete an instructional unit. While working on the unit, the student usually receives feedback from a teacher or advisor and from conferences providing reactions to draft plans and preliminary reports. Students may also receive direct instruction on a variety of subordinate capabilities, such as writing skills, techniques for finding sources, and others. Usually the evaluation is based on the appropriateness of procedures employed, competence in reporting and interpreting data, and ability to defend a rationale for the product or study that has been completed.

Functions of Modules

Modules may specify activities for groups, small or large. In such instances, a class chart shows the progress of each pupil and is used to form groups that are at the same point of progress.

Modules can also be designed as directions for laboratory or field exercises or for independent learning not based on instructional materials. In one industrial course for adults, the learners were given the entire set of course objectives and shown where they could go to take tests. They were then free to visit employees in appropriate departments to observe, ask questions, or see other ways to learn.

It should be noted that modules need not be restricted to cognitive objectives. Objectives in the affective or motor domain can be devised equally well. In shop courses, where machine time must be carefully planned, all needed cognitive learning can be made to take place before a skill is practiced with the equipment. This not only saves money for duplicate machines, but also avoids injury to persons and damage to equipment by assuring that the trainee knows safety precautions and correct procedures before having access to the machine. In many cases, a simulator (Chapter 11) of the actual machine also brings benefits in cost, safety, and efficiency. Simple training devices can be used for parts of the total task, reserving the more expensive, complex simulator for consolidation of skills and practice of emergency procedures in a safe environment.

USE OF MATERIALS IN INDIVIDUALIZED INSTRUCTION

A particular set of instructional materials has been developed for use with the individualized system called PLAN (described in Weisgerber, 1971). This particular system will be described to provide a concrete illustration of how such materials can be employed in individualizing instruction. PLAN was used by a number of schools throughout the United States in the mid-1970s.

The instructional objectives of PLAN formed the basis of a curriculum in language arts, social studies, science, and mathematics for grades 1–12. Within each grade and subject, these objectives were organized into *modules of study* for use by students. Usually, five or six objectives constituted a module. A program of studies was developed by the student and teacher, and this program guided the student in selecting modules appropriate to his needs and interests.

A central feature of the PLAN system involved the use of a computer, a terminal for which was usually located in each school. PLAN was a "computer-supported" system. The computer received and stored records about each student's previous study, progress, and performance record. On a daily basis, information was printed out for the teacher indicating (1) which lesson objectives were completed by which students and (2) what activities were begun or completed by each student. In addition, the computer furnished periodic reports of progress on each student. In general, the information stored in the computer constituted a base of essential information for planning individual student programs and guiding student learning activities.

Modules and TLUs

A PLAN module was a unit of study lasting two weeks, on the average. Sometimes modules dealt with single topics, sometimes not. They were collections of activites representing closely related objectives, such as those in writing, speaking, and spelling. In any case, modules were composed of several *teaching–learning units* (TLUs), each of which had a single objective.

The TLU began with a learning objective that told the student what was to be learned. Following this was a list of a number of learning activities. A typical TLU pertaining to a social module for the seventh grade is shown in Figure 15-1.

As will be seen, the TLU described the learning activities to be undertaken by the student and the references to be studied. Self-test questions and discussion questions were also included. In the early grades, pictorial techniques were used to communicate to pupils the objective and the learning activities. An accompanying sheet, called the activity sheet, described additional activities for the student to do in learning about the topic of the TLU. Once the activites given in the TLU and the activity sheet were completed, the student should have been able to do what was called for in the objective and was then ready to take a performance test. If the performance was satisfactory, the student moved forward to a new TLU; if not, additional work was suggested by the teacher.

Teacher Directions

The teacher directions that accompanied each TLU were designed to communicate the objective, the plan for student activities, materials needed, and test directions. Using this sheet, the teacher was able to see at a glance what kinds of activities needed planning—whether discussions, game playing, field trips, or self-study by the student. The teacher directions made evident which modes of instruction might be needed, such as small-group work, partners working together, tutoring, or other modes. Thus it was possible for the teacher to advise the student about options for learning acitivites.

Instructional Guides

Other materials that sometime accompanied TLUs were instructional guides. Such guides provided direct instruction to the student when it was unavailable in published sources. They often made it possible for the student to develop an intellectual skill necessary for further progress in a TLU. Figure 15-2 is an example of an instructional guide in fourth-grade language arts, related to the objective, "given a root word, change its form-class by adding suffixes."

Performance Measures

When the student had completed a TLU, he took a test designed to assess his performance on the stated objectives. In some instances, the test had a multiple-choice format that could be scored by the computer. In others, his performance was observed and evaluated by the teacher in accordance with definite standards. The teacher then transmitted this evaluation to the computer for record-keeping purposes. Computer printouts of performance records for all students were ready for the teacher's use on the next day.

Materials Handling

One distinctive feature of materials in modular form is that they tend to be arranged in smaller "chunks" than is the case with traditional instruction. The materials for a single objective must be either physically separate from the materials for other objectives, or they must be clearly identified and indexed to match the objective and the test for the module.

Whether the materials for a module represent one chapter in a book, several chapters from different books, or a specially designed programmed instruction sequence, there must be a system by which the student, the teacher, and the teacher's aides can locate the materials quickly. This requires either an indexing system or the separate physical packaging of all materials for each module. The materials may be collected into a folder, which is properly stored for easy retrieval. Some kind of numbering system is convenient to use, both for planning and record keeping for each pupil and for locating and storing materials. It is handy, for example, to have "Module No. 1, converting fractions to decimals," listed on planning sheets, record sheets, and on the materials themselves when filed on shelves or in cabinets.

A planning sheet of some kind is needed for each student, particularly when some freedom in choosing objectives is given to the student. In such a case, the student may confer with the teacher at intervals to plan in advance for taking one or more modules. If, on the other hand, all students begin with Module No. 1 and complete as many modules as time permits, a single sheet

Patriots and Politicians   4712-1

OBJECTIVE

Identify reasons for the development of political parties in the United States.

---

1   When the founders of our country were writing the Constitution, there were
    many different opinions about what should be done. Read *The Promise of
    America*, pp. 140-143, and *Promise of America: The Starting Line*, pp. 129-134.
    Make a list of at least four issues on which the authors of the Constitution
    disagreed. Were these the first differences of opinion among Americans?

---

2   When George Washington became President, there were no political parties.
    Read *The Promise of America*, pp. 153-155. With a partner, look at the
    filmstrip, **The Beginning of Political Parties.** If you were in George Washington's
    place, what problems would you have had to deal with? Discuss this question
    with your partner.

---

3   Raising money was a big problem for President Washington and his Secretary
    of the Treasury, Alexander Hamilton. Read *The Promise of America*, pp.
    157-162. Which groups in the colonies supported Hamilton's policies? Why?
    Which groups opposed them? Why?

---

4   Americans also differed on how to treat foreign countries. Read about these
    differences in *The Promise of America*, pp. 164-167, and *History of Our United
    States*, pp. 206-208. Why did some Americans favor France and some favor
    England?

---

5   Not long after Washington became President it became apparent that there
were two major groups with different solutions to our problems. One of these
groups was called the Federalists; the other was called the Anti-Federalists, or
the Republicans. These two groups became the first two political parties.
Political parties are organizations of men with similar views who work together
for the same goals. Read about the beginnings of political parties in *History of
Our United States*, pp. 205-206, and *The Promise of America*, pp. 162-164.
Now do the Activity Sheet.

6   Have a debate with a partner. Pretend that you are a farmer and a supporter
of Jefferson. Your partner is a merchant and a supporter of Hamilton. Try to
convince your partner that your party's programs are best for the United States.

7   George Washington was very disappointed by the development of political
parties. He believed that everyone should be able to agree on policies that
were good for everyone in the country. Discuss the following questions with a
partner.

a. Were Hamilton's programs "better" for everyone in the country than
Jefferson's programs?

b. Can we say that there really is one program that is best for the *whole*
country?

c. Why do some people favor one program rather than another?

d. Today's political leaders also say their programs will be good for
everyone in the country. Is it possible that these programs might be
good for some people and bad for others?

Look through the newspaper. Can you find examples of politicians who
disagree about what is good for the whole country?

8   In the filmstrip there is a statement that "the formation of our first political
parties was an important development in the democratic process of
government." Would Washington have agreed with this statement? Do you
agree? Do you think that it is possible to have a democratic government
without competing parties? Can you think of any alternatives to the two party
system?

OBJECTIVE

Identify reasons for the development of political parties in the United States.

FIGURE 15-1   An example of a TLU from PLAN, with an objective in seventh-grade
social studies.
(Reprinted by permission of the copyright owner, Westinghouse Learning Corporation.)

Food for Thought   1413-1

INSTRUCTIONAL GUIDE

Every pet needs pet food, **EVEN** pet words like the little feller you see here.

The best pet food for pet words happens to be Suffix Leaves like those on the bush here.

A suffix is a group of letters which when attached can change the meaning and the form-class of the word.

Let's feed our hungry word and see what happens!

Now let's find out if our pet word has changed form-class. Remember the test sentences:

Noun: I have one **noise**. I have many **noises**.

                 noiseless   **noiseless**.
Adjective: The **noisy** boy seemed very **noisy**.

                           noisily.   **Noisily**
Adverb: The boy ate the cake **noiselessly**. **Noiselessly** he ate the cake.

Can the pet word **noise** change form-class to become a verb? Try the verb test to find out! Write your answer below. Discuss your answer with a partner.

FIGURE 15-2   An example of an Instructional Guide from PLAN.
(Reprinted by permission of the copyright owner, Westinghouse Learning Corporation.)

for the entire class may be used for planning, monitoring, and record keeping.

It is no small task, even after the modules themselves have been designed and developed, to be sure there is a sufficient supply of each module and that the supply is stored for ready access, selection of material, and return of material. If some of the materials are expendable, someone (perhaps an aide) must be sure that after each use, the expendable portion is restocked and made ready for use again. A further problem is updating module material. As new material becomes available it must be cross-referenced to the objectives and test items and put into modular form. It is this maintenance that sometimes appears to spell the downfall of individualized instruction. Although there tended to be sufficient funding for the development of individualized systems like PLAN, a lack of maintenance may have been a major factor leading to their discontinuance (Reiser, 1987).

TEACHER TRAINING FOR INDIVIDUALIZED INSTRUCTION

At first glance, the task of storing, arranging, and using modules for instruction may lead one to believe it is all more trouble than it is worth. Indeed, teachers need training in how to manage individualized instruction. At first such training may lead the teacher to feel that his most cherished functions are being usurped by the system and that he is being asked to perform only the tasks of a librarian or clerk. This is because some of the teacher's tasks are new and strange compared with those required under a conventional method of teaching. All teachers need special training for conducting and managing individualized instruction, and they cannot be expected to function adequately, let alone enthusiastically, without such training. In the future, as such training is more frequently included as a regular part of preservice teacher training programs, the problem will likely be handled at that point in the teacher's education rather than later on.

Even with appropriate training, not all teachers will like the individualized approach. Some experienced teachers will not wish to relinquish their familiar role. Others, who are tired of saying approximately the same thing year after year in class, will welcome the change in role. Once teachers are trained and experienced in the new role, most come to prefer it, usually after one year (Briggs and Aronson, 1975).

Monitoring Student Progress

Monitoring the progress made by students consists of two related functions: (1) knowing what each student is undertaking to learn and (2) knowing how fast and how well each is progressing. A glance at the class chart can show which modules a student has finished and which one is being currently attempted. For a module to be recorded as finished, the student must have met some minimum standard of performance on a test or other evaluation of achievement of the objective. Sometimes this standard is stated in the objectives, as "by solving correctly 8 of 10 linear equations." At other times a product is to be evaluated: a laboratory report, a work of art, or an analysis of an editorial with respect to evidence of bias. To make such evaluations as objective and reliable as possible, a "grading sheet" or "criterion sheet" may be used; this sheet lists the features to be looked for in the student product and delineates some system for deciding whether it meets the standard. For example, points may be assigned on each separate feature to be evaluated; or the number of features present and satisfactory may be counted. Either of these techniques is preferable to making a single overall judgment, not only because it improves the evaluation, but also because it can serve a diagnostic function—it can show the student where improvement is needed. This same criterion sheet can be given to the student at the onset of instruction, and thereby can function to inform about what is expected, suggesting how it may be done and stating how the product will be evaluated.

Some evaluations can be done orally. By discussing the module and the work done on it by the learner, the teacher can often test in a more probing fashion than can be done in written form. The assessment can also involve the planning of the next work to be undertaken or the remedial work needed. Although oral tests may be less highly standardized than written ones, they are often convenient and effective when conducted with an individual student.

Regardless of how progress is monitored, the teacher usually knows more about the progress of each student in a well-designed individual plan than when group instruction is used. One reason is that in individualized instruction every student responds to every question. Even if all students work on the same objectives, this is a desirable feature. Of course, when students work on unique objectives or unique clusters of them, evaluation must be done individually.

Assessment of Student Performance

The performance of students is assessed throughout the conduct of an individualized instruction program to achieve a number of purposes: (1) initial placement of students in an approximate level with respect to first assignments in each subject; (2) assessment of mastery of each module and of the completion of instructional objectives on enrichment of tasks; (3) diagnosing learning difficulties in order to identify needed assignments; and (4) measuring student progress in areas of the curriculum over a yearly period.

Assessment of Mastery

The assessment of student performance is of particular importance to an individualized system, particularly in the areas of intellectual skills where new assignments are made on the basis of mastery of prerequisite skills. Such day-to-day assessment should not be considered a matter of formal testing, but instead likened to the informal probing typically done by every teacher in the classroom. It differs from the latter, not in its formality of administration, but in its provision of preestablished standards (criteria) used by the teacher or by the student to judge when mastery has been achieved. Criteria for mastery are specified in programs designed for individualized instruction, along with procedures and items used for the observation of individual student performance.

Diagnostic Testing

When a student encounters a difficulty with an assignment, a brief diagnostic test is typically administered by the teacher or aide. Diagnostic procedures provide an indication of prerequisite skills and information that have been inadequately learned or forgotten by the student. They therefore provide an indication to the teacher of a desirable next assignment or review assignment that will reestablish the necessary competence in the individual student.

Attitude Assessment

The assessment of student attitudes in areas such as cooperation, helping, control of aggressive acts, and others may be done by the teacher by means of checklists completed at periodic intervals. Other socially desirable attitudes of citizenship, which may be prominent objectives in social studies instruction, can be assessed in other ways, as by questionnaires.

Typical Daily Activities

The various activities of managing an individualized system suggest typical daily activities of students, teachers, teachers' aides, and (when applicable) student tutors. As all these participants gain experience with individualized methods, things go more smoothly. At first there may be quite a bit of "slack time" while a pupil is waiting for help or to be told what to do next. Gradually the pupil becomes more skilled as an independent learner, and he finds his way about within the system and with the resources available. Progressing from a rather harassed feeling at first, to a calm, easy pacing, the teacher also becomes both more skilled and more at ease. Initially it may seem that there are too many things to keep track of, but this changes with time.

Student activities are often quite varied over a relatively short time. Especially when an entire school, rather than just one course or classroom is engaged in an individualized system, the concept of flexible scheduling is combined with individualized instruction. Then a student may spend most of one day on one subject, but the next day he may be engaged in brief activities ranging over many subjects. This freedom to concentrate heavily at one point but to diversify at other points relieves schooling of much of the boredom resulting from the same schedule every day.

A student may thus move from a study carrel, to a slide-tape area, to a small-group activity, to a conference with one or more teachers, to a test station, and on to band practice or basketball, all in less than one complete school day. At other times, a three-hour laboratory and writing session may complete the work in chemistry for a week.

Teachers usually spend some time each day for advance planning sessions with one to six students; they may review progress and test results and give next assignments to other students. On still other occasions, the teacher may do individual diagnosis and remedial instruction or confer about a change in a planned schedule. Usually the teacher arranges and conducts small-group sessions for groups of students who are at approximately the same point in their learning process.

To a great extent, students and aides keep the materials files straight, once they are taught the filing system. Teacher's aides often administer and score tests, help keep records, and help students find materials they need. They may also serve as tutors, and generally provide back-up assistance when the teacher is especially busy. In general, their role is to help implement the plans agreed upon between teacher and student.

Classroom Control

Although in general the principles of classroom control and discipline are the same for individualized as for group instruction, several factors usually tend to minimize discipline problems in the individualized method. First, the personal attention and consideration given to individual students and to their plans, ambitions, and interests all tend to motivate them positively toward achievement of success. Second, the method is designed to promote success in learning, and this becomes rewarding in itself and motivating for continued effort. Third, the teacher spends less time teaching the class by a group procedure; this leaves fewer opportunities for a student to engage the attention of the entire class with his attention-seeking behavior. The teacher's dealing with a disturbance or lack of attention on the part of a student is less likely to be noticed by the entire group. All these factors help lessen the traditional adversary relationship that tends to grow between teachers and students.

Since a system of individualized instruction is clearly designed to help each pupil succeed, fair-minded youngsters usually respond favorably. Just as the system discourages "baiting" of the teacher, it also discourages public confrontations in which neither side wishes to "back down." Finally, it removes temptation for a teacher to employ sarcasm or ridicule of poor work. It quietly reminds the teacher that the goal is learning, not platform performance or crowd psychology. Most important of all, an individualized system emphasizes learning and achievement, privately attained and privately evaluated, by a student who has accepted major responsibility for his learning.

Contingency Management

Techniques of contingency management are of great usefulness in the administration of a system of individualized instruction. In simply stated form, these are techniques the teacher uses to arrange successions of student activities in such a way that an initially nonpreferred activity will be followed by a preferred activity, thus providing reinforcement for the former. The concept of reinforcement contingencies has been developed and applied to the activities of teaching by Skinner (1968). Application of the techniques of contingency management to school situations has been described by a number of writers (Homme, Czanyi, Gonzales, and Rechs, 1969; Buckley and Walker, 1970; Madsen and Madsen, 1970).

When used properly, contingency management aids in accomplishment of three objectives that form a part of successful instruction:

1. Establishing and maintaining orderly student behavior, freeing the classroom from disruption and distraction, and aiming students toward productive learning activities.
2. Managing learning so as to instill in students a positive liking for learning and for the accomplishments to which it leads.
3. Capturing the interest of students in desirable problem-solving activities as sources of satisfaction for mastery of the intellectual skills involved in them.

In general, the teacher needs to learn to identify differences in the interests, likes, and dislikes of individual students, and to employ these in selecting specific contingencies to achieve a task-oriented learning environment.

MEDIA DEVELOPMENT AND INDIVIDUALIZED INSTRUCTION

One of the primary roles of the teacher in individualized instruction is management of the system. Because the materials largely take over the role of instruction, the role of the instructor changes to that of manager and maintainer of the system. Initially, teachers resist this role, but, as previously mentioned, this resistance diminishes with continued experience. After becoming accustomed to the new role, teachers usually prefer individualized systems over traditional teaching. However, experience has indicated that satisfactory performance of individualized systems depends upon maintenance of appropriate procedures. It is owing to a lack of continuing support for these routines that individualized systems typically fail.

What are some possible solutions? The technology of media has changed drastically since the mid 1970s. Microcomputers have made it possible for teachers to have a personal computer in their classrooms for the purposes of instruction and management. These small computers can provide the processing capability that previously could only be provided by expensive mainframes. Of particular importance is the fact that it is possible to use microcomputers as instructional delivery systems. Most research on computer-assisted instruction (CAI) has shown it to be at least as effective as other modes.

The main problem facing individualized instruction continues to be the development of materials and the maintenance of the system. Perhaps these new technologies will demand an even more novel role for the teacher, that of instructional designer–developer. This would be an entirely new role for the teacher who is trained to deliver instruction in a conventional lecture-and-discussion classroom format.

Computers are not the only new technology that can be employed in instruction. Video technology, which was in the past very expensive and not particularly well used, has undergone a revolutionary change. Formerly bulky and hard to use, equipment is now light, compact, and relatively inexpensive. A teacher can learn to operate a videocamera-recorder unit in 15 minutes. The problem lies in training teachers how best to use this equipment to support and maintain individualized instruction.

Other new technology combines the use of the computer and laser video-disc libraries in the form of information retrieval systems. A related technology is electronic "desk top" publishing. Using computers, it is possible to "down load" text and graphics from data bases, and to reconfigure and print this material on high quality laser printers. It is conceivable that teachers could use this technology to produce a customized text for each student (or selected groups of students) in their classes. Information technology (the computer management of information and communication) will certainly have an effect on our individual lives. In what ways it will be used to support individualized instruction are matters yet to be determined.

A number of researchers such as Bork (1985), Park and Tennyson (1983), Ross (1984), and others continue to ask questions regarding the role of computers in instructional systems. From this modern work, one can predict the simplification of computer software to the point where teachers can quickly and easily develop lessons to meet individual student needs. Many of these advanced conceptions of the uses of technology to solve the 2-sigma problem are exciting to contemplate but will require reconstruction of teacher training curricula if they are to have lasting effects on instruction.

The history of technological solutions to educational problems does not provide grounds for unusual optimism. What needs to be emphasized is that technology has been proved to be effective (Suppes and Machen, 1978). Perhaps even newer technology can be found to alleviate some of the problems related to costs associated with access and maintenance.

SUMMARY

Individualized instruction is designed by the same processes of planning that apply to design of individual lessons for conventional group instruction. Our previous descriptions of performance objectives, learning hierarchies, sequencing, and employment of appropriate instructional events and conditions of learning apply to the design of modules for individualized instruction.

It is the *delivery system* that primarily distinguishes the design of modules from the design of lessons. The characteristics of materials for individualized instruction include the following:

1. Modules are usually more distinctly self-instructional than are conventional lessons. More of the needed instructional events and conditions of learning are designed into the materials making up the module than is the case for conventional materials.
2. The materials incorporated into modules do more of the direct teaching, whereas in conventional methods the teacher presents more of the necessary information. Thus the role of the teacher changes somewhat. Individualized instruction depends to a lesser degree on the teacher's function as provider of information; more stress is placed on counseling, evaluating, monitoring, and diagnosing.
3. Some systems provide alternative materials and media for each objective, thus letting the selection vary according to the learner's preferences as to style of learning.

Modules for individualized instruction sometimes contain all the materials, exercises, and tests needed. In other instances, they refer the learner to external materials and activities at appropriate times. A single module usually includes as a minimum:

1. a performance objective
2. a set of materials and learning activities either self-contained in the module or external to the module
3. a method for self-evaluation of mastery of the objective
4. a provision for verification of the learning outcome by the teacher.

As a consequence of its nature, individualized instruction typically provides more frequent feedback and more frequent progress checks than is the case for conventional instruction. It may permit more freedom of choice on the part of the learner, depending on the extent to which objectives are optional or required. Usually, as a minimum, the learner sets his own pace in learning activities.

Management of individualized instruction requires a way to index and store modules, a way to schedule modules to be used by each learner, a way to monitor pupil progress, and a way to assess performance. Sometimes "contracts" are arranged to provide for required work, for enrichment work, and for earned free time for activities the learner prefers.

Classroom control problems are usually less in individualized instruction than in conventional instruction. Teachers usually need special training in the management of such systems. Once they master the necessary routines, they often prefer individualized to conventional methods.

New technologies offer a variety of opportunities as vehicles for the delivery of instruction in individualized form. Computer-aided instruction appears to be used with increasing frequency in schools and workplaces. Computers also make possible rapid access to knowledge banks and the development of instructional materials designed for individual student use. Video technology has also advanced to a degree that makes the fabrication of video lessons or modules relatively inexpensive. It is possible that new technological advances of these and other varieties will make individualized instruction more readily available and maintainable in the years to come.

REFERENCES

Bloom, B. (1984). The 2-sigma problem: The search for methods of group instruction as effective as one-to-one tutoring. *Educational Researcher, 13(6),* 4–16.

Bork, A. (1985). *Personal computers for education.* New York: Harper & Row.

Briggs, L. J. (1947). Intensive classes for superior students. *Journal of Educational Psychology, 39,* 207–215.

Briggs, L. J. (1948). *The development and appraisal of special procedures for superior students, and an analysis of the effects of knowing of results.* Unpublished doctoral dissertation, Ohio State University, Columbus, OH.

Briggs, L. J. (1968). Learner variables and educational media. *Review of Educational Research, 38,* 160–176.

Briggs, L. J., & Aronson, D. (1975). *An interpretive study of individualized instruction in the schools: Procedures, problems, and prospects.* (Final Report, National Institute of Education, Grant No. NIE-G-740065). Tallahassee, FL: Florida State University.

Buckley, N. K. & Walker, H. M. (1970). *Modifying classroom behavior: A manual of procedures for classroom teachers.* Champaign, IL: Research Press.

Edling, J. V. (1970). *Individualized instruction: A manual for adminstrators.* Corvallis, OR: DCE Publications.

EPIE (1974, January). Evaluating instructional systems. *Educational product report: An in-depth report* (No. 58).

Frase, L. T. (1970). Boundary conditions for mathermagenic behavior. *Review of Educational Reesearch, 40,* 337–348.

Hiller, J.H. (1974). Learning from prose text: Effects of readability level, inserted questions difficulty, and individual differences. *Journal of Educational Psychology, 66,* 202–211.

Homme, L., Czanyi, A. P., Gonzales, M. A., & Rechs, J. R. (1969). *How to use contingency contracting in the classroom.* Champaign, IL: Research Press.

Keller, F. S. (1966). A personal course in psychology. In R. Ulrich, R. Stachnik, & J. Mabry (Eds.), *The control of behavior.* Glenview, IL: Scott, Foresman.

Madsen, C. H., Jr., & Madsen, C. K. (1970). *Teaching/discipline: Behavioral principles toward a positive approach.* Boston: Allyn & Bacon.

Park, D. & Tennyson, R. D. (1983). Computer-based instructional systems for adaptive education: A review. *Contemporary Education Review, 2(2),* 121–135.

Postlethwait, S. N., Novak, J., & Murray, H. T., Jr. (1969). *The audio tutorial approach to learning* (2nd ed.). Minneapolis: Burgess.

Pressey, S. L. (1950). Development and appraisal of devices providing immediate automatic scoring of objective tests and concomitant self-instruction. *Journal of Psychology, 29,* 417–447.

Reiser, Robert A. (1987). Instructional technology: A history. In R. M. Gagné (Ed.), *Instructional technology: Foundations.* Hillsdale, NJ: Erlbaum.

Rohwer, W. D., Jr. (1971). Prime time for education: Early childhood or adolescence? *Harvard Educational Review, 41,* 316–341.

Ross, S. M. (1984). Matching the lesson to the student: Alternative adaptive designs for individualized learning systems. *Journal of Computer-Based Instruction, 11,* 41–48.

Rothkopf, E. Z., & Bisbicos, E. E. (1967). Selective facilitative effects of interspersed questions in learning from written materials. *Journal of Educational Psychology, 58,* 56–61.

Ryan, B. A. (1974). *PSI: Keller's personalized system of instruction, An appraisal.* Washington, DC: American Psychological Association.

Skinner, B. F. (1968). *The technology of teaching.* New York: Appleton.

Suppes, P. & Machen, E. (1978). The historical path from research and development to operational use of CAI. *Educational Technology, 18* (4), 9–12.

Talmage, H. (Ed.). (1975). *Systems of individualized education.* Berkeley, CA: McCutchan.

Weisgerber, R. A. (1971). *Developmental efforts in individualized instruction.* Itasca, IL: Peacock.

16 EVALUATING INSTRUCTION

Every designer of instruction wants to have assurance that his topic, or course, or total system of instruction is valuable for learning in the schools. This means that he wishes to at least know whether his newly designed course or system works in the sense of achieving its objectives. More importantly, perhaps, he is interested in finding out whether his product works better than some other system it is designed to supplant.

Indications of how well an instructional product or system performs are best obtained from systematically gathered evidence. The means of gathering, analyzing, and interpreting such evidence are collectively called methods of *evaluation*, which is the subject of this final chapter. The placement of this chapter, by the way, should not be taken to indicate that the planning of evaluation for instruction should be undertaken as a final step. Quite the opposite is true; as will be shown, the design of evaluation requires principles of instructional planning that have been described in every chapter of this book.

Evidence sought in an enterprise whose purpose is the evaluation of instruction should be designed to answer at least the following specific questions concerning a lesson, topic, course, or instructional system:

1. To what extent have the stated objectives of instruction been met?
2. In what ways and to what degree is it better than the unit it will supplant?
3. What additional, possibly unanticipated, effects has it had, and to what extent are these better or worse than the supplanted unit?

These are but a small subset of the questions that are posed in the field of educational evaluation in general (cf. Popham, 1975). These three questions may best be considered critical ones for the evaluation of an instructional *product* or *procedure*. Before discussing them further, we attempt in the next section to provide a brief review of the larger context to which they belong.

EDUCATIONAL EVALUATION

In its most general sense, evaluation in education is to *assess the worth* of a variety of states or events, from small to large, from the specific to the very general. One can speak legitimately of the evaluation of students, of teachers, of administrators. Evaluation can be undertaken of educational *products,* the *producers* of such projects, or even of evaluation *proposals* (Scriven, 1974). Methods of evaluation applicable to many different aspects of educational systems and institutions have developed rapidly over the past several years. The subject of educational evaluation requires a book of its own. Here, we shall be able to indicate only the main ideas of some prominent methods.

Scriven's Evaluation Procedures

Scriven has proposed and tried out evaluation procedures that he considers applicable to educational products, courses, curricula, and projects proposing educational change (Scriven, 1967, 1974). One of the outstanding conceptions proposed by Scriven is called *goal-free evaluation.* In essence, this means that an evaluation undertakes to examine the effects of an educational innovation and to assess the worth of these effects, whatever they are. The evaluator does not confine himself to the stated objectives of a new product or procedure, but rather seeks to assess and evaluate outcomes of any sort. Thus, changes in teacher attitude might occur with the introduction of a scheme for using parent volunteers to tutor children in arithmetic. Such a change in attitude would be assessed in a goal-free evaluation, not simply as an "unanticipated outcome," but as one of a number of effects that a new procedure might produce.

The total scope of educational evaluation, as Scriven sees it, extends from the establishment of a need through the assessment of effects to a determination of cost-effectiveness and the likelihood of continued support. The following list summarizes suggestions of the assessments of worth that need to be made in evaluating a new educational program or product (Scriven, 1974):

1. *Need:* establishing that the proposed product will contribute to the health or survival of a system.
2. *Market:* determining the existence of a plan for getting the product used.
3. *Performance in field trials:* evidence of performance of the product or program under typical conditions of use.

4. *Consumer performance:* the appropriateness with which the product is addressed to, and likely to be used by, true consumers (teachers, principals, students).
5. *Performance—Comparison:* performance of the product compared with critically competitive products.
6. *Performance—Long-term:* data indicating performance over a period extending beyond initial field trials.
7. *Performance—Side effects:* outcomes other than the primary objective, revealed by goal-free assessment.
8. *Performance—Process:* indication that the processes of instruction are as proposed in the product.
9. *Performance—Causation:* demonstration that the effects observed are caused by the product or program.
10. *Performance—Statistical significance:* a quantitative indicator of effect.
11. *Performance—Educational significance:* in view of the achievement of the product or program as indicated by items 3 through 10, evaluating the importance of the gains thus identified for the educational institutions concerned.
12. *Costs and cost-effectiveness:* estimation of inclusive costs of a new program, and comparison with competitors.
13. *Extended support:* continued monitoring and updating of the product.

According to this system, judging the worth of a new educational product or procedure is a complex matter, based upon various kinds of information. The judgments made about each single factor may be recorded on a profile graph, which can then be used to make a systematic judgment of appropriateness and general worth of the product.

Stufflebeam's Evaluation Methods

The model of evaluation developed by Stufflebeam and his associates (1971) was originally designed to apply to any of a variety of educational improvements that might be considered for adoption, or actually adopted, by a school or school system. The model is called CIPP, in which the letters stand for context, input, process, and product.

The CIPP model considers evaluation as a continuing process. The information to be dealt with in evaluation of an education program must first be *delineated,* then *obtained,* and finally *provided.* The information produced by evaluation has the primary purpose of guiding decision making. Typically, Stufflebeam's procedures for evaluation can be thought of as having the purpose of guiding the decisions of a school superintendent who is faced with a proposal to institute a new curriculum program.

The kinds of decisions toward which evaluation may be oriented are four: planning, structuring, implementing, and recycling. Planning decisions are guided by *context* evaluation, which involves the determination of problems and unmet needs. Evaluation of the *input* includes consideration of alternative solutions (programs, products), their relative strengths and weaknesses, and their respective feasibilities (structuring decision). *Process* evaluation deals with information about the educational processes set in motion by the new program. These are implementing decisions, an example of which would be a decision about the appropriate use of tutoring as specified by a new mathematics program. Finally, there is *product* evaluation, which serves to guide decisions about recycling. An example of this type of evaluation would be identifying and assessing how well a new course of study is working, leading to a decision to continue the program, to drop it, or to modify it.

In general, this conception of evaluation procedures requires planning of the operation in a manner that seeks answers to questions in the following areas:

1. *Program*   What program is to be evaluated, and what population does it serve?
2. *Audiences*   What are the main concerns of the audience to be served by the evaluation?
3. *Purposes*   How will the information yielded by the evaluation be used?
4. *Approach*   What type of study will be conducted (context, input, process, output)?
5. *Questions*   What questions will be addressed?

Stufflebeam (1974) accepts Scriven's suggestion of goal-free evaluation, not as a substitute for, but as a valuable supplement to, goal-based evaluation. The distinction between formative and summative evaluation is also maintained in the CIPP model. Formative evaluation is seen as serving the needs of decision making about program development, whereas summative evaluation provides a basis for accountability.

It is apparent that there are few, if any, points of actual conflict in ideas between the evaluation models of Scriven and Stufflebeam. We note that "continuous planning" is a major emphasis in the Stufflebeam model, whereas "verified performance" is the emphasized concern of the Scriven system. Both models appear to view evaluation in a highly comprehensive manner, and both are most obviously relevant to the "large" decisions about educational procedures that must be made by the people responsible for the management of total school systems, on the one hand, or for the support of programs of widespread educational innovation on the other.

EVALUATION OF INSTRUCTION: TWO MAJOR ROLES

The view of evaluation assumed in the remainder of this chapter is more circumscribed than those of the general models already mentioned, but is otherwise not in conflict with them. Here we shall examine the logic and the procedures of evaluation as they apply to a single course of instruction. Such a course, we assume, would have been designed in accordance with the principles described in previous chapters of this volume. The questions to be addressed are: How does one tell whether the design process is working so as to achieve a worthwhile result? How does one tell whether the product designed has made a desirable difference in educational outcome?

The account of instructional evaluation to be given here is based upon the premise that a course (or smaller unit) of instruction is being designed, or has been designed, to meet certain specified objectives. Thus, the evaluation procedures to be described are concerned primarily with the *performance* aspects of Scriven's model, and with *process* and *product* evaluation as these terms are used by Stufflebeam. At the same time, we employ the customary distinction between *formative* evaluation and *summative* evaluation, as defined in an article by Scriven (1967). These two roles of evaluation lead to decisions about program revision, in the former case, and about program adoption and continuation, in the latter.

Formative Evaluation

Evidence of an instructional program's worth is sought for use in making decisions about how to revise the program while it is being developed. In other words, the evidence collected and interpreted during the phase of development is used to *form* the instructional program itself. If one discovers, by means of an evaluation effort, that a lesson is not feasible, or that the newly designed topic falls short of meeting its objectives, this information is used to revise the lesson or to replace portions of the topic, in the attempt to overcome the defects that have been revealed.

The decisions made possible by formative evaluation may be illustrated in a number of ways. For example, suppose that a lesson in elementary science has called for the employment of a particular organism found in fresh-water ponds. But when the lesson is tried in a school, it is found that without taking some elaborate precautions, this particular organism cannot be kept alive for more than two hours when transplanted to a jar of ordinary water. Such an instance calls into question the practical *feasibility* of the lesson as designed. Since evaluation has in this instance revealed the specific difficulty, it may be possible to revise the lesson by simply substituting another organism and changing the instructions for student activities appropriately. Alternatively, the lesson may have to be rewritten completely, or even abandoned.

Another type of example, illustrating *effectiveness,* may be provided by an instance in which a topic such as the "use of the definite article with German nouns" fails to meet its objective. Evidence from a formative evaluation study indicates that students use the definite article correctly in a large proportion of instances, but not in all. Further examination of the evidence reveals that the mistakes students are making center on the identification of the gender of the nouns. The designer of instruction for the topic is consequently led to consider how the lesson, or lessons, on the gender of nouns can be improved. He finds, perhaps, that some necessary concept has been omitted, or inadequately presented. This discovery in turn leads him to revise the lesson or possibly to introduce an additional lesson, designed to ensure the attainment of this subordinate objective. A detailed description of such procedures is given by Dick (1977a).

As described by Dick and Carey (1985), the procedures of formative evaluation involve three stages. Each stage consists of a tryout of the instructional material or program with a different sample of potential students, representative of the targeted student audience.

One-to-One Testing

In this stage, each student is presented with the instructions one at a time while the evaluator watches closely the student's performance. If the content of the instruction is presented via a computer screen, for example, the evaluator sits with the student while he works through the lesson or module. Another participant in this stage of evaluation is the *subject-matter expert*. This person is made thoroughly acquainted with the performance objectives of the instruction, and also with the test items or observations employed as performance indicators. The questions being asked pertain to the validity of objectives and to the accuracy and clarity of materials and test items.

The types of information obtained in one-to-one testing include evidence of the following features: (1) errors in estimates of the entry capabilities of students; (2) lack of clarity in the presentation of instruction; (3) unclear test questions and directions; and (4) inappropriate expectations of learning outcomes. On the basis of such information, systematic revisions of the instructional content can be made.

Small-Group Testing

A second stage of formative evaluation uses a small group of students who represent the target population. Typically, such testing begins with a pretest of the skills and knowledge to be taught during instruction. The instruction is then presented, followed by the administration of a posttest. Additionally, an attitude questionnaire seeks to assess student attitudes toward various aspects of the instructions. Students may also be asked to discuss the instruction, the pretest, and the posttest.

Information obtained from small-group testing begins to answer questions about the occurrence of learning and its amount, based on comparison of pretest and posttest scores. Other results may provide indications about the clarity of presentations and questions, which will be used to guide revision.

Field Trial

The instructional program is next tried out with an appropriate sample of the population intended as its audience. With this larger group, a pretest and a posttest (revised on the basis of small-group testing) are given, framing the presentation of the instruction itself. Attitude surveys are administered to learners and to participating instructors. Observations are made during this trial regarding the adequacy of the presentation of materials and their directions. In addition, information is collected on the quality and adequacy of instructors' performances in using the materials.

The field trial is designed to be a critical test of the instruction, its feasibility of use, and its effectiveness. Student and instructor behavior and attitudes yield valuable information that can make possible a near-final revision and improvement of the lessons and modules. Regarding effectiveness, the test scores and gains in achievement of students in this representative group, under near-typical conditions of use, are of course of crucial interest and importance.

Interpretation of the Evidence

These various kinds of evidence, collected by means of observational records, questionnaires, and tests, are employed throughout the stages of formative evaluation to draw conclusions as to whether a lesson needs to be kept as is, revised, reformulated, or discarded.

The question of *feasibility* may be decided, for example, by considering reports of the difficulties experienced by instructors or students in the conduct of the lesson. The question of *effectiveness* is a somewhat more complex judgment. It may depend, in part, on the reports of an observer that the materials could not be used in the manner intended or that the instructor did not carry out the intended procedures. It may also depend, in part, on student attitudes incidentally established by the lesson, as revealed by answers to questionnaires by both instructors and students. And, of course, it may depend to a most important degree upon the extent to which the performance of students, as revealed by tests, are successful.

Referring to the three questions at the beginning of this chapter, it will be evident that formative evaluation is most cogently concerned with Question 1—To what extent have the stated objectives of instruction been met? This is one of the principal kinds of evidence that may be brought to bear on the revision and improvement of the designed instruction. On occasion, evidence may also become available that permits comparison with an alternative or supplanted instructional entity (Question 2), and such evidence may also be utilized for formative purposes. Similarly, observations that reveal unanticipated effects (Question 3), good or bad, may surely have an effect on decisions about revision or refinement of instruction. However useful these additional pieces of evidence may be, it remains true that Question 1 defines an essential kind of evidence leading to decisions about revising and improving the instructional unit that is being developed.

Summative Evaluation

Summative evaluation is usually undertaken when development of an instructional entity is in some sense completed, rather than on-going. Its purpose is to permit conclusions to be drawn about how well the instruction has worked. Such findings permit schools to make decisions about adopting and using the instructional entity (cf. Dick, 1977b; Dick and Carey, 1985).

In general, summative evaluation concerns itself with the effectiveness of an instructional system, course, or topic. Individual lessons may of course be evaluated as components of these larger units, but rarely as separate entities. The evaluation is called summative because it is intended to obtain evidence about the *summed* effects of a set of lessons making up a larger unit of instruction. Naturally, though, such evidence may include information pointing to defects or positive accomplishments of particular lessons, and this can be used in a formative sense for the *next* development or the *next* revision.

The main kind of decision for which the evidence of a summative evaluation is useful is whether a new course (or other unit) is better than one it has replaced and, therefore, should be adopted for continued use. Conceivably, it may be no better, in which case considerations other than effectiveness (such as cost) will come to determine the choice. Also conceivably, it might be worse than what it has replaced, in which case the decision would likely be an easy one to reach.

Suppose that a newly designed course in American government has replaced one of the same title and has been adopted by a school. A summative evaluation finds that student enthusiasm for the new course is little changed compared with that for the old; that 137 of the 150 defined objectives of the new course are adequately met by students (the previous course did not have defined objectives nor means of assessing them); and that a test on American government given at the end of the semester yields an average score of 87 as opposed to 62 on the same test in the previous year. The new course is liked by teachers, for the specific reason that it permits them to take more time for individual student conferences. Now, provided that the new course does not cost more than the old, this set of evidence would very likely lead to a decision to adopt and continue the new course and to abandon the old.

In contrast to formative evaluation, summative evaluation usually has many formal features, some of which are indicated by this example. Measures of student attitudes, for example, are likely to be based upon carefully constructed questionnaires, so that they can be directly and validly compared with those of last year's students. The assessment of mastery of each objective is also systematically done, in order that there will be a quantitative indication of the accomplishments of the entire course. In addition, measures of achievement are taken from a test serving as a semester examination. As is true of formative evaluation, each of these summative measures needs to be obtained with the use of methods that make possible the collection of convincing evidence of effectiveness.

Evidence Sought

Summative evaluation of a topic, course, or instructional system is primarily concerned with evidence of learning outcomes. As will be discussed in the next section of this chapter, obtaining such evidence requires the collection of data on "input measures" and "process measures," as well as on those measures that directly assess outcomes. Learning outcomes are assessed by means of observations or tests of human capabilities, as reflected in the objectives specified for the instruction. Accordingly, the measures of outcomes might consist of any or all of the following types:

1. Measures indicating the mastery of *intellectual skills,* assessing whether or not particular skills have been acquired. Example: A test requiring solutions for designated variables in linear algebraic equations.
2. Measures of *problem-solving ability,* assessing the quality or efficiency of the student's thinking. Example: Exercises requiring the design of a scientific experiment to test the effect of a particular factor on some natural phenomenon, in a situation novel to the student.
3. Tests of *information,* assessing whether or not a specified set of facts or generalizations has been learned. Example: A test requiring the student to state the names and roles of the principal characters in a work of literature. Alternatively, tests assessing the breadth of knowledge attained by the student. Example: A test that asks the student to describe the major antecedents of a historical event.
4. Observations or other measures of the adequacy of *motor skills,* usually with reference to a specified standard of performance. Example: An exercise in which the child is asked to print the alphabet in capital letters.
5. Self-report questionnaires assessing *attitude.* Example: A questionnaire asking the student to indicate "probability of choice" for actions concerned with the disposal of personal trash.

Interpretation of Summative Evidence

The various measures appropriate for the outcomes of learning are interpreted primarily in comparison with similar measures obtained on an instructional entity representing an alternative mode of instruction. Referring again to the questions at the beginning of this chapter, the primary emphases of evidence obtained for summative purposes is on the answer to Question 2—. In what ways, and to what degree, is this unit better than some other? Usually, the comparison to be made is with a topic or course that the newly designed unit is intended to replace. Sometimes, two different newly designed instructional entities may be compared with one another. In either case, such comparisons require methods of data collection that can demonstrate that "all other things are equal," which is by no means an easy thing to do.

Answers to Questions 1 and 3 are also desirable outcomes of a summative evaluation. One wants to determine, as a minimal condition, whether the objectives of the new instructional unit have been met (Question 1). Should it turn out that they have not, this result will obviously affect the possible conclusions to be drawn from comparison with an alternative unit. In addition, it is always of some importance to explore whether the newly designed instruction has had some unanticipated effects (Question 3). A topic designed to teach basic concepts of weather, for example, might turn out to have some unexpected effects on attitudes toward disposal of personal trash.

EVALUATION OF INSTRUCTIONAL PROGRAMS

Evaluation methods may be applied to lessons or courses, and also to entire *programs* of instruction. A program may consist of a number of different courses, which can be viewed as contributing to a common purpose, and may have a duration of months or years. The various measures of outcomes of instruction that were described in Chapter 12 may be employed in program evaluation studies. Often the measures of outcome must be separately developed to meet the needs of an evaluation effort. In some instances, however, the required measures, tests, observation schedules, or questionnaires are commercially available or can be adapted from commercially available instruments.

Besides the development of tests or other types of measures, the enterprise of evaluation requires careful, scientifically based methods that serve to ensure that the evidence obtained is truly convincing. To describe these methods in full detail would require at least a separate volume; in fact, a number of books are available that deal with the design of evaluation studies (e.g., Fitz-Gibbon and Morris, 1978; Popham, 1975; Thorndike and Hagen, 1986). In this chapter, we can deal only with the *logic* of evaluation studies, beginning with the logic of data collection and interpretation already introduced. Beyond this, however, is the rationale for identifying and controlling variables in evaluation efforts so that valid conclusions can be drawn about instructional outcomes.

The Variables of Evaluation Studies

The intention of studies to evaluate an instructional program is to draw conclusions about the effects of the instruction on learning outcomes—on the human capabilities the instruction has been designed to establish or improve.

But these capabilities are affected by other factors in the educational setting, not only by the instruction itself. It is therefore necessary to *control or otherwise account for these other variables* in order to draw valid conclusions about instructional effectiveness. Considered as a whole, the educational situation into which instruction is introduced contains the classes of variables described in the following paragraphs.

Outcome Variables

We begin to list the variables of the educational situation with outcome variables, the dependent or measured variables that are the primary focus of interest. These have already been described as measures of the human capabilities intended to be affected by instruction. The classes of variables that influence educational outcomes, and their various sources, are shown in Figure 16-1.

Process Variables

What factors in the school situation might influence the outcomes, given the existence of an instructional program? Obviously, there may be some effects on how the instructional entity (topic, course, system) is conducted. Outcomes may be influenced, in other words, by the *operations* carried out to put the instruction into effect, typically by the teacher (cf. Astin and Panos, 1971). For example, the instruction as designed may call for a particular type and frequency of teacher questioning. To what extent has this been done? Or, the designed course may call for a particular sequence of intellectual skills, some to be mastered before others are undertaken. To what extent has this operation been carried out? As still another example, the designed instruction

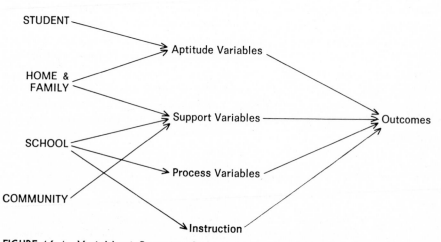

FIGURE 16-1   Variables influencing the outcomes of an instructional program.

may specify that a particular sort of feedback is to be incorporated in each lesson (see Chapter 9). Has this been systematically and consistently done?

One cannot simply assume that process variables of the sort specified by the designed instruction or intended by the designer will inevitably occur in the way they are expected to. Of course, well-designed instruction provides for whatever action may be required to ensure that the program operates as planned; for example, provisions are often made to train teachers in these operations. Nevertheless, such efforts are not always fully successful—teachers are no more free of human inadequacies than are the members of any other professional group. Designers of new programs of individualized instruction, for example, have rather frequently found that the operations specified for these programs are not being executed in the manner originally intended. As a consequence, it is essential that assessments of process variables be made, and this is particularly so when the newly designed instructional entity is being tried out for the first time.

Process variables comprise many factors in the instructional situation that may directly affect student learning. Such factors, then, may concern matters of *sequence* or matters of the institution and arrangement of the *events of instruction,* both of which are described in Chapter 10. Another factor is the *amount of time* devoted by students to particular lessons or portions of a course. Naturally, one of the major variations to be found in topics or courses of instruction is the degree to which these classes of process variables are specified. A textbook, for example, may imply a sequence for instruction in its organization of chapters, but may leave the arrangement of events of instruction entirely to the teacher or to the learner himself (as does this book). In contrast, a topic designed for instruction in language skills for the sixth grade may not only specify a sequence of subordinate skills but also particular events, such as informing the learner of objectives, stimulating recall of prerequisite learning, providing learning guidance, and providing feedback to the learner, among others. Regardless of the extent to which process variables are prescribed by the designed entity, it is necessary to take them into account in a well-planned evaluation study. After all, the outcomes observed may be substantially affected by the ways a new instructional program is *operated,* whatever its designed intentions.

Typically, process variables are assessed by means of systematic observations in the classroom (or another educational setting). This is the function of the observer, not the teacher, in the conduct of the evaluation. The observer may employ a checklist or observation schedule as an aid in recording his observations. Such instruments normally have to be specially designed to meet the purposes of each particular evaluation study.

Support Variables

Still another class of variables, occurring partly in the student's home and community, has to be considered as potentially influential in the outcomes of an instructional program. These include such factors as the presence of adequate materials (in the classroom and the school library), the availability of a quiet place for study, the "climate" of the classroom with reference to its encouragement of good achievement, the actions of parents in reinforcing favorable attitudes toward homework and other learning activities, and many others. The number of different variables in this class is quite large, and not enough is known about them to make possible a confident differentiation among them with regard to their relative importance.

The general nature of this class of variables is to be seen in their effects on the *opportunities for learning.* Materials in the classroom, for example, may present greater or fewer opportunities for learning, depending on their availability; parents may make opportunities for adequate attention to homework more or less available; and so on. In contrast to process variables, *support variables* do not directly influence the process of learning, as the former set of factors is expected to do. Instead, they tend to determine the more general environmental conditions of those times during which process variables may exert their effects. For example, the designed instruction may call for a period of independent study on the part of the student. In operation, the teacher may make suitable time provisions for this independent study, thus ensuring that the process variable has been accounted for. But what will be the difference in the outcome, for (1) a student who has a relatively quiet place in which to pursue his learning uninterrupted and for (2) a student who must perform his independent study in an open corner of a noisy classroom? This contrast describes a difference in a *support* variable. The opportunities for learning are presumably less in the second case, although the actual effects of this variable on the outcome cannot be stated for this hypothetical example.

Support variables require various means of assessment. What parents do in encouraging the completion of homework may be assessed by means of a questionnaire. The availability of materials relevant to a topic or course may be assessed by counting books, pamphlets, and other reference sources. The climate of a classroom may be found by the use of a systematic schedule of observations. Other measure of this class, such as number of students or the pupil-teacher ratio, may be readily available at the outset of the study. For any of a number of support variables, it is likely to be necessary to select or develop the technique of assessment best suited to the particular situation.

Aptitude Variables

It is of great importance to note that of all the variables likely to determine the outcomes of learning, the most influential is probably the students *aptitude for learning.* Such aptitude is usually measured by means of an intelligence test or a test of scholastic aptitude. This kind of intelligence, sometimes called *crystallized intelligence* (Cattell, 1963; see also Corno and Snow, 1986), is found to be highly correlated with achievement in school subjects. Whatever may be accomplished by improved methods of instruction, by arrangements of process variables, and by assuring the best possible support for learning, this entire set of favorable circumstances cannot influence learning outcomes as much as can students' aptitude for learning.

Aptitude for learning is undoubtedly determined in part by genetic inheritance, as well as by environmental influences occurring before birth (such as nutrition). An individual's aptitude is partly determined also by the kinds of prior learning accomplished and by opportunities to learn. It should be clear, then, that aptitude is a variable that has its own multiple determinants. As it enters an evaluation study, however, aptitude is usually an *input* variable (Astin and Panos, 1971). In this role, aptitude is not subject to alteration by the evaluation; it can only be measured, not manipulated. Many studies have shown that aptitude for learning, as measured by intelligence tests, may account for as much as 50 percent of the variations in learning outcome, measured as student achievement in capabilities falling in the categories of verbal information, intellectual skills, and cognitive strategies.

The aptitude for learning that a student brings to the instructional situation is likely to have a very great effect on learning, when learning is assessed in terms of its outcomes. Thus, if the effectiveness of an instructional program is to be assessed, the effect of instruction itself must be demonstrated by instituting controls that make possible the separation of the influence of the students' entering aptitudes for learning. Evaluation studies having particular purposes may require that learner characteristics identified as prerequisites be treated as input variables. For example, an instructional program devoted to teaching adult basic skills might be concerned with the question of the effectiveness of the newly designed instruction independent from the entering skills of the students. In such a case, an evaluation study would take account of prerequisite skills (such as are described in Chapter 6) as input variables.

Although measures of learning aptitude are often most conveniently identified by scores on intelligence tests, other measures are sometimes employed. A *combination* of several aptitude tests may be used to yield a combined score to assess learning aptitude. (Actually, most intelligence tests are themselves collections of subtests sampling several different aptitudes.) Another procedure involves the use of measures that are known to *correlate* with intelligence scores to a fairly high degree. Previous school grades exhibit such high correlations, particularly in subjects such as reading comprehension and mathematics. Still another correlated measure is family income, or family socioeconomic status (SES). It seems reasonable, though, that although correlated measures are sometimes useful, they are not to be preferred in evaluation studies over measures that attempt to assess learning aptitude in the most direct manner possible.

INTERPRETING EVALUATIVE EVIDENCE

We have pointed out that measures of the outcomes of an instructional program—that is, measures of learned intellectual skills, cognitive strategies, information, attitudes, and motor skills—are influenced by a number of variables in the educational situation, besides the program itself. Process variables in the operation of the instructional program may directly affect learning and thus also affect its outcomes. Support variables in the school or in the home determine the opportunites for learning and thus influence the outcomes of learning that are observed. And most prominently of all, the learning aptitude of students strongly influences the outcomes measured in an evaluation study.

If the effectiveness of the designed instruction is to be evaluated, certain *controls* must be instituted over process, support, and aptitude variables to ensure that the "net effect" of the instruction is revealed. Procedures for accomplishing this control are described in this section. Again it may be necessary to point out that only the basic logic of these procedures can be accounted for here. However, such logic is of critical importance in the design of evaluation studies.

Controlling for Aptitude Effects

The assessment of outcomes of instruction in terms of Question 1 (To what extent have objectives been met?) needs to take account of the effects of aptitude variables. In the context of this question, it is mainly desirable to state what *is* the level of intelligence of the students being instructed. This may be done most simply by giving the average score and some measure of dispersion of the distribution of scores (such as the standard deviation) on a standard test of intelligence. However, correlated measures such as SES are frequently used for this purpose. Supposing that 117 out of 130 objectives of a designed course are found to have been met; it is of some importance to know whether the average IQ of the students is 115 (as might be true in a suburban school) or 102 (as might occur in some sections of a city, or in a rural area). It is possible that, in the former setting, the number of objectives achieved might be 117 out of 130, whereas in the latter, this might drop to 98 out of 130. The aims of evaluation may best be accomplished by trying out the instructional entity in several different schools, each having a somewhat different range of student learning aptitude.

When the purposes of Question 2 (To what degree is it better?) are being served in evaluation, one must go beyond simply reporting the nature and amount of the aptitude variable. In this case, the concern is to show whether any difference exists between the new instructional program and some other—in other words, to make a *comparison*. Simply stated, making a comparison requires the demonstration that the two groups of students were *equivalent* to begin with. Equivalence of students in aptitude is most likely to occur when successive classes of students in the same school, coming from the same neighborhood, are employed as comparison groups. This is the case when a newly designed course is introduced in a classroom or school and is to be compared with a different course given the previous year.

Other methods of establishing equivalence of initial aptitudes are often employed. Sometimes, it is possible to assign students *randomly* to different classrooms within a single school, half of which receive the newly designed instruction and half of which do not. When such a design is used, definite administrative arrangements must be made to ensure randomness—it cannot be assumed. Another procedure is to select a set of schools that are "matched," insofar as possible, in the aptitudes of their students and to try out the new instruction in half of these, making a comparison with the outcomes obtained in those schools not receiving the new instruction. All of these methods contain certain complexities of design that necessitate careful management if valid comparisons are to be made.

There are also statistical methods of control for aptitude variables—methods that "partial out" the effects of aptitude variables and thus reveal the net effect of the instruction. In general, these methods follow this logic: If the measured outcome is produced by A and I, where A is aptitude and I is instruction, what would be the effect of I alone if A were assumed to have a constant value, rather than a variable one? Such methods are of considerable value in revealing instructional effectiveness, bearing in mind particularly the prominent influence the A variable is likely to have.

Whatever particular procedure is employed, it should be clear that any valid comparison of the effectiveness of instruction in two or more groups of students requires that equivalence of initial aptitudes be established. Measures of intelligence, or other correlated measures, may be employed in the comparison. Students may be randomly assigned to the different groups, or their aptitudes may be compared when assignment has been made on other grounds (such as school location). Statistical means may be employed to make possible the assumption of equivalence. Any or all of these means are aimed at making a convincing case for equivalence of learning aptitudes among groups of students whose capabilities following instruction are being compared. No study evaluating learning outcomes can provide valid evidence of instructional effectiveness without having a way of controlling this important variable.

Controlling for the Effects of Support Variables

For many purposes of evaluation, support variables may be treated as input variables, and thus controlled in ways similar to those used for learning aptitude. Thus, when interest is centered upon the attainment of objectives (Question 1), the measures made of support variables can be reported along with outcome measures so they can be considered in interpreting the outcomes. Here again, a useful procedure is to try out the instruction in a variety of schools displaying different characteristics (or different amounts) of support.

Similarly, the comparisons implied by Question 2 and part of Question 3 require the demonstration of *equivalence* among the classes or schools whose learning outcomes are being compared. Suppose that outcome measures are obtained from two different aptitude-equivalent groups of students in a school, one of which has been trying out a newly designed course in English composition, while the other continues with a different course. Assume that, despite differences in the instruction, the objectives of the two courses are largely the same and that assessment of outcomes is based on these common objectives. Class M is found to show significantly better performance, on the average, than does Class N. Before the evidence that the new instruction is "better" can be truly convincing, it must be shown that no differences exist in support variables. Since the school is the same, many variables of this sort can be shown to be equivalent, such as the library, the kinds of materials available, and others of this nature. Where might differences in support variables be found? One possibility is the climate of the two classrooms—one may be more encouraging to achievement than the other. Two different teachers are involved—one may be disliked, the other liked. Student attitudes may be different—more students in one class may seek new opportunities for learning than do students in the other. Variables of this sort that affect opportunities for learning may accordingly affect outcomes. Therefore, it is quite essential that equivalence of groups with respect to these variables be demonstrated or taken into account by statistical means.

Controlling for the Effects of Process Variables

The assessment and control of process variables is of particular concern in seeking evidence bearing on the attainment of stated objectives (Question 1). Quite evidently, an instructional entity may work either better or worse depending upon how the operations it specifies are carried out. Suppose, for example, that a new course in elementary science presumes that teachers will treat the directing of students' activities as something left almost entirely to the students themselves (guided by an exercise booklet). Teachers find that under these circumstances, the students tend to raise questions to which they (the teachers) don't always know the answers. One teacher may deal with this circumstance by encouraging students to see if they can invent a way of finding the answer. Another teacher may require that students do only what their exercise book describes. Thus the same instructional program may lead to quite different operations. The process variable differs markedly in these two instances, and equally marked effects may show up in measures of outcome. If the evaluation is of the formative type, the designer may interpret such evidence as showing the need for additional teacher instructions or training. If summative evaluation is being conducted, results from the two groups of students must be treated separately to disclose the effects of the process variable.

In comparison studies (Question 2), process variables are equally important. As in the case of aptitude or support variables, they must be controlled in one way or another in order for valid evidence of the effectiveness of instruction to be obtained. Equivalence of groups in terms of process variables must be shown, either by exercising direct control over them by a randomizing approach, or by statistical means. It may be noted that process variables are more amenable to direct control than are either support or aptitude variables. If a school or class is conducted in a noisy environment (a support variable), the means of changing the noise level may not be readily at hand. If, however, a formative evaluation study shows that some teachers have failed to use the operations specified by the new instructional program (a process variable), instruction of these teachers can be undertaken, so that the next trial starts off with a desirable set of process variables.

Unanticipated outcomes (Question 3) are equally likely to be influenced by process variables and accordingly require similar control procedures. A set of positive attitudes on the part of students of a newly designed program *could* result from the human modeling of a particular teacher and, thus, contrast with less favorable attitudes in another group of students who have otherwise had the same instruction. It is necessary in this case also, to demonstrate equivalence of process variables before drawing conclusions about effects of the instructional entity.

Controlling Variables by Randomization

It is generally agreed that the best possible way to control variables in an evaluation study is to ensure that their effects occur in a random fashion. This is the case when students can be assigned to *control* and *experimental* groups in a truly random manner, or when an entire set of classes or schools can be divided into such groups randomly. In the simplest case, if the outcomes of Group A (the new instructional entity) are compared with those of Group B (the previously employed instruction), and students drawn from a given population have been assigned to these groups in equal numbers at random, the comparison of the outcomes may be assumed to be equally influenced by aptitude variables. Similar reasoning applies to the effects of randomizing the assignment of classrooms, teachers, and schools to experimental and control groups in order to equalize process and support variables.

Randomization has the effect of controlling not only the specific variables that have been identified, but also other variables that may not have been singled out for measurement because their potential influence is unknown. Although ideal for purposes of control, in practice randomizing procedures are usually difficult to arrange. Schools do not customarily draw their students randomly from a community, nor assign them randomly to classes or teachers. Accordingly, the identification and measurement of aptitude, support, and process variables must usually be undertaken as described in the preceding sections. When random assignment of students, teachers, or classes is possible, evaluation studies achieve a degree of elegance they do not otherwise possess.

EXAMPLES OF EVALUATION STUDIES

The four kinds of variables in evaluation studies—aptitude, support, process, and outcome—are typically given careful consideration and measurement in any evaluation study, whether formative or summative. Interpretation of these measures differs for the two evaluation roles, as will be seen in the following examples.

Evaluation of a Program in Reading for Beginners

A varied set of lessons in reading readiness and beginning reading was developed and evaluated over a two-year period by the Educational Development Laboratories of McGraw-Hill, Inc., and by the L. W. Singer Company of Random House, Inc. This system of instruction is called *Listen Look Learn*. In brief, the instructional materials include: (1) a set of filmstrips accompanied by sound, designed to develop listening comprehension and oral recounting; (2) an eye-hand coordination workbook dealing with the identification and printing of letters and numerals; (3) a set of filmstrips providing letter-writing tasks, accompanying the workbook; (4) letter charts for kinesthetic letter identification; (5) picture sequence cards, and other cards for "hear and read" practice; (6) a set of colored filmstrips for the analysis of word sounds and the presentation of words in story contexts.

As reported by Heflin and Scheier (1968), a systematic formative evaluation of this instructional system was undertaken, which at the same time obtained some initial data for summative purposes. Table 16-1 summarizes some of the main points of the study, abstracted from this report. The purpose of the table is to illustrate how the major classes of variables were treated and interpreted; naturally, many details of the study covered in the report cannot be reported in the brief space of such a table.

Classes of first-grade pupils from schools located in 11 states were included in the evaluation study. A group of 40 classes comprising 917 pupils were given instruction provided by the *Listen Look Learn* system, and a group of 1000 pupils in 42 classes was constituted as a control group. Control-group classes used the "basal reading" instructional system. Each school district was asked to provide classes for the experimental and control groups that were as equivalent as possible in terms of characteristics of teachers and pupils.

Aptitude Variables

Owing to differences in the availability of aptitude scores in the various schools, no initial measures of aptitude were employed. Instead, information was obtained concerning the socioeconomic status of the pupils' families, as indicated in Table 16-1. When aptitude measures were administered during the second year of the study (Metropolitan Readiness, Pintner Primary IQ),

TABLE 16-1. Variables Measured and Their Interpretation for Formative and Summative Evaluation in a Study of a System of Instruction for Beginning Reading (*Listen Look Learn*)*

Type of Variable	How Measured	Interpretation
Aptitude	Initially, by means of socioeconomic status (SES), a correlated measure. During second year, standardized test scores for IQ and Reading Readiness.	*Formative:* A variety of classes providing a range of SES from high to low. *Summative:* Equivalence of SES, and later of aptitude, shown for experimental and control groups.
Support	1. Level of formal education of teachers. 2. Amount of teacher education in reading methods. 3. Years of teaching experience.	*Formative:* Range of these variables typical of most elementary schools. *Summative:* Reasonable equivalence of experimental and control groups on these variables.
Process	1. Appropriateness of lessons as judged by teachers. 2. Success of program components as judged by teachers. 3. Strength and weaknesses of individual lessons judged by teachers.	*Formative:* Judgments of appropriateness used to test feasibility. Indirect indications of effectiveness of pupil learning, based on teachers' estimates.
Outcome	Metropolitan Primary I Achievement word knowledge means: *LLL* group—25.5 Control group—24.1 Word discrimination means: *LLL* group—25.9 Control group—24.7 Reading means: *LLL* group—27.3 Control group—25.2	*Summative:* Achievement scores on standardized test indicate scores on component reading skills significantly higher than those of an equivalent control group.

*Information and results abstracted from Heflin and Scheier (1968), *The Formative Period of Listen Look Learn, a Multi-Media Communication Skills System.* Huntington, NY.: Educational Development Laboratories, Inc.

verification was obtained of a broad range of aptitude, as well as of equivalence of the experimental and control groups.

For purposes of formative evaluation, it is necessary to know that the classes selected for instruction included a range of student aptitudes that is representative of schools in the country as a whole, since that is the intended usage for the system being evaluated. From the report (Heflin and Scheier,

1968), it would appear that the schools taking part in the study represented a great majority of U. S. elementary schools, although by no means all of them. For example, inner-city schools were apparently not included. Nevertheless, the study offers reasonably good evidence that a broad range of pupil aptitudes was represented. In addition, it is clear from the reported data that the two groups of pupils were reasonably equivalent in aptitude.

Support Variables

The range of SES of pupils' families provides the additional indication that support for learning, insofar as it may be assumed to originate in the home environment, exhibited a suitable range of variation for the study. Other evidences of learning support are inferred from measures of the characteristics of teachers, as indicated in Table 16-1. The inference is that teachers having a typical range of educational backgrounds will conduct themselves in ways that provide a range of differential opportunities for learning. A reasonable degree of equivalence is also demonstrated on these variables between experimental and control groups.

Other measures of support for learning, not systematically obtained in this study, are perhaps of greater relevance to summative evaluation. Such variables as "availability of reading materials," "encouragment of independent reading," and others of this general nature would be examples. In the *Listen Look Learn* study, incomplete evidence was obtained of the number of books read by individual children, and this number was found to vary from 0 to 132 (Heflin and Scheier, 1968, p. 45).

Process Variables

As Table 16-1 indicates, a measure of the feasibility of the various parts of the program was obtained by asking teachers to judge the appropriateness of the materials for groups of fast, medium, and slow learners. Various features of the individual lessons might have contributed to appropriateness, such as the familiarity of the subject of a story or the difficulty of the words employed. Teacher's judgments led to conclusions about feasibility that resulted in elimination or revision of a number of elements of the program.

Teachers' estimates also formed the bases for evidence of the success of the various activities constituting the *Listen Look Learn* program. Such measures are of course indirect evidence bearing on process variables, as contrasted with such indicators as how many exercises were attempted by each student, how long a time was spent on each, what feedback was provided for correct or incorrect responses, and other factors of this nature. The materials of this program do not make immediately evident what the desired process variables may have been. Consequently, teachers' reports about "how effective the lesson was" were probably as good indicators of these variables as could be obtained in this instance.

Outcome Variables

Learning outcomes for this program were assessed by means of standardized tests of word knowledge, word discrimination, and reading (portions of the Metropolitan Primary I Achievement Test). As can be seen from Table 16-1, mean scores on these three kinds of activities were higher for the experimental group than for the control group, which had been shown to be reasonably equivalent so far as the operation of aptitude and support variables were concerned. Statistical tests of the difference between the various pairs of means indicated that these differences were significant at an acceptable level of probability.

It should be pointed out that the evidences of learning outcome obtained in this study were considered by its authors as no more than initial indications of the success of the *Listen Look Learn* program. Further studies were subsequently conducted to evaluate learning outcomes in a summative sense (Brickner and Scheier, 1968, 1970; Kennard and Scheier, 1971). In general, these studies have yielded data and conclusions that show improvements in early reading achievement considerably greater than are produced by other instructional programs they are designed to supplant (usually basal reading approaches).

Evaluation of an Individualized Arithmetic Program

A second example of an evaluation study, summative in character, is provided by an investigation of an individualized instruction system developed by the Learning Research and Development Center, University of Pittsburgh (Cooley, 1971). In this study, a program of individualized instruction in arithmetic for the second grade of the Frick School was compared with the previously used program. The new program had undergone several years of formative evaluation and development. It provided for individual progress of pupils in attaining arithmetic skills, based upon mastery of prerequisite skills.

Table 16-2 summarizes the treatment of variables in this evaluation study, and presents the major outcome findings.

Aptitude Variables

First, it will be seen from the table that aptitude variables were measured from year to year at the time the children first entered the school. Over a period of several years, the aptitude of entering classes was found to be essentially the same. In addition, the correlated variable of socioeconomic status (SES) was found to remain stable. Accordingly, it was considered a reasonable assumption in this study that successive classes of pupils would have the same initial aptitudes. An experimental group (individualized instruction) in the second grade in 1971 could be compared with a control group (regular instruction) who were in the second grade in 1970.

TABLE 16-2.   Variables and Their Interpretation in an Evaluation of an Individualized Program in Arithmetic for the Second Grade, Frick School

Variable	How Controlled or Measured	Interpretation
Aptitude	Classes of pupils used in control and experimental groups equivalent in aptitude when they entered school. SES of pupils in both groups shown to be equivalent.	Aptitude of classes of pupil remains unchanged in this school from year to year.
Support	Same school facilities present for both groups, and same teachers involved. SES of pupils equivalent.	Specific support variables of the school and the home are equivalent.
Process	Contrasting process in individualized and regular instruction. Same teachers involved in both groups.	Effects of process variables in individualized instruction to be examined; other specific process variables equivalent in both groups.
Outcome	Wide Range Achievement Test— *Arithmetic* Mean scores in second grade: Experimental group (1971)— 25.22 Control group (1970)—23.40	Significant differences obtained in outcome scores for equivalent groups.

Information and results abstracted from Cooley (1971), *Methods of Evaluating School Innovations.* Pittsburgh, PA: Learning Research and Development Center, University of Pittsburgh.

Support Variables

Support variables were not specifically singled out and measured individually. Instead, there was a demonstrated equivalence of classrooms and teachers. Under these circumstances, particular support variables were assumed to be equivalent for both groups. Similarly, those support variables originating in the home could be assumed equivalent in view of the demonstrated absence of differences in SES variables for the two classes.

Process Variables

The most important process variables, those associated with the specific technique of individualized instruction, were deliberately contrasted in the two groups, and this variation was verified by classroom observations. Other process variables (such as the encouragement provided by teachers to pupils) could be assumed to be equivalent because the same teachers were involved for both experimental and control groups.

Outcome

As a consequence of this study design, certain influencing variables in the categories of aptitude, support, and process are either shown to be, or reasonably assumed to be, equivalent in their effects on both groups of pupils. Outcome variables are therefore expected to reflect the effects of the changes in instruction in an unbiased manner. Measures of arithmetic achievement, as shown in the final row of the table, indicate a significant improvement when the new (individualized) instruction is compared with the previously used instruction.

A Generalized Example

Every evaluation study presents the evaluator with a different set of circumstances to which he must apply the logic we have described. In practice, compromise must sometimes be made because of the existence of inadequate measures of learning outcomes, the difficulties of achieving equivalence in groups to be compared, the occurrence of particular events affecting one school or class without affecting others, and many other possibilities too numerous to mention. Part of the evaluator's job, of course, is to judge the severity of these occurrences and the ways in which they must be taken into account to arrive at convincing evidence.

A reference set of representative evaluation situations is shown in Table 16-3, together with their most likely interpretations. These situations serve as one kind of summary of our previous discussion of the types of variables affecting learning outcomes.

TABLE 16-3. Comparison of Learning Outcomes in School A (Using Course A) and School B (Using Course B), and Their Interpretation

Situation	Outcome Comparison	Most Likely Interpretation
1. Aptitude variable: A>B Support variables: A=B Process variables: A=B	A>B	Most of the outcome differences, if not all, attributable to aptitude differences.
2. Aptitude variable: A=B Support variables: A>B Process variables: A=B	A>B	Differences may be caused by instruction, by support, or both.
3. Aptitude variable: A=B Support variables: A=B Process variables: A>B	A>B	Differences may be caused by instruction, by process differences, or both.
4. Aptitude variable: A=B Support variables: A=B Process variables: A=B	A>B	Difference is attributable to effects of instruction.

The hypothetical comparisons of Table 16-3 suppose that School A has been trying out a newly developed course (also labeled A), and that its outcomes are being compared with those from School B, which has been using a different course (B). In all cases, it is further supposed, the measures of outcome have been found superior in School A to what they are in School B.

Situation 1 is that in which support variables and process variables have been controlled, that is, shown to be equivalent. Aptitude variables indicate higher intelligence, on the average, in School A than in School B. Since this variable is such a powerful one, the effects of instruction cannot be expected to show up, and the likely interpretation is as shown in the final column. Situation 2 is one in which all influencing variables have been shown equivalent except for the support variables. Differences in outcomes may be caused by these variables, by the instruction, or by both in some unknown proportion. Similarly, Situation 3, in which process variables differ, can lead only to the conclusion that either process or instruction, or both, have produced the observed differences in outcome.

Situation 4 is what is aimed for in studies of summative evaluation. Here all the influencing variables have been shown to be equivalent by one method or another. This situation is one that makes possible the interpretation that outcome differences are attributable to the instruction itself.

SUMMARY

Evaluation of courses, programs, and instructional programs usually has at least the following questions in view: (1) have the objectives of instruction been met; (2) is the new program better than one it is expected to supplant; and (3) what additional effects does the new program produce?

Formative evaluation is undertaken while the new unit is being developed. Its purpose is to provide evidence on feasibility and effectiveness, so that revisions and improvements can be made. It seeks evidence from observers, teachers, and students.

Summative evaluation is concerned with the effectiveness of the course or program once it has been developed. Mainly, the evidence sought is in terms of student performance. Measures are taken of the kinds of student capabilities the program is intended to establish, as described in Chapters 3 through 6.

Summative evaluations may be undertaken to compare an entire instructional program with another. A number of kinds of variables must be taken into account. The outcomes of the program are influenced by variables whose effects must be controlled (or factored out) in order to test the effects of instruction. These variables include.

1. Aptitude variables, reflecting the students' aptitude for learning.
2. Process variables, arising from the manner of operation of instruction in the class or school.

3. Support variables—conditions in the home, school, workplace, and community that affect opportunities for learning.

Evaluation studies use various means to control these influencing variables, to demonstrate the effects of the newly designed instruction. Sometimes, the operation of these variables can be made equivalent by assigning students, schools, or communities in a randomized way to different groups to be instructed. More frequently, statistical means must be employed to establish the equivalence of the groups to be compared. If two courses or programs of instruction are to be evaluated to determine which is better, evaluation logic requires that control be exercised over these other variables. Ideally, everything should be equivalent except the instructional programs themselves.

REFERENCES

Astin, A. W., & Panos, R. J. (1971). The evaluation of educational programs. In R. L. Thorndike (Ed.), *Educational measurement* (2d ed.). Washington, DC: American Council on Education.

Brickner, A., & Scheier, E. (1968). *Summative evaluation of Listen Look Learn, Cycles R-40, 1967–68.* Huntington, NY: Educational Development Laboratories, Inc.

Brickner, A., & Scheier, E. (1970). *Summative evaluation of Listen Look Learn 2nd year students, Cycles R-70, 1968–69.* Huntington, NY: Educational Development Laboratories, Inc.

Cattell, R. B. (1963). Theory of fluid and crystallized intelligence: A critical experiment. *Journal of Educational Psychology, 54,* 1–22.

Cooley, W. W. (1971). *Methods of evaluating school innovations.* Pittsburgh, PA: Learning Research and Development Center, University of Pittsburgh.

Corno, L. & Snow, R. E. (1986). Adapting teaching to individual differences among learners. In M. C. Wittrock (Ed.), *Handbook of research on teaching* (3rd. ed.). New York: Macmillan.

Dick, W. (1977a). Formative evaluation. In L. J. Briggs (Ed.), *Instructional design: Principles and applications.* Englewood Cliffs, NJ: Educational Technology Publications.

Dick, W. (1977b). Summative evaluation. In L. J. Briggs (Ed.), *Instructional design: Principles and applications.* Englewood Cliffs, NJ: Educational Technology Publications.

Dick, W. & Carey, L. (1985). *The systematic design of instruction* (2nd ed.). Glenview, IL: Scott, Foresman.

Fitz-Gibbon, C. T., & Morris, L. L. (1978). *How to design a program evaluation.* Beverly Hills, CA: Sage.

Heflin, V. B., & Scheier, E. (1968). *The formative period of Listen Look Learn, and multi-media communication skills sytems.* Huntington, NY: Educational Development Laboratories.

Kennard, A. D. & Scheier, E. (1971). *An investigation to compare the effect of three different reading programs on first-grade students in Elk Grove Village, Illinois, 1969–1970.* Huntington, NY: Educational Development Laboratories.

Popham, W. J. (1975). *Educational evaluation.* Englewood Cliffs, NJ: Prentice-Hall.

Scriven, M. (1967). The methodology of evaluation. In R. Tyler, R. M. Gagné, and M. Scriven, *Perspectives of curriculum evaluation*. (AERA Monograph Series on Curriculum Evaluation, No. 1). Chicago: Rand McNally.

Scriven, M. (1974). Evaluation perspectives and procedures. In W. J. Popham (Ed.), *Evaluation in education*. Berkeley, CA: McCutchan.

Stufflebeam, D. L. (1974). Alternative approaches to educational evaluation: A self-study guide for educators. In W. J. Popham (Ed.), *Evaluation in education*. Berkeley, CA: McCutchan.

Stufflebeam, D. L., Foley, W. J., Gephart, W. R., Guba, E. G., Hammond, R. L., Merriman, H. O., & Provus, M. M. (1971). *Educational evaluation and decision making*. Itasca, IL: Peacock.

Thorndike, R. L., & Hagen, E. (1986). *Measurement and evaluation in psychology and education* (5th ed.), New York: Wiley.

NAME INDEX

Rubinstein, M. F., 71, *75*
Rumelhart, D. E., 104, *117*
Ryan, B. A., 296, *317*

Salomon, G., 79, *94*, 204, *216*
Schank, R. C., 104, *117*
Scheier, E., 336, 337, 338, 339, *343*
Schramm, W., 205, *216*
Schulman, L. S., 65, *75*
Scriven, M., 319, 322, *344*
Sharan, S., 271, *288*
Simmon, H. A., 70, *75*
Singer, R. N., 47, *53*, 92, 93, *95*
Skinner, B. F., 45, *53*, 87, *95*, 313, *317*
Snow, R. E., 106, 110, 111, 112, *116*, *117*, 330, *343*
Spielberger, C. D., 67, *75*
Spiro, R. J., 83, *95*
Stein, F. S., 161, 166, *176*
Stufflebeam, D. L., 320, 321, *344*
Suber, J. R., *287*
Suppes, P., 314, *317*

Tallmadge, G. K., 31, *36*
Talmage, H., 291, *317*
Tennyson, R. D., 188, *197*, 314, *316*
Thorndike, E. L., 8, *19*

Thorndike, R. L., *117*, 261, *264*, 327, *344*
Thurstone, L. L., 106, *117*
Tobias, S., 104, 106, 107, *117*
Torrance, E. P., 254, *264*
Triandis, H. C., 86, *96*, 257, *264*
Tyler, L. E., 69, *75*, 261, *264*
Tyler, R. W., 166, *176*

Wager, W. W., *19*, 24, *35*, 155, *159*, 164, *176*, 193, *197*, 204, 206, *216*, 222, 237, 238, *239*, 245, *263*
Walberg, H. J., 268, *288*
Walbesser, H. H., 148, *159*
Wald, A., 250, *264*
Walker, H. M., 313, *316*
Watson, J. R., 87, *96*
Weil, M., 278, *287*
Weiner, B., 106, *117*
Weinstein, C. E., 68, *75*
Weisgerber, R. A., 21, 30, *36*, 290, 291, *317*
White, R. T., 108, 109, *117*, 150, *160*
Wickelgren, W. A., 71, *75*
Wietecha, E. J., *287*
Wittrock, M. C., 187, *197*
Womer, F. G., 40, *53*
Worthen, B. R., 65, *75*

SUBJECT INDEX